COMBINED ARMS WARFARE

MODERN

WAR

STUDIES

Theodore A. Wilson

General Editor

Raymond A. Callahan

J. Garry Clifford

Jacob W. Kipp

Jay Luvaas

Allan R. Millett

Carol Reardon

Dennis Showalter

Series Editors

COMBINED ARMS

WARFARE IN

THE TWENTIETH

CENTURY

JONATHAN M. HOUSE

Maps by George Skoch

University Press of Kansas

© 2001 by the University Press of Kansas
All rights reserved
Published by the University Press of Kansas
(Lawrence, Kansas 66049), which was
organized by the Kansas Board of Regents
and is operated and funded by Emporia State
University, Fort Hays State University, Kansas
State University, Pittsburg State University,
the University of Kansas, and Wichita State
University
The paper used in this publication meets the
minimum requirements of the American
National Standard for Permanence of Paper
for Printed Library Materials Z39.48-1984.

Library of Congress
Cataloging-in-Publication Data
House, Jonathan M. (Jonathan Mallory),
1950–
Combined arms warfare in the twentieth
century / Jonathan M. House ; maps by
George Skoch.
p. cm.—(Modern war studies)
Rev. ed. of: Toward combined arms warfare /
Jonathan M. House. 1985.
Includes bibliographical references and index.
ISBN 0-7006-1081-2 (cloth : alk. paper) —
ISBN 0-7006-1098-7 (pbk. : alk. paper)
1. Tactics—History—20th century.
2. Armies—Organization—History—20th
century. 3. Military art and science—
History—20th century. 4. Military history,
Modern—20th century. I. House, Jonathan
M. (Jonathan Mallory), 1950– Toward
combined arms warfare. II. Title.
U165 .H8 2001
355.4'2'0904—dc21   00-068652
British Library Cataloguing in Publication
Data is available.
Printed in the United States of America
10 9 8 7 6 5 4 3 2

To Major General William A. Stofft,
the man who made it possible
and who revived the study of military history
when the U.S. Army had discarded it.
"Take risks, not gambles."

We have gotten into the fashion of
talking of cavalry tactics, artillery tactics,
and infantry tactics. This distinction is
nothing but a mere abstraction. There is
but one art, and that is the tactics of
the combined arms.

> *Major Gerald Gilbert*
> British Army, 1907

CONTENTS

• • • • • • • • • • • • • •

FIGURES, MAPS, AND PHOTOGRAPHS

• • • • • • • • • • • • • • •

Maps

Photographs

PREFACE

• • • • • • • • • • • • • • . •

In 1981, while teaching at the Command and General Staff College in Fort Leavenworth, Kansas, I inherited the extra duty of helping the U.S. Army service schools to include the study of military history in their curricula. Every first lieutenant or junior captain was supposed to study combined arms warfare, but there was no textbook for the class. The conventional wisdom was that technology changed so frequently that it was virtually impossible to reach any useful or enduring conclusions about the evolution of tactical combat. So I began to write out my own ideas on the subject, based on having taught it at the Armor School a few years earlier. How did the major armies of the world intend to fight on the battlefield, and what concepts, weapons, and organizations had they evolved to accomplish that battle? How did the different armies influence each other in their development? My department chair, then-colonel William A. Stofft, encouraged me to continue in this effort and later gave me a reduced teaching load to finish writing. The result was Research Survey no. 2, *Toward Combined Arms Warfare: A Survey of 20th-Century Tactics, Doctrine, and Organization* (Ft. Leavenworth, KS, 1984), which became a textbook in a variety of schools for the U.S. Army and other services.

The current revision has two aims. First, I have endeavored to bring the story of combined arms warfare forward from the late 1970s to the end of the century and in the process to incorporate developments that were not clearly evident in 1984. Second, the original text was written for serving officers by a serving officer; it inevitably included terminology and assumptions that might distract or mystify the general reader. This new edition includes both an appendix with terms and acronyms and a conscious attempt to explain concepts in nontechnical terms.

As with my other books, I am indebted to the staffs of the Combined Arms Research Library at Ft. Leavenworth and the Military History Institute at Carlisle Barracks, Pennsylvania. Ed Drea assisted in every step of the first edition, and he reinforced Ray Callahan in providing encouragement of the second. My mentor and sometime coauthor, David Glantz, taught me about the Soviet Army, supervised the writing of the first edition, and provided invaluable comments and resources for the revision. Steve Zaloga was once again generous with information and rare photographs. George Skoch (for maps) and Jackie Johnson (for the diagrams) responded superbly, despite my frequent changes and impenetrable handwriting. Finally, I must thank by family for tolerating my frequent absences, both physical and mental, while I was writing this edition.

SYMBOLS AND TERMS

• • • • • • • • • • • • • • •

Like other complex human endeavors, the profession of arms has its own terminology, abbreviations, and symbols. To assist the general reader, I have explained or avoided many specialized terms; a list of acronyms and common terms is also included as an appendix. No special knowledge is required to understand the various battles and military organizations described here. If, however, the reader is unfamiliar with the standard symbols used to depict such battles and organizations, a brief explanation is provided in the following chart. When a unit symbol appears on a map, the reader must understand that it represents only the location of the unit headquarters, or the center of a force that may extend over a much larger area.

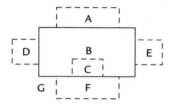

AREA A: SIZE OF UNIT

XXXXX = Army Group or Front
XXXX = Field Army
XXX = Corps
XX = Division
X = Brigade
III = Regiment
II = Battalion/Squadron
I = Company/Battery/Troop
••• = Platoon
•• = Section or Detachment
• = Squad

AREA B: PRIMARY TYPE OF UNIT

✕ = Infantry
╱ = Cavalry/ Reconnaissance
• = Artillery
◯ = Armor or Mechanized
+ = Medical
⋈ = Maintenance

˟ = Chemical
⌒ = Air Defense/Anti-Aircraft
⊓ = Engineer
∧ = Anti-tank
•–• = Aviation
⟍ = Signal/Communication
▬ = Supply

AREA C: ADDITIONAL QUALIFICATION OR TYPE OF UNIT

⌒ = Airborne
v = Air Assault
LT = Light

AREA D: Designation of the Actual unit depicted

AREA E: Designation of the parent unit/ higher headquarters of this unit

AREA F: Special qualifications/type; for artillery units, this space indicates the type and caliber of weapon

AREA G: If a line extends from the left edge of the rectangle, this indicates that the headquarters (not necessarily the entire unit) is located where this line terminates on the map

Two small circles under the square indicate the trains–the logistical element of the unit depicted.

EXAMPLES:

 17TH Airborne Infantry Division, XVIII Corps

 Battery A (155mm self-propelled), 2d Battalion, 17th Armored Field Artillery

2d Squadron, 11th Armored Cavalry Regiment

 19th Engineer Brigade (Combat), 5th Army

Standard Military Unit Symbols

INTRODUCTION

• • • • • • • • • • • • • • • • •

The Combat Arms

The first warriors fought on foot, armed with stones, clubs, and spears. In time, the domestication of draft animals permitted some men to ride on horseback or in chariots. These mounted warriors not only moved more swiftly but could also use the size and momentum of their steeds to frighten or crush their opponents on foot.

To counter this advantage, foot soldiers learned to fight shoulder to shoulder, their shields overlapping one another and a hedge of spears projecting from the shield wall. When this formation became deeper, with six or more rows of men standing one behind the other, the resulting phalanx could withstand the charge of mounted opponents and in fact frighten off any but the most determined attackers. With enough practice and discipline, the phalanx could move forward at a fast walk while retaining its unity, gaining enough momentum to crash through any opponent except an equally well-drilled phalanx. Yet to achieve this advantage, the members of a phalanx ceased to be individual warriors and became parts of a well-drilled unit. The individual warrior, however heroic or physically skilled, lost out to ordinary men working together within a military organization.

Of course, the phalanx had its vulnerabilities; it could not adjust its shield wall to uneven ground, nor could it change its orientation to face attack from unexpected directions. The individual member of a phalanx was hampered by the weight of shield, sword, spear, and helmet and could not move quickly without sacrificing the safety of his formation. As a result, in the fourth century B.C. the Greek phalanx lost its military dominance to the Macedonian army of Philip II and Alexander the Great. The Macedonians combined traditional phalanxes of foot soldiers with two other forces: horse-mounted cavalrymen and dismounted runners, armed only with small shields and short swords. Alexander's cavalry outmaneuvered opposing infantrymen by charging through gaps that appeared in enemy formations, and the lightly armed, running infantrymen linked the phalanx to the cavalry and blocked or delayed attacks from unexpected directions.

In turn, the Macedonian system gave way to the Roman legion. Instead of operating as a single, ponderous phalanx, each legion deployed in a checkerboard formation of up to thirty small phalanxes, each consisting of about 100 to 120 men. Thus the legion could adjust to the shape of the battlefield and of the enemy. The legion was more than an infantry unit, however—for most of its history, it included a small number of horsemen to scout for the enemy as well as a number of mechanical devices to fling rocks, spears, and other projectiles at enemy units or fortresses. Moreover, each legion carried the tools to construct its own temporary earthen fortress, within which it rested at the end of each day's march. Thus each legion possessed the four basic branches, traditionally known as the combat arms—infantry, cavalry, artillery, and engineers.

During the Middle Ages, the invention of the stirrup gave horse-mounted warriors a much more stable platform from which to fight, and strenuous training and a culture of individual courage made the armored knight a formidable foe. Yet the foot soldier never disappeared entirely. Beginning in the fourteenth century, a series of innovative commanders rediscovered the ancient formations of the phalanx and legion. When these formations were married to projectile weapons—the English longbow and later primitive firearms—they proved more than a match for armored cavalry, however brave or skillful.

In short, the development of military tactics and organization followed the pattern later seen in the industrial revolution. Individual craftsmen, individual types of warrior, could not compete effectively with an opponent who used specialization of labor to combine many different capabilities into an efficient, tightly controlled organization. On the battlefield, what mattered was how a commander orchestrated these different specialties—foot soldiers, horse-mounted soldiers, soldiers firing projectiles, and soldiers building or destroying fortifications. Tactics meant the deployment, maneuver, and interaction of these different specialists, known as the combat arms.

Combined Arms

We have gotten into the fashion of talking of cavalry tactics, artillery tactics, and infantry tactics. This distinction is nothing but a mere abstraction. There is but one art, and that is the tactics of the combined arms. (Major Gerald Gilbert. British Army, 1907[1])

The concept of combined arms in ground combat has existed for centuries, but the nature of that combination and the organizational level at which it occurred have varied greatly. Prior to the seventeenth century, the combat arms of infantry, artillery, and cavalry usually operated in separate units of a few hundred to a few thousand men. Engineers worked in even smaller groups, building bridges or advising commanders on the construction of trenches. Each branch served a specific function on the battlefield, and only the senior commanders in the field needed to coordinate the effects of the different arms. In succeeding centuries, the trend has been to combine the arms at progressively lower levels of the military hierarchy. The concern of commanders has gone from coordinating the separate actions of separate arms, to achieving greater cooperation among those arms, to finally combining their actions to maximize the effect of all components of an armed force.

At the time that Maj. Gerald Gilbert made his plea for the "tactics of the combined arms," many officers paid lip service to the idea but few understood the need to achieve such cooperation or combination among the branches at levels below that of, say, a brigade of approximately 2,000 men. Since then twentieth-century warfare and especially mechanized warfare have developed to the point where some form of combined arms is essential for survival, let alone victory, even in an infantry squad of six to ten soldiers. No matter how powerful a single arm—tanks, attack helicopters, or whatever—may be, that arm has many of the same strengths and weaknesses as its counterpart in the opposing army. As a practical matter, therefore, a carefully adjusted mixture of different weapons will almost always prove superior to a single type of weapon.

Given this necessity for combined arms, many observers have criticized the fact that most armies remain organized, at least in peacetime, into battalions composed of one or at most three major weapons systems. A mechanized infantry battalion of 500 to 800 men, for example, normally includes direct-fire infantry weapons (rifles, machine guns, small-caliber cannon), antiarmor guns and missiles, and short-range indirect-fire support (mortars and grenade launchers). This battalion has little or no organic capability in the areas of armor, air defense, engineers, long-range artillery, or air support. A tank, artillery, combat engineer, air defense, or aviation battalion is even more specialized and restricted in its weapons and capabilities.

Historical conservatism and unit identity are not the only reasons for segregating the combat arms in this manner. In garrison, it is more efficient to concentrate all vehicles and weapons of one type in a single unit for ease of repair and maintenance. Similarly, individual soldiers or small crews with the same duties and weapons can be trained more effectively if they are concentrated in a single unit. In any event, no permanent combined arms organization, however well trained, would be equally prepared to deal with all different circumstances. Fighting in a desert environment requires a preponderance of tanks and attack helicopters because of the ease of maneuvering and shooting over long distances, but fighting in an urban area necessitates more infantry and engineers to clear the enemy from buildings. Every army must tailor its forces to the specific situation.

The very term "combined arms" means different things to different people, or it is left undefined and vague. For our purposes, this term has at least three related meanings. First, the combined arms concept is the basic idea that different combat arms and weapons systems must be used in concert to maximize the survival and combat effectiveness of the others. The strengths of one system must be used to compensate for the weaknesses of others. The specific arms and weapons included in this concept have varied greatly among national armies and over time. Today, however, the list of combined arms could include at least infantry (mechanized, motorized, airborne, air assault, light, and special operations units), armor, reconnaissance (often called cavalry), artillery, antitank weapons, air defense or antiaircraft forces, combat engineers (often called sappers), attack helicopters, and ground-attack aircraft. Under certain circumstances, this list may also include electronic warfare to jam or deceive the enemy's communications, as well as guerrilla forces and nuclear and chemical weapons. Beyond this long list of combat forces is a host of combat support (intelligence, communications, chemical decontamination crews, smoke generators, military police, and construction engineers) and service or logistical support elements (transportation, maintenance, supply, medical, personnel, and finance). In many instances, logistical support is key to sustained combat and hence to victory. In the interests of brevity, however, the logistical aspects of combined arms will be examined only as they are incidental to tactical considerations.

A second meaning of combined arms is combined arms organization, the command and communications structure that brings the different weapons together for combat. This may include both permanent, peacetime organizations and ad hoc or "task-organized" combinations of the different elements in wartime. Any such temporary grouping of two or more combat arms is usually described as a "combat team" or a "task force." (see appendix).

Third, combined arms tactics and operations are the actual roles performed and techniques applied by these different arms and weapons in supporting one another in battle. This is the area that is of most concern to professional soldiers, yet it is precisely this area where historical records and tactical manuals often neglect important details. Moreover, combined arms tactics and techniques at the level of battalion (300 to 800 troops) or company (50 to 200 troops) are the most difficult aspects about which to generalize historically, because they are most subject to frequent changes in technology.

Tactics

Before we examine the historical development of combined arms, some basic terms and explanations are provided in order to assist the general reader. First, war can be planned and analyzed at three different levels, each with its own focus and objectives. *Tactics,* as we have seen, is the process of combining different arms and services to win a battle. At the other end of the analytical spectrum, *strategy* is concerned with the plans and actions that directly contribute to the outcome of the war. Of necessity, therefore, military strategists must take into consideration the political objectives and limitations of governments.

The American Civil War and World War I both demonstrated the need for some intermediate level of planning and analysis, some means of connecting individual, tactical battles with overall strategic intentions. The great complexity of modern warfare meant that victory required more than a single, Napoleonic battle of annihilation. Commanders entered the twentieth century without a clear concept or terminology for this intermediate level, which is today called *operational art* or the *operational level of war.* Operational art involves sequencing steps, coordinating actions and battles to achieve the strategic goal. It is tempting, but simplistic, to equate the tactical, operational, and strategic levels to different levels of military command, from a single company through a theater headquarters. In fact, the three levels must be defined in terms of their purposes and objectives rather than by any specific organizational level. Consider, for example, a hypothetical action in which a small team of special operations troops penetrates the enemy rear area in order to ambush an enemy commander or leader. If that ambush disrupts enemy command and control sufficiently to cause a military defeat, the action of this handful of soldiers might be described as an operational or even as a strategic endeavor rather than just as a tactical skirmish.

This book is focused on the tactical combination of arms and services at the level of division and below. However, in order to place those tactics within their larger, operational or even strategic context of warfare, it will be necessary to digress periodically to higher levels of organization and planning.

What, then, is the essence of tactics? In the abstract, tactical warfare may be considered as a combination of three elements or functions: mobility, protection, and offensive power.[2] *Mobility* means not only the capacity to maneuver and concentrate forces over varied terrain but also the ability to move men and units when exposed to the enemy's weapons. Mobility is not an absolute, but it must be measured relative to the mobility of other friendly or enemy forces as well as to the difficulty of the terrain. For a combined arms team, the least mobile element may determine the mobility of the entire force. Without mobility, an army would be unable to implement the military principles of massing and maneuvering forces to seize the initiative in battle, and the principle of surprise would also be difficult to achieve.

Protection includes both security against enemy surprise attack and safeguards to allow troops to attack or defend on the battlefield. Such battlefield protection may be accomplished by using defensive fortifications and the folds of the ground to block enemy fire or by employing artificial means such as armor. Last, *offensive power* is necessary to impose one's will on the enemy. Offensive power usually relies on the firepower of modern weapons to overcome the enemy's protection. Moreover, the psychological ascendancy of troops aggressively advancing toward the enemy is also part of offensive power, although such an advance may result in massive casualties for the attackers.

These three elements have interacted continually throughout military history. The past century has been characterized by a vast increase in firepower, an increase that can be overcome only with great difficulty by a carefully designed combination of protected mobility and other firepower. The most obvious example of such a combination was the defensive system of World War I. That combination of firepower and protection had to be countered by close coordination of infantry (mobility), fire support (offensive power), and armor (which theoretically combined the three elements). Even this explanation of trench warfare is simplistic, but the three elements of mobility, protection, and offensive power are present in all tactical equations.

At a more practical level, these three elements are combined technically in the design and employment of individual weapons and tactically in the combination of different arms and units. The resulting combination may be either supplementary or complementary in its effects on the battle.[3] As the term implies, supplementary combined arms increase the effect of one weapon system or arm by adding the similar effects of other weapons and arms. For example, the effects of mortars and artillery may reinforce or supplement one another in an integrated fire plan. Engineers may enhance the protection of armored vehicles by digging earthen revetments around those vehicles and may increase the mobility of vehicles by building bridges and breaching enemy obstacles. Complementary combined arms, by contrast, combine different effects or characteristics, so that together they pose a more complicated threat, a dilemma for the enemy. By itself, a minefield will slow but not halt the advance of an attacking force. However, the defender may place that minefield so that it halts the attacker in the open, where artillery and antitank fires can engage the attacking force while it has stopped to clear the minefield. This dilemma forces the attacker to accept casualties while clearing the mines or to seek passage elsewhere. The defender has thus integrated different weapons to provide a much greater effect than any one weapon by itself could achieve.

Orchestrating the Battle

An army needs to accomplish many tasks in order to produce effective battlefield interaction. Orchestrating a battle requires organization and doctrine; training; command, control, and communications; and motivation. No army achieves perfection in all these areas; victory often goes to the side that is less inefficient and that makes fewer mistakes.

Organization and doctrine together reflect how an army expects to fight. The wartime organization of land forces must be sufficiently flexible to respond to the ever-changing confusion of battle with an appropriate mixture of weapons and soldiers. Yet the best military structure is useless if its members lack some shared terminology and concepts of how that organization will function. This shared understanding is called doctrine. Doctrine in the military sense of the term is not rigid or dogmatic, however; it simply provides a common set of procedures and a frame of reference for dealing with the unique nature of each tactical situation. Indeed, doctrine cannot be dictated from the top of the military hierarchy—soldiers at every level must both understand the doctrine and believe that they can apply it with the personnel and equipment available.

Outside observers frequently criticize armies, and especially officer corps, for a perceived resistance to technological change. In truth, this is not simply social conservatism but the painful process of adjusting organizations and doctrine to make the best use of new weapons. This adjustment takes time, not only to determine how the new weapons should be used but also to modify the doctrinal understanding of the officers involved. Senior officers tend to view matters through the prism of their personal experience, even if it may be ten to thirty years old. As a young platoon leader, an officer may have struggled with unreliable, vacuum-tube radios; twenty years later, that same officer might find it difficult to accept a communications structure based on satellite relays and laptop personal computers.

In the twentieth century, armies have had to make this adjustment for a host of new weapons, including machine guns, tanks, ground-attack aircraft, and helicopters. In each case, a three-stage pattern of organizational behavior appears. First, the army cannot find an organizational or doctrinal home for the new weapon. As a result, commanders tend to view the new technology as a specialized adjunct, useful only under certain conditions where the existing combination of arms and services has proved inadequate. Next, a group of enthusiasts seeks to make the new weapon into its own separate combat arm, in the process asserting exaggerated claims about its ability to achieve victory on its own, or perhaps with only one other existing combat arm. Because the enemy army is also having difficulty adjusting, the new weapon may, in fact, achieve a brief success as an independent arm. Ultimately, however, each side develops countermeasures to reduce and limit its effectiveness. Thereafter, the new weapon can no longer achieve victory by itself but must become a full-fledged member of the combined arms team. Within this expanded team, professional soldiers eventually reach a doctrinal solution, a shared concept of how to integrate the new weapon into the complementary effects of the other arms and services.[4]

For example, the tank was originally viewed as a slow, unreliable specialized weapon that might assist the infantry and artillery to deal with the unusual phenomenon of trench warfare but that would never accomplish much on its own. Early leaders in the new tank units chafed at their limited, subordinate status and sought to develop their new weapon into an entirely separate, war-winning force. Forces composed largely of tanks had some initial success against unprepared defenders in World War II, but eventually most armies developed techniques both for employing and for defeating tanks. Thereafter, the tank remained an important weapon system but only if used in conjunction with the other combat arms. A similar evolution occurred for ground-attack aircraft.

Implementing doctrine requires constant training at every level, from the individual soldier and weapons crew to the largest military command. Soldiers must practice to perform complex tasks under any conditions of darkness, adverse weather, or enemy fire. A recurring theme of this book is that professional soldiers tend to overestimate the amount and quality of training necessary for the rank and file to perform effectively in war. There is no substitute for realistic training, but historically leaders with high standards have rejected or modified doctrine that their troops seemed incapable of executing.

On the other hand, training levels may in fact force modifications in the existing military structure and doctrine. If, for example, company commanders are capable of coordinating only eighty men and two types of weapons systems, it would be futile to design 170-man companies with ten different weapons systems. Training leaders to handle these larger, more complex units may be prohibitively expensive in peacetime or may require too much time in war.

This example highlights another trend in warfare. The growing complexity of combat has forced armies to depend increasingly on the judgment and abilities of junior leaders. Before 1914, a battalion commander could maneuver his 300 to 800 soldiers almost as a unit. His troops were close enough that everyone could hear his voice and obey orders immediately. The increased lethality of weapons soon forced troops to disperse over wider areas, beyond the range of one man's voice. Eventually, reliable portable radios again permitted one officer to control a large unit, but by that time each platoon and company had acquired so many different weapons that the battalion commander had to rely on his subordinate officers to make many of the detailed decisions in execution of his intentions. Thus, although a nineteenth-century battalion was controlled by one officer with fifteen or more years' experience, twentieth-century combat is often decided by more junior, less experienced leaders.

Command, control, and communications are necessary to train in peacetime and to coordinate the battle in wartime. Good commanders at any level not only make the ultimate decisions in battle but also encourage and inspire their subordinates. As a result of this dual role, commanders are torn between two extremes. On the one hand, the commander may remain in the rear, where he has the communications and information necessary to coordinate the entire battle. On the other hand, aggressive commanders at every level wish to be out in front, where they can see with their own eyes and lead by personal example. Unfortunately, such a forward role not only endangers the commander personally but often tempts him to neglect his overall duties as he becomes preoccupied with the portion of the battlefield that he can see. Alternatively, a particular level of command may be organized to perform most of the logistical and administrative functions for subordinate levels of command, freeing that lower commander to focus on the tactical aspects of battle.

Whatever compromise a commander may choose, the mechanisms by which he directs the battle may be termed control and communications. Control includes standardized procedures and trained staff officers to relieve the commander of innumerable burdens by monitoring events, recommending courses of action, and implementing the commander's intent. The growing complexity and scale of warfare has often resulted in the growing complexity and scale of headquarters staffs. One frequent solution to this problem is to assign different parts of the control function to different headquarters. Thus, for example, the commander may have a small, mobile command post that allows him to control current operations while his main headquarters, farther to the rear, plans future actions and coordinates the logistical support necessary for sustained operations.

Command and control take time—time to gather information and to assess the situation, time to develop and choose the best course of action, time to disseminate the decision and to coordinate the different elements of the unit, and finally time to actually launch the operation. These time delays give rise to two problems. First, the battle situation may have changed, making the commander's plan inappropriate. Second, if one force can accelerate its decision process, it may take action more quickly than its opponent can react. This acceleration, often called "turning inside the enemy's decision cycle," is one of the factors that permitted the German Army to be so successful in both world wars.

Regardless of how commanders and staffs function, communications are vital to gather information, disseminate orders, and coordinate their execution. Inflexible communications may tie commanders to their headquarters and impede the decision cycle; disrupted communications may result in defeat before the commander is even aware of a threat.

The final factor in battle is *motivation,* or more simply, intelligent courage. Doctrine, organization, and staffwork are only inadequate attempts to impose order on the chaos of battle. Often plans that made perfect sense in one environment or situation will be suicidally useless under different circumstances. No battle ever unfolds strictly according to plan, nor is any battle without personal risk to the participants. The ability of junior leaders to make rapid decisions, the willingness of leaders and followers to risk their lives for the sake of their friends and units—those qualities translate the dry concepts of this book into action. In the words of Field Marshal William Slim,

> In the end every important battle develops to a point where there is no real control by senior commanders. Each soldier feels himself to be alone. Discipline may have got him to the place where he is, and discipline may hold him there—for a time. Co-operation with other men in the same situation can help him move forward. Self-preservation will make him defend himself to the death, if there is no other way. But what makes him go on, alone, determined to break the will of the enemy opposite him, is morale.[5]

In short, armies must constantly review their doctrine and organization to accommodate changes in technology. Commanders must train and co-ordinate the resulting military units in order to implement that doctrine intelligently. Finally, each individual in the organization must function smoothly and rapidly, despite the unpredictable and deadly nature of the battle.

THE TRIUMPH OF FIREPOWER, 1871–1939

The Mexican Punitive Expedition, 1916

Late on the morning of 15 March 1916, the 6th U.S. Infantry Regiment crossed the Rio Grande River near Columbus, New Mexico, as the first element of Brig. Gen. John J. Pershing's punitive expedition into Mexico. In the khaki uniforms and broad-brimmed campaign hats of the era, the soldiers splashed across the shallow river with their rifles held high to keep them dry. On the far shore, the first company of men spread out briefly to protect the main body of the regiment crossing behind them.

The 6th had been selected to lead the advance to commemorate its similar role in the Mexican War of 1846–1848, but once the river was behind them, the horsemen of Troop K, 13th Cavalry Regiment, clattered past to become the advance guard. The mission of this expedition was to destroy the guerrilla forces of Pancho Villa while respecting the sovereignty of the Mexican government. However, the local commander of Mexican forces had threatened to resist any American intervention by force, so the cavalrymen approached the hamlet of Palomas cautiously. The town proved to be deserted, save for an old couple too crippled to flee.*

*Clarence C. Clendenen, *Blood on the Border: The United States Army and the Mexican Irregulars* (New York, 1969), 220.

For months, Pershing and his men engaged in a fruitless effort to catch or at least to disperse Villa's guerrillas. The tactics involved—multiple small columns of soldiers on foot or on horseback, moving quickly in an attempt to surround their quarry—varied little from those used against Native Americans in the campaigns of the later 1800s. A truck-mounted infantry company and a few primitive radios did not materially change the situation. When part of the army's fledgling air service arrived to join in the search, the unreliable aircraft crashed with alarming frequency. Frustrated but hardened by the experience, the U.S. Army finally withdrew from Mexico in early 1917.

While a few thousand Americans vainly pursued Mexican guerrillas, millions of Europeans were locked in the massive stalemate of World War I. By European standards, the U.S. Army's tactical organization in 1916 was almost nonexistent. As an experiment, in 1911 the War Department had ordered the assembly of a "maneuver division" in San Antonio, Texas. After a ninety-day struggle, a hodgepodge of understrength infantry, artillery, and cavalry units gathered at San Antonio to experiment with the most basic concepts of communications and maneuver. Based on this limited success, most of the army's units received assignments, at least on paper, as parts of specific brigades and divisions, a plan that was tested in 1913 in response to the continuing crisis of the Mexican Revolution. The National Defense Act of 1916, passed just before Pershing's operation, recognized the existence of tactical divisions and brigades. Yet the act still authorized the organization of the army in terms of regiments of infantry, artillery, cavalry, engineers, and specialized supporting units. Each such regiment, nominally consisting of about 1,000 soldiers, was armed with only one type of weapon.* The infantry regiment, for example, had only four machine guns to support a force that was otherwise equipped exclusively with bolt-action, single-shot rifles.

Moreover, politics had perpetuated the tiny forts of the Indian-fighting era. Pershing's force had to assemble from a large number of garrisons, most of which had only a few hundred soldiers in any one location. Few of the officers had any training or experience in coordinating different types of soldiers in large formations, nor had the troops practiced together as a large unit. Even finding and trapping Villa proved to be an impossible task, and the effort, so reminiscent of previous Indian campaigns, encouraged American officers to look to the past rather than to the future. Yet only one year after crossing the Rio Grande, General Pershing found himself trying to bring the U.S. Army from the primitive state of the Mexican Punitive Expedition to the modern, complex standards of trench warfare.

---

*Russell F. Weigley, *History of the United States Army* (New York, 1967), 334–35, 348. See also James L. Abrahamson, *America Arms for a New Century: The Making of a Great Military Power* (New York, 1981), 45, 110–13.

CHAPTER 1

. . . .

Prologue to 1914

In the 1690s, European armies developed and fielded the socket bayonet, a long, triangular-shaped blade that could be fixed on the end of a musket without obstructing the bore of the weapon during loading and firing.[1] This simple device allowed well-disciplined infantry to withstand horse cavalry charges without the aid of specialized weapons such as the pike. However, the infantryman had to remain standing in order to ram gunpowder and musket balls down the muzzle of his weapon. For the next 150 years, infantry units armed solely with smoothbore firearms and bayonets were the backbone of all Western armies. Skilled senior commanders understood how to coordinate this infantry with cavalry and with direct-fire smoothbore artillery. Yet combined arms coordination was rarely important at the level of a regiment (600 to 1,000 men) or below, because those units were basically armed with a single type of weapon. The need to maximize the firepower of inaccurate smoothbore weapons led to extremely linear deployments on the battlefield. The infantry maneuvered from marching columns into long lines that were only two or three men deep, bringing as many muskets as possible into play. The artillery was usually located between or slightly behind the infantry battalions. In the absence of accurate guns and reliable exploding artillery shells, the effect of even such carefully arrayed firepower was limited. In turn, this made it possible, if dangerous, for dense masses of cavalry and infantry to attack at a specific point and break the thin lines of the defender. Fire-support coordination was simple, because the infantry and artillery unit commanders had face-to-face contact or used hand signals to designate targets.

The fundamentals of weaponry, technology, and small-unit tactics were refined but remained basically unchanged until the mid-1800s. In the later 1700s, for example, the French Army led Europe in developing means to make that artillery as mobile and flexible as the infantry it supported. The stability of tactics made professional soldiers skeptical of innovations even when such changes come from serious students of combat.

Technological Change

During the period 1827 to 1870, the first of two waves of technological change revolutionized the battlefield. The most important innovation of this first wave was the development of mass-produced, rifled, breech-loading firearms. In the 1840s, the muzzle-loading rifle with a bullet-shaped projectile initially replaced the smoothbore musket. Rifling and an improved seal between the bullet and the barrel of the gun increased both the accuracy and the velocity of small-arms fire to an effective range of nearly 500 meters.[2] During the American Civil War (1861–1865), the dense infantry formations of the preceding two centuries provided lucrative targets for defenders armed with such rifles. Both sides learned to spread out into skirmish lines, with one or more paces between soldiers, when attacking. Defenders, for their part, had to dig trenches to reduce their own vulnerability to the attackers' rifle fire.

The muzzle-loading rifles used by most soldiers during the Civil War were obsolescent, as the Prussian Army had already fielded a breech-loading rifle.[3] Unlike muzzle-loaders, breech-loaders could be reloaded in a prone position, allowing infantry to remain prone while firing repeatedly. Soon metallic-cased ammunition made loading even faster, because the soldier could insert a single, neat package of bullet, powder, and primer into the rifle's breech. By the time of the Franco-Prussian War in 1870–1871, most armies had adopted breech-loading artillery as well as rifles.

This first wave of technological change also included the introduction of the railroad and the telegraph, which greatly increased the speed of communication, mobilization, and troop movement at the strategic and operational levels. These inventions also made it possible to provide sustained logistical support for large armies. At the tactical level, however, troops still maneuvered on foot or on horseback, moving no faster than the armies of Napoleon.

The second wave of technological change occurred in the 1880s and 1890s. Smokeless powder, magazine-fed repeating rifles, recoiling and quick-firing artillery, improved artillery fuses, machine guns, and internal combustion engines appeared in rapid succession. With the exception of the gasoline engine, these developments favored the defensive. They increased the volume, range, and accuracy of fire, and smokeless powder made it difficult to identify the location of the shooter. This placed a soldier advancing over open ground at a tremendous disadvantage, compared to the soldier defending prepared positions. Artillery, in particular, was able to fire much more rapidly, spewing as many as thirty shells per minute. The new recoiling mechanisms meant that the field gun no longer rolled backward when it fired; this in turn not only increased accuracy but permitted the gunners to take cover behind an armored shield mounted on the gun carriage.[4] Communications, however, like mobility, lagged behind firepower. Although radiotelegraphs existed in the armies of 1914, the radio was not yet sufficiently mobile and reliable to permit commanders to follow and direct events on the expanded battlefields of the day.

The cumulative effect of these two waves was to make cooperation and coordination between different units and arms absolutely essential. Anything less than total coordination in the attack might well result in defeat by superior firepower. Conversely, an uncoordinated defense invited disaster.

The American Civil War and the Wars of German Unification (1864–1871) gave professional soldiers many opportunities to adjust to the first wave of technological change. That technology, combined with an effective system of mobilizing reservists, provided the tools for Prussia's struggles to unite Germany. When World War I began, however, professional soldiers had not yet digested and agreed upon the effects of the second wave of change. Thus most tactical doctrines in 1914 showed a healthy respect for the effects of firepower, but such doctrines had not solved the resulting problems of the battlefield.

Reservists and Regulars

Quite apart from changes in technology, the late 1800s saw major changes in both the enlisted ranks and the officer corps of most major armies. Most nations followed the example set by the Franco-Prussian War, where German armies composed largely of reservists had overwhelmed the better-trained but smaller professional armies of France. For the next four decades, a large portion of the young manpower of Europe was conscripted for two or three years of training, then released into civilian life until war broke out. General staffs developed elaborate plans to mobilize and deploy these reservists by railroad at the outbreak of war. By 1900 Germany had only 545,000 men on active duty in peacetime but a total wartime strength of 3,013,000; France had 544,450 men in peacetime but 4,660,000 in war; and Russia could mobilize well over 4 million from a peacetime strength of 896,000.[5] Even Britain, which retained a volunteer army, relied on reservists in the event of general war because so many of its professional soldiers were deployed in the colonies. In 1914, 61.8 percent of the British Expeditionary Force was composed of reservists, although some of these had long years of experience on active duty.[6]

The Prussian reserve formations of the 1860s were successful in part because they were filled with veterans of previous Prussian wars. After 1871, however, a long period of peace deprived most armies of such experienced reservists. Every army had to develop its own system of reserve training and organization, and every army had to decide what percentage of reservists could be absorbed into an active-duty unit on mobilization. In many cases, this meant that the officers and men who had to fight together were strangers to one another.

Many officers distrusted the competence of their citizen-soldiers and compared them unfavorably to the long-term volunteers of previous eras. The absence of reservists from active army formations during most of the year meant that units were well below authorized strength and were in effect skeleton formations, which made realistic training difficult for both officers and conscripts.

Though the rank and file were composed of conscripts and reservists, the officer corps were evolving into bodies of trained professionals. The pace and success of this evolution varied widely between armies, however. In Germany, the prestige of the Wars of German Unification resulted in fierce competition even for commissions as reserve officers, and the unsuccessful candidates for these posts gave the Germans a remarkably well-trained cadre of noncommissioned leaders. In France, by contrast, the army was tainted by an image of political conservatism; although there were many able French commanders, the officer corps did not always attract the best young men from all walks of life. The British, American, and Russian officer corps lagged behind their competitors; despite the intellectual achievements of individual officers, the social origins and habits of the majority of officers often discouraged the development of true professionalism. In many armies, career officers tended to come from rural and small-town backgrounds and therefore had little experience with the industrial, technical portions of society. Under the influence of the Boer War (1899–1902), the Spanish-American War (1898–1899), and the Russo-Japanese War (1904–1905), these attitudes began to change, although conventional wisdom dismissed these wars as having few lessons for the future European conflict. Thus, the typical British, U.S., or Russian officer tended to be less technically competent, especially concerning large-unit operations, than was his French or German counterpart.[7]

This disparity in the development of officer corps was even more apparent with regard to the general staffs, the officers responsible for planning, mobilizing, and directing the forthcoming wars. All European armies recognized the great ability of the Prussian (later German) General Staff, which had been a key component for victory in the 1860s. Throughout the nineteenth century, rigorous training and selection processes produced an elite group of German officers who shared a detailed understanding of how war should be conducted. These general staff officers were not infallible, of course; for example, they tended to neglect the logistical aspects of their plans. Yet by 1914 the German General Staff was a fully developed institution. At the other end of the spectrum, the British and American General Staffs did not really exist until the last decade before World War I and were still in the process of development during that conflict. Senior officers had not yet risen through the staff system, and older commanders often distrusted or ignored the young general staff officers who served under them.[8]

Organization and Doctrine

Pre-1914 armies organized the different combat arms into divisions and corps that bore a superficial resemblance to those of today. The most obvious difference was the absence of the vehicles and electronics associated with more modern combat. By the end of the Napoleonic Wars, European armies had accepted the division as the wartime unit for combining infantry and artillery, although most cavalry was concentrated into separate brigades, divisions, or even corps.[9] As in so many other areas, the Prussian example had produced considerable agreement by 1914 on the basic organization of an infantry division. Most divisions contained twelve battalions of infantry, each with two machine guns either assigned or in direct support (see Figs. 1 and 2).[10] These battalions were usually grouped under four regimental headquarters that in turn belonged to two brigades. (The British had three brigades and lacked the regimental level, because the British regimental headquarters remained in garrison, training replacements.) Divisional cavalry was universally restricted to a few hundred men since most functions of reconnaissance and security screening were assigned to the separate cavalry brigades or divisions. These large cavalry formations were almost purely horse-mounted troops, with a few horse artillery batteries attached. Not until 1913–1914, for example, did the Germans add small numbers (one company

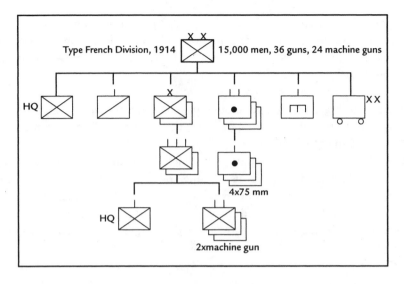

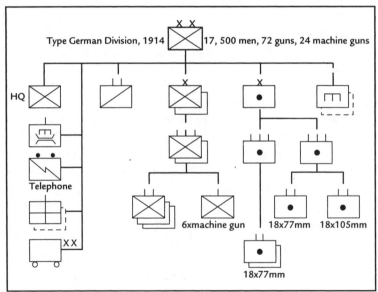

Figure 1. Type French and German Divisions, 1914

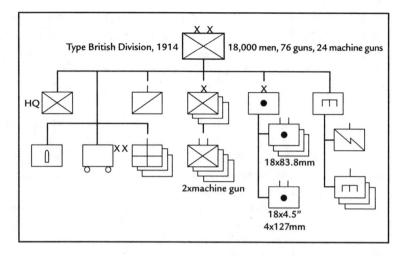

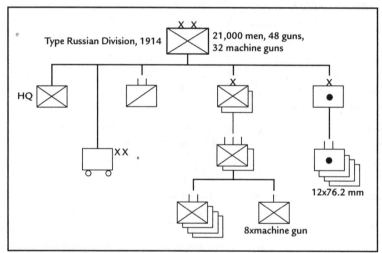

Figure 2. Type British and Russian Divisions, 1914

each) of mounted engineers and bicycle-equipped infantry to their cavalry divisions.[11]

The armies differed most markedly in the proportion and size (caliber) of artillery included in the infantry divisions. Divisional artillery varied from as few as thirty-six light guns of 75mm in the French division to as many as seventy-six artillery pieces, including eighteen 4.5-inch (114.5mm) howitzers and four 127mm guns in the British division. These variations in structure reflected profound confusion and disagreement over the role of artillery and the importance of combined arms.

In order to understand the doctrinal interrelationships of the different arms before World War I, some consideration of each arm is in order. Cavalry and engineers may be discussed briefly, but infantry and artillery deserve a more detailed explanation. Because the U.S. division was only just developing during the period 1911 to 1917, it is omitted from this discussion.

Cavalry had the greatest mobility in the days before motor vehicles and was therefore closely associated with functions requiring speed and movement. Traditionally, cavalry had three missions: reconnaissance and security before the battle, shock action charges on the battlefield, and pursuit after the battle. The increases in firepower during the later 1800s led many tacticians to suggest that shock action was no longer a feasible role except under rare circumstances. They argued that, because charging against rifled weapons seemed suicidal, cavalry should be reequipped as dragoons or mounted infantry. The mounted arm could then continue its reconnaissance and security mission while still functioning as highly mobile infantry that dismounted to fight as soon as it made contact with the enemy. Cavalry actually operated in this manner during the American Civil War, the Boer War, and the Russo-Japanese War. By 1914 the British and German Armies had equipped their cavalry with machine guns and trained them to fight dismounted when necessary.

Yet the desire to retain cavalry's mobility in reconnaissance, security, and pursuit caused many cavalrymen to prefer mounted fighting whenever possible, despite the large target a horse and rider presented to the enemy. In colonial warfare, the British Army found that cavalry was still useful in its traditional role, charging at the battle of Omdurman in 1898. A thundering mass of horses and riders could still demoralize and scatter undisciplined troops. Another factor, social conservatism, helped preserve the traditional cavalry of lances and sabers in most armies. In addition, defenders of cavalry shock action justified their views by citing one charge of the Franco-Prussian War, an action appropriately known as "von Bredow's death ride." At the battle of Vionville-Mars-la-Tour on 16 August 1870, Major General von Bredow led a Prussian cavalry brigade down a depression until it was within a few hundred meters of the left flank of the French VI Corps. The French had already suffered from artillery fire and were not entrenched when Bredow charged out of the smoke. The charge achieved its objective, overrunning the French guns. Yet during an attack that took less than five minutes and produced only a momentary tactical advantage, 380 of the 800 Prussian cavalry were killed or wounded.[12]

Of the four combat arms, engineers were the most neglected in doctrine. They generally operated in very small units, dedicated to assisting an army's mobility through bridges and other projects or in countering enemy mobility by constructing obstacles. Because of these missions, engineers were often the only troops trained in the detailed construction and destruction of obstacles and field fortifications. Moreover, as the technical experts of the army, engineers often performed specialized tasks such as maintaining weapons or equipment.[13]

With respect to infantry, a rifle battalion before 1914 was just that—four companies of rifle-armed infantry plus, in most cases, two heavy machine guns. The rifles in question—the British Lee-Enfield, Austrian Mannlicher (fielded in 1886), French Lebel (1886), and German Mauser (1889)—seem primitive by today's standards. Yet in the hands of trained soldiers, these bolt-action, magazine-fed rifles could put out six or more aimed shots per minute, an unprecedented volume of fire that seemed to obviate the need for other weapons. As a result, these battalions lacked the many indirect-fire weapons, such as grenades and mortars, that we today associate with the word "infantry."

To some extent, armies neglected specialized weapons because of the additional training they required and because machine guns and mortars were too heavy to keep pace with riflemen advancing on foot. Machine guns were usually cast in an economy-of-force role, covering open flanks or other areas where riflemen could not be spared. On the eve of war, the Germans began to double the number of machine guns in their highest priority divisions, but they did not display any unusual understanding of how to integrate such guns with the riflemen. In Britain, the first light, portable Lewis automatic rifles were tested for aircraft mounting but not to accompany the infantry.[14]

The Offensive

Once an infantry battalion left the railroad line and advanced to contact the enemy, it was neither more mobile nor more protected than infantry in the eighteenth or nineteenth century. The firepower of rifles and machine guns had outstripped the mobility and survivability of infantry. As everyone discovered in the fall of 1914, the only immediate remedy was to entrench, using earth to stop enemy bullets. All professional soldiers had been aware of this problem before the war, but they regarded defensive firepower as a costly obstacle that had to be overcome by a highly motivated attacker. Attacking infantry was expected to forego protection in order to maximize its own firepower and mobility.

In order to explain this belief, we must consider the war that professional soldiers expected to fight in 1914. The Wars of German Unification had provided models of short conflicts won by decisive offensive action. Over and over during the summer of 1870, the better-trained and better-armed French infantry had taken up carefully selected defensive positions, only to be outflanked and driven back by determined and costly German attacks.[15] Thus, many professional soldiers concluded that standing on the defensive was a sure road to defeat. In any event, no one believed that a war that mobilized the entire manpower of a nation could endure for more than a few months. War in 1914 meant that an entire economy halted while the reserves mobilized and fought; even the workers in the munitions factories were recalled to the colors. Under such circumstances, societies and economies would collapse if the war dragged on.

This belief in a short offensive war determined many of the tactical expectations of European soldiers. With few exceptions, they did not anticipate assaulting prepared fortifications across open fields. Instead, most soldiers envisaged a series of encounter battles in which neither side had time to entrench.[16] Each commander hoped that his cavalry reconnaissance screen or his infantry advance guard would locate an enemy weak point, a place that he would attack immediately to develop the situation and force his opponent onto the defensive. The attacker's artillery would then act to pin down and isolate the defender, preventing reinforcement or serious defensive entrenchment.

Meanwhile, the attacking infantry would approach the hastily entrenched enemy, preferably by maneuvering to an open flank. The goal was to infiltrate to within 400 meters of the defender by using all available cover and concealment. During the Balkan Wars of 1912–1913, Serbian and Bulgarian infantry had infiltrated as close as 200 meters from the enemy before opening fire. Most soldiers considered this to be an exceptionally successful movement, however.[17] Once the defender engaged the advancing infantry, the attacker would reduce the target he presented by spreading out into a series of skirmish lines. The desired density of these lines varied between armies and over time, but soldiers generally moved in thin lines, with one to three meters between each man. Because of the recognized strength of the defender's firepower, skirmishers would advance by fire and movement, one group lying prone and providing covering fire while another group rushed forward for a short distance and then threw themselves onto the ground. The covering fire would (supposedly) keep the outnumbered defender hugging the earth, unwilling to aim his own rifle. Then the roles of the different groups would be reversed, leapfrogging toward the enemy. The size of each group and the distance covered by one rush would both become smaller as the attacker closed with his opponent. Enemy fire would intensify as the attacker found cover more sparse. Casualties were expected, but supporting troops would fill the resulting gaps in the skirmish line. Prewar machine guns were too heavy to accompany the advancing skirmishers, so these guns were usually assigned to provide fire support from the rear. The defender would be outnumbered and isolated and would gradually feel overwhelmed by the threat advancing toward him. Eventually, the attacker expected to arrive within a short distance of the defender, keep the defenders' heads down by a superior volume of rifle fire, and assault with the bayonet.

With certain variations, most armies before 1914 shared this conception of warfare. It had a number of problems that are obvious in retrospect but that were not so evident at the time. First, the attacker assumed that he would have numerical superiority over the defender. Yet the numbers of troops fielded in 1914 were so similar that numerical superiority, even at specific points, was difficult to achieve.

Second, this scenario assumed, perhaps unconsciously, that the enemy and friendly forces were operating in a vacuum, moving to contact against each other with their flanks open for envelopment. In practice, however, the density of forces along the French, German, and Belgian frontiers in 1914 was so great that anyone seeking to maneuver to the flank was likely to encounter another unit, either friendly or enemy. Open flanks did occur, notably in the battles of the Marne and Tannenberg at the end of August, but these were exceptions caused by faulty command decisions on a battlefield that was still fluid. It is also worth noting that, in each case, the attacker was able to exploit these mistakes because of superior mobility *off* the battlefield—Parisian taxicabs moving the French infantry to the attack, and East Prussian railroads switching the German Eighth Army to outflank the Russians.[18]

Third, professional soldiers had not yet fully assimilated the new realities of firepower. In reaction to the new lethality of the battlefield, many soldiers tended to overemphasize the importance of morale, of the attackers' offensive spirit. The goal of the attack was not so much to destroy the enemy by fire but to break the morale of the opposing commander or of his troops. Some of this emphasis was appropriate, because the attacker needed great confidence and discipline to advance in the face of withering enemy fire. The danger, however, was that commanders might believe that morale was more important than firepower and might insist on discipline at the expense of individual initiative on the battlefield. In the context of nineteenth-century Social Darwinism, the ability to advance despite massive casualties was subconsciously considered to be a demonstration of national toughness and superiority. This overemphasis on the morale factors of warfare, a fallacy Tim Travers has termed "the Psychological Battlefield," was widely held by British, French, Russian, and even German officers.[19]

In turn, the psychological approach to battle led to a fourth problem with prewar doctrine: many professional soldiers considered their subordinates incapable of executing the official tactics of the day. The kind of battle envisioned seemed to depend on three aspects: excellent training, high morale, and firm control. Unfortunately, the system of short-term conscription followed by long service in the reserves meant that a majority of troops in many units would be untrained or unprepared for the rigors of battle. A French first-line infantry company, for example, had a wartime authorized strength of 225 enlisted personnel, of which 65 percent were reservists or first-year conscripts.[20] According to many observers of peacetime maneuvers, these reservists and conscripts demonstrated that they lacked the training, discipline, and physical stamina necessary to conduct dispersed fire-and-movement tactics under heavy enemy fire. Lying prone in the weeds, a soldier would feel alone and would require great self-motivation to continue the attack. Professional soldiers argued that, once these reservists and trainees took cover for the first time, they would never stand back up and advance again. This belief, correct or not, led French, Russian, Austrian, and other officers to attack standing up in relatively dense formations. Such formations were not only easier to control, but they also gave the troops a sense of unity, an illusion of strength in numbers. The officers recognized the risks of attacking in this manner but felt that there was no other way to achieve the necessary rapid victory with undertrained personnel.[21]

Although German officers also tended to think in terms of rigid discipline, the German Army of 1914 had minimized the reservist problem that faced its opponents. Just as the German officers and sergeants were trained to a high standard, German conscript training was also extremely demanding, aiming to make a trained soldier during the first year of service so that he was fully prepared for the annual fall maneuvers. Moreover, the Germans organized their troops into a three-tiered system of units, consisting of twenty regular army corps, fourteen reserve corps composed of regular cadres with large numbers of reservists, and numerous smaller Landwehr or militia formations. In the regular army corps, most reservists had left active duty only one year before and knew they would be recalled to the same regiment for annual maneuvers in the fall of 1914. Such units had a considerable advantage in training and coherence over their opponents, because officers and men knew each other well. Yet even the German reserve and Landwehr units were well trained. Indeed, one of the great surprises for France in 1914 was the German willingness to use such "low priority" cadred formations in the line of battle immediately. Prewar French intelligence estimates of enemy strength had ignored these reserve units.[22]

British officers were just as prone to this mistake as were their continental counterparts. However, at least some of the reservists in the British Expeditionary Force (BEF) had extensive active military experience. Moreover, the BEF fought its first battles on the defensive, where there was less demand for such suicidal bravery.

Both the German and British Armies, however, suffered heavy casualties in the initial campaigns. During the fall of 1914 and spring of 1915, they formed new divisions from half-trained patriotic volunteers; they were then used in rigid attacks that repeated the suicidal French tactics of August–September 1914.

Given the emphasis in all armies on the meeting engagement and the hasty attack, prewar training often neglected the defense. The Germans constructed field fortifications during their annual maneuvers, but their defensive doctrine focused on rigidly holding a single trench by filling it with large numbers of troops that would soon become the targets for enemy artillery. French defensive doctrine, as reflected in prewar engineer manuals, called for a defense-in-depth, with a forward position to delay the enemy, a main line of resistance, and a second position to limit a successful enemy penetration.[23] Ironically, these doctrines became reversed by 1915. The same belief in a psychological battlefield motivated the French and British to defend their forward-most positions in a rigid structure, refusing to give ground without massive casualties. By contrast, the outnumbered Germans were beginning to develop a defense-in-depth (see chapter 2).

Artillery

If infantry had difficulty adjusting to the requirements of the new firepower, artillery was even slower to react. The traditional tactic for artillery, as perfected by Napoleon, was to concentrate the guns in a direct-fire role, placing them between or a few hundred meters behind the infantry units they were supporting. In this position, the gunners could see clearly to hit their targets, and the massed fire had a psychological effect on friend and foe alike. This tradition of direct-fire support meant that by 1905 all armies had standardized on relatively small, highly maneuverable field guns with flat trajectories, designed to hit targets that were in view of the guns rather than to shoot over intervening hills. The French 75mm-diameter field gun, the German 77mm, the American and Russian three-inch (76.2mm), and the British eighteen-pounder (83.8mm) were designed for this role. Larger weapons were too heavy for a standard team of six horses to pull across country. These guns were too small to have much effect against even hasty field fortifications, and they lacked the higher trajectory necessary for indirect fire, that is, fire over the horizon to a target that was not in direct view of the guns. This was perfectly satisfactory to the French. In preparation for an infantry attack, French commanders relied on an extremely rapid rate of direct artillery fire to suppress the enemy's defensive fire momentarily— causing the defenders to take cover while the attacking infantry advanced— rather than use it to destroy the enemy troops themselves.[24] The volume of such fire was intended to force the defender to remain under cover, unable to provide effective aimed fire from his own weapons even if he were not wounded by the French shells. The colonial wars of the nineteenth century had encouraged the British to believe in a similar suppressive function. That same experience had also prompted the British Army to maintain a much higher proportion of artillery than that in French divisions, because British infantry had discovered the value of such fire support.[25]

The Boer War, and especially the Russo-Japanese War, provided a glimpse into the future, with trench systems and the skillful use, particularly by the Japanese, of indirect-fire artillery. Many professional soldiers dismissed these conflicts as minor wars fought at the end of long supply lines and having no useful lessons for a future war in Europe. Yet observers of the Russo-Japanese War, especially those from the German Army and the British Royal Artillery, were impressed by the necessity for indirect fire. Rather than locate the guns in an open field where they were exposed to enemy counter-fire, these forward-thinking artillerymen wanted to locate the guns behind hills, shooting over them to hit targets designated by forward observers. The rest of the British Army, however, insisted on having close direct-fire support and believed simplistically that massed firepower was accomplished only by concentrating the guns forward on the ground. Thus, the British in 1914 fell between two chairs: they possessed an assortment of weapons but no clear doctrine on how to employ them.[26]

The German Army, by contrast, conducted a serious study of indirect-fire techniques and equipment. Beginning in 1909, the Germans increased their indirect-fire capability by converting one battalion in each division to larger, 105mm howitzers and by adding a battalion of 150mm howitzers to the artillery assigned to each corps of two divisions. Frequently, two batteries of these larger howitzers would be subordinated to each division artillery commander. Such howitzers could be concealed behind hills because they had a higher trajectory, a greater angle of fire than conventional artillery guns, with an effective range of 7.5 kilometers. By contrast, the French 75mm field gun had a range of 4.0 kilometers. Howitzers were also, by their design, much lighter in weight and more accurate in indirect fire than were field guns of the same caliber.

Nineteen fourteen found the various armies in varying stages of response to the changes in artillery. Russian artillery officers had learned the need for indirect fire to protect their batteries from enemy counterbattery fire. Unfortunately, the limited Russian defense budget and weapons industry meant that Russian army divisions began the war still armed with the 76.2mm gun; the artillery assigned to a typical Russian corps of two to three divisions included only a single battalion of 122mm howitzers.[27]

The French artillery also recognized the need to place their guns in defilade, protected from direct enemy observation and fire. Yet the French, like the British, subordinated their artillery to infantry commanders, who continued to believe in direct-fire support of the assaulting infantry. Moreover, the French were so convinced of the inherent superiority of the 75mm gun that they were slow to obtain larger pieces and even then produced flat-trajectory guns instead of the more accurate, higher angle, howitzers. By 1914 the German Army had 3,500 artillery pieces with calibers of 105mm or larger, including many howitzers and large siege mortars; France had only 300 modern guns larger than 75mm. These larger guns were usually assigned to corps or higher level headquarters, who retained control of them instead of attaching them to the forward divisions.[28]

The British and French artillery paid the price for the continued insistence on supporting the troops in a forward, direct-fire mode. At the battle of Le Cateau, on 26 August 1914, a British artillery commander ordered seven batteries to deploy in the open, in clear view of the Germans, in order to support the British infantry. German artillery, firing from hidden positions, massacred the British gun crews and horses, with twenty-five out of forty-two guns being abandoned to the enemy. The French fared somewhat better, generally putting out a large volume of fire in 1914. Still, the French habit of subordinating these guns to infantry commanders meant that their artillery fire was often less flexible and less effective than that of their opponents.[29]

The small caliber and limited number of guns involved in most of the lesser wars at the end of the 1800s meant that no one was prepared for the devastating effects of massed, large-caliber artillery fire on the battlefield. To complicate matters further, in the nine years between the Russo-Japanese War and the start of World War I, a final technological change occurred in the explosive charges contained in artillery rounds. The experiments of Alfred Nobel and others gave all armies high explosive rounds that were much more destructive than the artillery shells of the nineteenth century.[30]

Thus, at the outbreak of World War I, cavalry in most armies had not yet adjusted to the realities of the new technology, and infantry and cavalry were in the midst of change. Infantry commanders had at least a partial understanding of the lethality of the battlefield but doubted their ability to execute the relatively sophisticated fire-and-movement tactics of the day. Perhaps most significantly, none of the combat arms had trained for truly close cooperation with the others, an oversight that proved disastrous in 1914. The most obvious example of this parochialism was the standard method of describing the size of the army in the field. Instead of counting combined arms divisions, or even single arm regiments, the average professional soldier described any force in terms of the numbers of rifles, sabers, and guns— the separate weapons of the three principal arms.

CHAPTER 2

• • • • • • • • • • • • • •

World War I

The defensive firepower of indirect artillery and machine guns dominated the battlefields of 1914. From the very first engagements, commanders had to restrain the "impetuosity" of their troops and insist upon careful preparation by the engineers in the defense and artillery support in the attack. All armies discovered the fallacy of their prewar beliefs in the need for officer control and the necessity of massed rifle fire.[1] The French and British were shocked by the vulnerability of their exposed infantry and artillery to carefully sited German machine guns and artillery. The Germans, in turn, were surprised by the accuracy and rapidity of British riflemen and British and French field guns. By the end of 1914 this firepower had forced the opponents to create a continuous line of foxholes and hasty trenches from Switzerland to the North Sea. Thereafter, every attack was of necessity a frontal attack, seeking to break into and through these trenches.

The stereotype of trench warfare did not appear overnight. On both the Eastern and Western fronts, the battles of August–September 1914 were characterized by considerable maneuver and fluidity. Under the right circumstances, prewar infantry tactics worked effectively. At 0430 on 8 September, for example, the infantry of the Prussian Guard Corps infiltrated forward and, in a surprise attack without artillery preparation, overran the positions of the French XI Corps.[2] On the Eastern Front, the German Eighth

Army surrounded and destroyed the entire Russian Second Army by a double envelopment. In fact, the Eastern Front was never as immobile as the Western, because of the greater frontages involved and the inability of Russian industry to produce enough weapons and obstacles to re-create fully the trenches of the west. Thus the German Army was to some extent able to continue its prewar tactics in the east. Yet this fluidity produced indecisive results until first the Russians and then the Austro-Hungarians became exhausted and demoralized by attrition.

Given these examples of maneuver, many commanders on the Western Front regarded the thin line of entrenchments in 1914 as an unnatural and temporary pause in the war. British and French commanders spent most of the war seeking the means to penetrate and disrupt the German defenses in order to restore the war of maneuver. Because the Germans concentrated most of their efforts on the Eastern Front during 1914–1916, they conducted an economy-of-force defense with relatively few attacks in the west. In order to understand the nature of World War I tactics, therefore, we need to examine the problem of Allied attacks and, then, the development of German doctrine to defeat such attacks. The solutions to both problems involved greater cooperation between arms than had previously existed on either side. Moreover, the nature of infantry changed fundamentally, and different combat arms made their appearance.

Artillery and Cooperation

Once the infantry attacks failed and trench warfare became the reality of combat, the opposing generals had three options, although this choice was not evident at the time. First, an army could increase the mobility and firepower of the infantry. Second, armies could greatly increase the volume of artillery fire to create a gap through which the infantry could advance. The final option would have been to increase the cooperation between the different combat arms. The British and French rapidly gave up any idea of combining artillery fire with infantry maneuver and instead concentrated on the second option, trying to achieve overwhelming destruction with artillery fires before the attack. In the mind of Douglas Haig, the British field commander for most of the war, the battle fell into three phases, each one to be fought by a separate combat arm—the artillery would prepare the battlefield, the infantry would seize the enemy trenches, and then the cavalry would exploit the resulting penetration. In practice, many British and French commanders interpreted the new techniques as "the artillery conquers, the infantry occupies."[3]

Artillery conquest was not easy. Everyone had expected a short war, and thus few armies had sufficient supplies of ammunition and heavy artillery to conduct the massive preparations necessary to demolish even temporary field fortifications. In both Britain and Russia, scandals arose over the long delays necessary to produce more ammunition and guns. Even when France began to produce more guns, the first models of medium and heavy field guns had extremely slow rates of fire, and the continued preference for guns over howitzers made accuracy at long ranges even more difficult than it might otherwise have been. Meanwhile, the more rapid 75mm gun, the prewar standard weapon, had such a short range that it had to move well forward and displace frequently behind the advancing troops in order to destroy or even disrupt German defenses-in-depth.[4]

After initial difficulties, the Germans continued to expand their inventory of divisional and nondivisional howitzers and thus were able to conduct short, whirlwind bombardments designed to disrupt and suppress the defenders before the infantry attack. Because of their shortages in artillery, however, the Allies had to fire smaller numbers of guns and howitzers for longer periods of time in order to achieve the same destructive effect. Such prolonged bombardments not only wore out the gun tubes but also deprived the ultimate attack of much of its surprise.[5]

Adding to the Allied problem was the fact that most gunners had little experience in precision indirect fire. Many of the procedures that are commonplace to artillerymen today were developed painfully, against much resistance, during the period 1914–1917: establishing forward observer techniques, measuring and compensating for the effects of weather and worn gun barrels, and using ammunition from the same production lot to ensure that successive volleys fell in the same general location. The first French regulation describing such procedures was not published until November 1915. The British Royal Field Artillery needed new, carefully measured maps of the entire area of northeastern France before it could establish a grid system for surveying battery locations and adjusting indirect fire. The fledgling air services of the belligerents had to provide aircraft for photographic mapping and both aircraft and balloons to tell the artillery batteries where their shells actually landed. Eventually, improved radiotelegraphs allowed aerial observers to talk to the artillery-fire controllers.[6] Such developments took most of the war to reach perfection.

Quite apart from the technical problems of indirect fire, there were numerous problems of artillery command and tactics. First, the cavalry and infantry commanders of most armies regarded artillery as a supporting arm and therefore saw no reason for an artillery commander or fire-support coordinator above the level of a division. This attitude was particularly true of the British Army, but even the Germans did not habitually appoint artillery commanders above the division until 1916. Without such a senior gunner, artillery planning and tactics varied greatly among units, and innovative ideas were ignored.

Artillery tactics therefore developed only by trial and error. Senior commanders did not understand the importance of counterbattery fire, of consciously locating and attacking the enemy's artillery. As late as 1916, British staff officers and commanders continued to call for large numbers of shrapnel shells, in which an explosive charge threw pieces of metal outward from the point of detonation. Such shells were effective against troops in the open, but trench warfare required more high explosive shells to destroy barbed wire and fortifications.[7] Repeatedly, British and French commanders overestimated the effects of their massive artillery bombardments, confusing destruction of German trench positions with destruction of the German soldiers in those positions. Thus, the attacking infantry often encountered intact barbed wire obstacles defended by enough surviving German machine gunners to exact a considerable price.

Coordination between infantry and artillery was especially difficult. The two arms were physically separated on the battlefield, the infantry in trenches and the artillery deployed behind hills several kilometers to the rear. The first deliberate attacks conducted by the British and French during late 1914 and early 1915 were particularly difficult to control, because both artillerymen and commanders lacked experience in indirect fire. The easiest procedure seemed to be the establishment of a series of phase lines, with artillery aiming at the far side of a phase line while all infantry remained on the friendly side. The troops could not advance beyond that phase line until the commander, following a fixed timetable, directed the artillery fires to shift forward to a new phase line. Unfortunately, as soon as the artillery fire shifted, the surviving defenders hurried out of their bunkers to engage the attacking infantry.

Such phase lines encouraged commanders to ignore the terrain contours to their front and the possibility for maneuver and to favor instead simple advances by all units on line. This in turn discouraged massing of artillery or infantry at critical points, while the linear formations of the infantry blocked any fields of fire for supporting machine guns. More important, there were no effective communications procedures that would allow the leading infantry units to talk to their supporting artillery during the assault. In the Champagne campaign of 1915, the French went to the extreme of sewing white cloths on the backs of their soldiers to help observers determine the forward progress of troops, but casualties from friendly fire still occurred. The Germans experimented with colored flares and signal lamps to communicate between infantry and artillery, but such signals were often difficult to recognize amid the destruction of battle.[8]

Beginning with the battle of the Somme in July 1916, Allied artillery was able to provide a rolling barrage of shrapnel that could advance at a steady rate of speed. The use of shrapnel instead of high explosive made it safer for the infantry to advance close (40 to 100 meters) behind the artillery barrage, because the destructive power of shrapnel was focused forward along the line of flight of the shell. Yet shrapnel had almost no effect against well-prepared positions—the best it could do was force the defender to stay under cover during the assault. Moreover, there was still no means for the infantry to adjust the rate at which the rolling barrage moved forward. The rigid forward movement of artillery barrages often outran the heavily laden infantrymen struggling across the shell-pocked battlefield, allowing the defender time to leave his shelter and engage the attacker after the barrage had passed over a trench.

Command, Control, and Communications

The problems of infantry-artillery coordination were only one aspect of the greater issues of command, control, and communications that plagued a World War I commander. The huge scope of offensives and the scarcity of trained, competent staff officers at lower echelons of command meant that most operations were planned at the level of corps or field army headquarters. Artillery timetables, attack times, objectives, phase lines, reserves, and so on were controlled at a high level, although within those constraints the division and lower headquarters still planned the details. This was particularly true in the British (and later American) Army, which underwent massive expansion during the war; in these armies qualified general staff officers, the essential experts on planning and coordination, were few and far between. Given the crude nature of artillery procedures for much of the war, artillery planning and control were also centralized at a high level. Thus, even when forward observer procedures had improved, commanders retained personal direction of the artillery fires. It was impossible for commanders in headquarters behind the lines to evaluate the flow of battle, hence the difficulties of adjusting artillery fires to support the infantry. The demand for staff officers also drained the British, American, and French combat units of some of their most innovative junior leaders, but their counterparts in the German Army tended to remain with the troops for longer periods of time.[9]

Compounding this problem of centralized control was the inflexible nature of tactical communications systems. Although primitive radiotelegraphs were used even in 1914, radios were bulky, unreliable, and generally suspect because of the possibility that enemy intelligence units would also intercept radio messages. Instead, the basic means of communication during

World War I was the field telephone. In rear areas, this system worked relatively well, except when, as in 1918, higher headquarters were forced to relocate themselves suddenly over long distances and had to restring their telephone wires. At the front, field telephone lines, however carefully buried, were vulnerable to artillery fire, causing frequent breakdowns in communications below the level of a division headquarters. Moreover, when the British and French infantry advanced, they had to leave their telephones behind in their trenches. Thus, senior commanders often insisted that brigade and even battalion commanders remain behind in the friendly trenches; these midranking officers maintained telephone contact with their superiors but were themselves out of touch with the realities of battle.[10] Meanwhile, the advancing infantry, often commanded only by junior officers, struggled forward across the battlefield. Each time these troops reached an objective or phase line, they had to stop and request permission to continue the advance or to commit reserve forces. A messenger had to hand-carry the request under fire back to the lowest headquarters (usually a brigade, regiment, or division) where the field telephone circuits had survived enemy counterfire. These circuits then relayed the report through different levels of organization in order to receive a decision from the corps, field army, or higher commander in charge of the offensive. Often the higher headquarters had to wait hours to receive detailed reports from various sectors and then decide whether it could safely reinforce a success in one sector or if it would have to wait until all the attacking units had reached the objective. Even after a staff estimate had been made and the commander's decision announced, this communications process had to operate in reverse before the troops could advance or reserve units could move up. For example, at the battle of Neuve Chapelle on 10 March 1915, one of the first concentrated artillery preparations of the war destroyed most of the shallow German defenses. The forward British troops, however, had to wait at a phase line for seven hours before they received authorization from their corps commanders to continue the advance. During this delay, the Germans were able to move in reserves and reestablish a defense in the very path of the British advance.[11] Once the momentum of an attack was lost in this manner, it was difficult to organize a renewed attack. Yet senior commanders, elated by initial success, would continue to launch follow-up attacks, often suffering much greater casualties than the initial advance had incurred.

To some extent, these communications problems were a product of the technology of the time. Senior commanders and staff officers could not command close to the front even if they wished to; the battlefield was simply too large for personal observation. Instead, higher headquarters were tied to the field telephone system that brought all information to them and transmitted all orders forward. Under these circumstances, subordinate initiative and rapid exploitation were risky, if not disastrous. If successful attackers did not coordinate with higher headquarters, they might well fall prey to their own artillery support.

By 1918 improvements in artillery techniques and communications made such initiative more practical and facilitated the efforts of higher commanders to track the flow of battle. The Australian general Sir John Monash, for example, developed an elaborate system to determine the forward progress of his forces. Advancing troops carried specially colored flares, and a detachment of aircraft did nothing but spot the location of these flares, write out reports based on the locations, and air-drop the results to Monash's headquarters. This system gave a corps commander the forward trace of his forces with a delay of twenty or fewer minutes, provided he had air superiority.[12]

Yet the problems of command, control, and communications created habits that lingered long after 1918. The British and French Armies, in particular, became accustomed to centralized control with methodical planning and long delays in reports from the front line. As late as 1940, French commanders were expected to remain in their headquarters, where they could control their subordinates through wire communications, grasping the "handle of the fan" rather than leading from the front.[13] In turn, this methodical, centralized approach to command and control later spread to the armies of former British colonies, including Egypt and Iraq. The result was a long decision cycle, a cycle that worked superbly unless the opponent acted more quickly, derailing the plan.

The Problem of Penetration

The problems of indirect artillery fire and of command, control, and communications were two aspects of the crucial tactical question of World War I—how to achieve and exploit a penetration more rapidly than the defender could redeploy to prevent or seal it off. In most instances before 1918, armies learned how to break into an opponent's defenses but were unsuccessful in breaking through and out of them to exploit into the enemy rear areas.

Consider the abstract diagram of a fully developed trench system (see Fig. 3). In order to advance, one side began by attempting to neutralize the defensive fire of the enemy's trenches and artillery batteries. As early as the battle of Neuve Chapelle the British had demonstrated the possibility of achieving such a penetration by concentrated or prolonged artillery fire. This artillery preparation could not, of course, kill all the defenders nor destroy all their positions. Eliminating the barbed wire and similar obstacles was

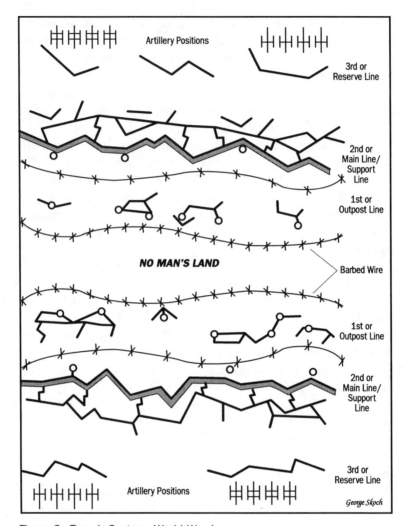

Figure 3. Trench System, World War I

especially difficult. The metal fragments of shrapnel had little effect against wire; prewar fuses for high explosive rounds were insufficiently sensitive to detonate when they encountered the slight resistance of barbed wire. By 1917 the British had developed the instantaneous Model 106 fuse that would detonate with sufficient rapidity to destroy wire.[14] Of course, some defenders survived almost every bombardment, so that the infantry still had to fight to take its objectives. Yet even the Germans conceded that artillery followed by infantry could always capture the first and even the second trench lines, especially if a short artillery bombardment and good operational security maintained surprise.

The problem came when the attacker tried to develop and exploit the resulting partial penetration. After massive artillery bombardments, the defending trenches often disappeared into a moonscape of muddy shell holes and debris. The physical exertion necessary to struggle across this terrain, coupled with the casualties inflicted by the surviving defenders, meant that the assaulting infantry was usually exhausted after seizing one or at most two lines of the enemy defense. Fresh battalions would have to continue the attack, while the attacking artillery would have to move forward in order to extend its range, aiming at the defender's third line and artillery positions. Even after the senior commander learned of success, decided to exploit, and communicated his orders, his troops, guns, and supplies had to move across the intervening No Man's Land and captured trenches. In most cases, by the time the attacker had completed this displacement, the defender had been able to bring up reserves and establish new trench lines in front of the attacker. The defender's role was much easier, because his reserves could move by railroad and motor truck while the attacker's forces toiled forward over the broken ground of the battlefield. Moreover, the defender could counterattack and pinch off any penetration that did not occur on a broad frontage, because the newly captured area would be exposed to defensive artillery fire from all the guns within range.

Even if the attacker moved faster than the defender and actually penetrated through existing trenches and gun positions, the second-echelon infantry would again be tired, out of range of artillery support and communications, and essentially restricted to foot mobility. Thus, another passage of lines to bring up still more reinforcements would be necessary. In theory, this was the stage when horse cavalry could use its greater mobility to exploit, although in practice a few defending machine guns could delay such exploitation significantly.

The timing of the decision to exploit and the problems of mobility across No Man's Land remained major obstacles for any attacker seeking to achieve a breakthrough. Various solutions were tried. Some artillery batteries secretly moved forward prior to the battle and camouflaged themselves just behind the friendly first-line trenches, allowing artillery support a slightly deeper range. Attacking brigades and regiments developed a system of leapfrogging, with second-echelon battalions passing through the attacking battalions to sustain the advance. Ultimately, however, the point would be reached where the attacker's advantages of artillery preparation and, if possible, surprise, were canceled by the defender's advantages of depth, terrain, and operational mobility.

Of course, these problems could be minimized if the attacker did not try to achieve a complete breakthrough in any one attack but settled for capturing a limited objective. Meticulous planning and preparation would allow such a surprise attack to succeed within the limits of artillery range and of command and control capabilities, after which a new defense would be organized to halt the inevitable counterattack. French commanders such as Philippe Pétain were particularly noted for using this technique during 1917–1918, after French morale had been shattered by too many blind frontal attacks. Such a set-piece battle certainly improved morale and could achieve a limited victory at low cost; it could not, however, break the stalemate and win the war. In practice, the commanders at different levels were often unclear as to whether the objective was breakthrough or limited success, causing further confusion. Ultimately, a combination of attrition, new weapons, and new infantry tactics was required to achieve the elusive victory.

Flexible Defense

While the British, French, and later the Americans sought to solve the mystery of the penetration, the Germans gradually perfected their defenses against it. This evolution of German defensive doctrine was by no means rapid or easy, but the result by 1917 was a system of flexible defense-in-depth that not only hindered enemy attacks but also developed the capabilities of the German infantry.

At the beginning of the war, senior commanders on both sides emphasized a rigid defense of forward trenches. As the cost of taking ground increased astronomically, it seemed treasonous to surrender voluntarily even one foot of precious soil to an enemy attack. Moreover, many commanders, inspired by the psychological approach to battle, believed that creating defenses-in-depth and allowing units to withdraw under pressure would encourage cowardice; troops expecting a retreat would defend their positions only half-heartedly.[15] Only gradually did German leaders realize that massing their forces in the forward trenches was suicidal; the artillery bombardment before a French or British offensive eliminated many of the defenders in them, increasing the possibility of enemy penetration. This was most obvious at the battle of Neuve Chapelle, when the single line of German trenches disappeared under the weight of a British bombardment, leaving nothing but a string of concrete pillboxes behind the lines to block the British advance until reinforcements arrived.

Beginning with the shock of Neuve Chapelle, Germany gradually evolved a system that by 1917 included up to five successive defensive positions, one behind the other, in critical sectors. The first two or three of these positions were located on reverse slopes, that is, on the far side of a ridgeline, hidden from the attacker's view. This not only complicated the task of adjusting enemy fire on those trenches, but it meant that the attacking British and French infantry were out of sight and therefore out of communication with their own headquarters when they reached the German defenses. Further, if a German trench on a reverse slope were captured, it would be fully exposed to fire and counterattack from the German rear positions. The forward trenches were beyond the range of enemy light and medium artillery, making them more difficult to reduce.

Quite apart from the choice of terrain, the German defensive system emphasized three principles: flexibility, decentralized control, and counterattack. In terms of flexibility, the forward German positions, those most exposed to enemy bombardment, contained few troops, with perhaps one battalion out of every four in the first two trenches. By contrast, the French put two-thirds of every regiment in these forward lines, with orders to hold at all costs. By 1916 the Germans had decided that trench lines were useful shelters only during quiet periods between battles. Once an enemy bombardment began, the rearward German troops moved into deep bunkers, and the forward outposts moved out of the trenches, taking cover in nearby shell holes. The British and French artillery bombarded the deserted trenches until their barrage lifted and their infantry began to advance. At that point the surviving Germans came out of their shelters and opened fire from the shell holes or from the remains of the trenches. These defenders might not halt the enemy completely, but they certainly broke up the enemy formations, with some parts of the advance moving more slowly than the others. Such an uneven advance presented the centralized British and French command with a dilemma as to whether and where to commit its reserves. Meanwhile, wherever the German strong points delayed parts of the British and French advance, the rolling artillery barrage continued to move forward, separated from the infantry force.

The second aspect of the German system was decentralized control. Squad and platoon leaders had considerable independence and might defend or delay anywhere forward of the third, or main, defensive position. The forward or "front battalion commander" frequently directed the entire defense of a regimental sector. In the mature system of 1917–1918, this battalion commander had the authority to commit the remaining two or three battalions of his regiment in a counterattack at the moment he judged most appropriate. Although the division's artillery remained independent of these local infantry commanders, in practice the artillery liaison officers assigned to them almost always obtained the fire support they requested. Similarly, the division commander in charge of a sector usually had the authority to commit part or all of a second division located in reserve behind that sector.

The German officer corps' spirit of intelligent cooperation allowed it to deal with the situation pragmatically, regardless of the normal chain of command. Thus the German defenses were controlled by two levels of commander— battalion and division—each of whom was familiar with the details of his sector. This system only exaggerated the difference in decision cycles: the British and French attackers had to seek orders and reinforcements from their corps or field army commander located miles to the rear, but the defending German battalion commander could direct a regimental counterattack on the spot.[16]

This, in fact, pertains to the third element of the German defensive tactics: counterattacks at every echelon to retake lost ground before the attacker could consolidate his hold on it. Whenever a major offensive began, the German defenders sought to contain the flanks of the penetration by blocking positions; hasty counterattacks would then pinch off or push back the resulting salient.

Firepower had already forced troops to disperse widely to avoid presenting targets; the German tactics took this tendency to an extreme, creating an "empty battlefield" in which most troops were concealed or located on reverse slopes. On average, there was only one German soldier for every fifty square meters of the battle zone. To support this dispersed defense, the German Army rapidly expanded the ratio of machine guns to rifles in infantry units. By the end of 1916, every infantry battalion had a heavy machine gun company. The Germans also formed separate machine-gun sharpshooter battalions. By January 1918 four light machine guns had been assigned to every rifle company, and their size had been reduced by one-third.[17]

Why did the German Army, rather than the British or French, develop these defensive tactics? To some extent, necessity was the mother of invention: the Germans were not only on the defensive along the Western Front, but they were usually outnumbered until 1918. Further, the German officer corps was somewhat more open to debate and experimentation about the best answers to practical problems. This debate was encouraged by the legendary Gen. Erich von Ludendorff, a longtime advocate of technology and efficiency who in 1916 became head of operations of the German High Command. In January 1917 Ludendorff created a full infantry division whose only function was to experiment with and train German units and commanders in the application of the new tactics. This same collegial, cooperative approach to warfare made the Germans willing to entrust decisions to junior officers and sergeants. By contrast, the British Army had no institutional system for developing tactics and training; many British commanders continued to emphasize centralized planning and strict discipline.[18]

The Germans were by no means infallible, of course. Many German commanders bitterly opposed the flexibility and decentralized control of the elastic defense. For example, at Passchendaele in July–August 1917, the local German commander ordered all outposts to hold in place while awaiting the counterattack. The British thwarted such tactics, however, by attacking for strictly limited objectives. After advancing approximately three kilometers into the depth of the German defenses, the British dug in and successfully blocked the counterattack.[19]

Still, the combination of flexibility, decentralized control, and counterattack at every echelon made the German defensive system almost invincible until attrition and demoralization gave the Allies an overwhelming superiority. Meanwhile, the British and French found it difficult to achieve the same results when they had to defend against German offensives in the spring of 1918. A committee of British generals attempted to adapt the German defensive tactics for British use, but the resulting instructions denied junior commanders any choice about where to defend or when to counterattack. Although the British defenses were laid out in a manner similar to those of the Germans, they failed due to the absence of local initiative and counterattack, coupled with the British determination to defend hard-won ground even when that ground placed them on forward, rather than reverse, slopes. The French fared little better. As early as 8 July 1915, a French order directed that the majority of troops be held in the rear for counterattacks, but the order was frequently ignored. Not until the five German offensives of 1918 did French field commanders learn to array their forces in depth and to accept the loss of lightly defended forward positions.[20]

Technological Change

Like all major conflicts, World War I accelerated the development of new technology. In addition to changes in artillery and communications, a number of new weapons appeared as the result of efforts to solve the penetration problem. None of these efforts was entirely successful, but they all represented additional weapons or tools to be combined with the traditional arms.

GAS. The first attempt to break the trench defense was gas warfare. Although the French had experimented with various noxious gases on a small scale at the end of 1914, it was the Germans who first conducted major gas attacks. The first German test of gas took place in January 1915, at Lodz on the Russian front. Much of the chemical, however, failed to vaporize because of low temperatures and thus had only limited effect. The first use on the Western Front was on 22 April 1915 at the Ypres salient. There a surprise attack routed French colonial troops on a five-mile front, but the Germans were not prepared to exploit their success. They had no significant reserves available to advance before the French sealed the breach. Thereafter, each side found that primitive gas masks and uncertain weather conditions made the existing nonpersistent (i.e., rapidly dissipating) and early persistent chemical agents difficult to employ. When the British first used gas at Loos on 25 September 1915, there was virtually no wind, and the gas moved too slowly or in the wrong direction along most of the front. The British troops advanced into their own gas clouds, suffering more casualties than their opponents. The Germans, for their part, had problems with chemical warfare on the Western Front because the prevailing winds came from the west, often blowing gases back in their faces. Gas warfare became only an adjunct, useful to degrade enemy effectiveness but not to achieve a penetration by itself. The main effect of chemical weapons was to increase the horror, discomfort, and complexity of warfare. By 1917–1918, the most common use of gas was to mix chemical and high-explosive artillery shells during a preparatory fire, in hopes that the low-lying gas would force the enemy out of his deep shelters.[21]

AIRCRAFT. World War I was also the first conflict to have significant air action.[22] Military aviation developed at a tremendous rate during the war but was still in its infancy in 1918. The publicity went to fighter pilots, whose primary mission was to achieve local air superiority. This condition allowed the pilots of the more primitive aircraft of the time to conduct their more basic functions of reconnaissance and artillery-fire adjustment. Even Germany, which established a separate air staff in 1916 and developed a full-blown concept of airpower, recognized the need for reconnaissance and air superiority as well as long-range bombing.

As early as 1914, the British Royal Flying Corps began to require its reconnaissance aircraft to carry bombs and to attack targets of opportunity. The bombs in question were tiny, weighing twenty-five pounds or less, but both sides rapidly realized the extraordinary psychological impact of air attacks on ground troops. Unlike artillery, where the guns were often out of sight, the airplane was a visible and personal enemy, an avenging monster thundering down on the ground troops at the then-incredible speed of 150 miles per hour. For the defense, the presence of friendly aircraft convinced the ground troops that higher headquarters was concerned about their situation. Slowly, the different European powers developed a concept not only of interdicting enemy supply lines but also of close air support, bombing and strafing the enemy's front lines in immediate proximity to friendly forces. Although this practice was still in its infancy at the end of the war, the successful attacks of 1917–1918 almost always included careful planning for air–ground cooperation. Thus, although most of the combatant air services sought to become independent of ground commanders, even the great American theorist of airpower, Billy Mitchell, recognized the need for such close air support.

The different air forces took various approaches to satisfying this need, however. The British Royal Air Force (RAF), and by extension the U.S. Army Air Service, regarded ground support as simply another task for ordinary pursuit or fighter aircraft. No specialized units or equipment was developed until the final months of the war. The result was often very expensive, because the pilots were not particularly well trained, and they suffered heavy casualties while flying low over enemy positions in fragile aircraft. During the last ten months of the war, for example, No. 80 Squadron of the RAF averaged losses of 75 percent of its pilots every month.[23] These casualty rates go far to explain the RAF's postwar rejection of close air support.

By contrast, the Germans, and to a lesser extent the French, developed specialized formations, giving their pilots training and, in Germany's case, modified aircraft for ground support. A limited amount of armor was placed around the cockpits of German J-1 aircraft to protect the pilots from small-arms fire as they flew over the enemy. At the time of the Ludendorff Offensives in the spring of 1918, Germany had thirty-eight *Schlachtstaffeln* (battle squadrons) as support. These aircraft were trained to bomb and strafe in methodical patterns at altitudes below 200 feet.

TANKS. The military motor vehicle also developed from a few primitive automobiles in 1914 to thousands of large trucks by 1918. Although not a battlefield weapon, the truck allowed the rapid movement of troops and supplies between widely separated points. As such, it increased operational mobility as significantly as had the railroad in previous generations. Trucks made it possible to mass suddenly and conduct a surprise attack at an unexpected point or to move reserves to blunt a penetration. They were also essential for stockpiling the ammunition and materiel needed for major offensives.

The tank was originally designed as a special weapon to solve an unusual tactical situation, the stalemate of the trenches. Potentially, the early tank could bring the firepower of artillery and machine guns across the morass of No Man's Land while providing more protection than an infantry unit could carry. Yet the intended purpose of this weapon was to assist the infantry in creating the breakthrough so that cavalry, which had been waiting for the opportunity since 1914, could exploit into the German rear area. It is true that, as early as April 1916, Ernest Swinton, one of the organizers of the British tank force, had considered the possibility of the tanks penetrating all the German defensive lines without waiting for infantry support. However, even Swinton assumed that the standard artillery barrage would still be fired to destroy enemy barbed wire and other defenses. The resulting destruction would inevitably hamper the already slow advance of the tanks. In practice, therefore, the tanks were limited to helping the infantry break into an enemy defensive belt, rather than breaking through and exploiting to the rear. Moreover, even if Swinton's penetration were possible, it would have disrupted the centrally controlled, phased advance of British infantry and artillery.[24] Tanks were simply one more piece of the offensive team instead of a separate arm in themselves.

This purpose must be remembered in order to understand the shortcomings of early tanks. British and especially French heavy tanks had slow speeds, poor mechanical reliability, and great vulnerability to direct-fire artillery once the initial surprise wore off. After all, these new weapons were designed and fielded in great haste and were intended only to advance a few miles and then turn the battle over to the cavalry. Moreover, the great secrecy surrounding tank development, coupled with the skepticism of infantry commanders, often meant that infantry had little opportunity to practice cooperation with tanks. Under such circumstances, the infantry would become separated from the tanks, allowing the German infantry to defeat the two arms separately. Generally speaking, infantry that had the opportunity to train with tanks before battle and to work with them in battle swore by them; infantry thrown into battle without prior tank training swore at them.

On 15 September 1916 tanks first entered combat when thirty-two vehicles, spread out across a seven-mile front, supported a British offensive at Flers-Courcelette. A series of such small attacks dissipated the initial surprise value of the tank, although even early tankers argued that such experiments were necessary to identify and correct design problems. The resulting success was sufficient for Gen. Douglas Haig, the British commander, to request 1,000 tanks for future use. Contrary to later accusations, Haig and other British commanders supported the development of tanks; the problem was to get enough of them and to find the circumstances under which they could best achieve success.[25]

The British Tank Corps did not get such conditions until 20 November 1917, at the battle of Cambrai (Map 1). Using new survey techniques, the British artillery moved into position without firing ranging shots prior to the offensive. The tanks then began to move forward at the start of a very short artillery bombardment, the infantry following in the lee of the tanks. The elimination of the usual long artillery preparation not only achieved surprise but also left the ground more trafficable for tanks and infantry. Four hun-

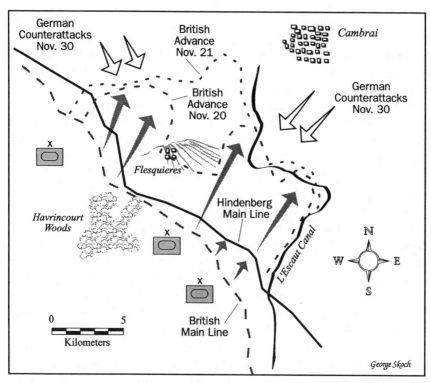

Map 1. Battle of Cambrai, 20–30 November 1917

dred seventy-four heavy tanks in three brigades had practiced extensively with five of the six infantry divisions they accompanied. Tanks operated in sections of three: one tank used machine-gun fire and its treads to suppress the defending infantry, and the other two tanks, accompanied by the British infantry, crossed the trenches. These tactics worked well except at Flesquières Ridge, in the center of the Cambrai sector. Here the commander of the 51st Highland Division, believing that German fire would be focused on the armor, had forbidden his infantry to come within 100 yards of their tanks. Furthermore, the Royal Flying Corps, which had received no advance notice to prepare for the offensive, erroneously reported that it had driven off the German artillery in the area. In fact, one German battery had moved onto the reverse slope of Flesquières Ridge. The British tanks were unsupported when they slowly climbed up the hill and, one by one, topped the ridge. Direct-fire German artillery knocked out sixteen unmaneuverable tanks in a few minutes.[26] This incident convinced many observers that armor could not survive when separated from infantry, an attitude that persisted after 1918, even when tank speed and maneuverability improved. In any event, the available tanks were distributed evenly across the Cambrai front, leaving no reserve to exploit the greatest success.

Moreover, the British had regarded Cambrai as a raid rather than as an attempt at breakthrough and had few infantry reserves available when they achieved an initial success. The usual problems of Allied generals commanding from the rear meant that the Germans rebuilt their defenses before the British cavalry moved forward to exploit the penetration. Ten days after the British offensive at Cambrai, the Germans counterattacked and restored the original front line. In its own way, this counterattack also reflected the latest developments of the war: surprise, colored flares to shift artillery at phase lines, and multiple attacking waves to clear out British strong points bypassed by the first wave.

Even before Cambrai, the Germans had begun to develop an antitank doctrine. In marked contrast to the beliefs of British armor commanders, the German doctrine was more concerned with the psychological effect of tank attacks than with the limited firepower and armor of the primitive models. In 1917–1918, however, the Germans lacked the material resources to compete in tank production. Instead, they relied on tank obstacles combined with existing light artillery pieces (the 77mm guns) and some armor-piercing ammunition for infantry weapons. This special ammunition was effective against early British tanks, and by 1918 the Germans had developed oversized antitank rifles to counter later British models. To combat the terror of tanks, German troops received training on how to defeat them. Where possible, German infantry would wait until the attacking tank had passed, engage the accompanying British infantry, and throw bundles of grenades underneath the tanks. The resulting explosions often disabled a tank by destroying one of its treads.[27]

By the end of the war, tanks were extremely vulnerable unless accompanied by infantry and ground-attack aircraft, both of which worked to locate and suppress antitank defenses. During the first three days of the battle of Bapaume in August 1918, German antitank defenses or mechanical breakdowns immobilized 81 percent of the attacking tanks.[28] Any tank that broke down on the battlefield was almost certain to be knocked out by antitank fire in a few minutes. Even if a German shell did not penetrate the armor, it often jarred loose sharp metal slivers from the inside of the armor plate. Again, such experiences shaped perceptions of its capabilities and roles long after technological change had restored the tank's initial advantage.

Tank tactics as well as weapons varied between armies. The British, who had fielded the first armored vehicles, designed huge heavy tanks intended to advance in front of the infantry, crushing wire and crossing trenches as they went. By contrast, the head of the French armored force, Gen. Jean-

Baptiste Estienne, regarded heavy tanks as essentially self-propelled artillery for destroying strong points. Estienne also fostered the development of tracked carriers to move supplies and fuel across No Man's Land in order to sustain the tank advance. Moreover, the French produced thousands of light tanks designed to escort the assaulting infantry and to suppress any bypassed centers of German resistance. By 1918 the British and Americans had embraced the concept of light as well as heavy tanks.[29] The British Whippet light tank was faster than most heavy tanks (7.5 versus 4.0 miles per hour) but was still hardly a vehicle for rapid exploitation. Light tanks did have one advantage, however; they were much easier to move in secret from one sector to another, because they could be loaded onto trucks instead of moved by rail.

Although the Royal Tank Corps experimented with special armored vehicles in which to transport radios, supplies, and even machine guns, all tank units in World War I were just that—pure tank formations of up to brigade size, intended for attachment to infantry units rather than for independent combined arms mechanized operations.

Gas warfare, aviation, motor transport, and tanks had two effects, apart from those derived from their individual characteristics, on the positional battlefield of 1914–1918. On the one hand, their development further complicated the problem of combining different weapons for attack or defense. This reinforced the tendency for detailed planning and centralized control at a time when infantry–artillery cooperation was still being developed. On the other hand, the army that succeeded in this orchestration had a much better chance of eventually defeating its opponent by attrition, even if true breakthrough was never achieved.

The Resurgence of Infantry

Many of the developments in artillery, gas warfare, aircraft, and armor were based on the supposed inability of 1914 infantry to advance under fire. Once the artillery was dispersed into hidden positions in the rear, infantrymen were even more isolated in the forward trenches. During the course of World War I, however, the infantry of several armies gradually evolved to a point where it had recovered some of its original ability to take and hold terrain on its own. In the process, modern infantry organization was born.

The 1914 infantry battalion was almost exclusively armed with rifles, plus a few heavy and almost immobile machine guns. Once the firepower of the latter became apparent, the tendency in the British, American, and some other armies was to segregate the heavy machine guns into separate battalions, leading to efforts to turn the machine guns into a separate combat arm.

The infantry of various armies sought to increase its own firepower. The first such effort was the trench mortar, an opportunity for the infantry commander to regain control of fire support after the artillery was segregated on the battlefield. Because of its high trajectory, the relatively light, simple mortar could lob a shell from one trench line to an opponent's positions without exposing the firers to direct enemy fire. Mortars had existed as a form of heavy artillery for centuries, but they were generally regarded as obsolete. In 1914 only the Germans had a limited number of large (250mm) mortars *(Minenwerfer);* the weapons belonged to specialized combat engineer units assigned to assault the Belgian fortresses. The popularity and convenience of such weapons soon led to the formation of additional engineer *Minenwerfer* units. By 1916 the typical German infantry division was supported by a *Minenwerfer* company equipped with three calibers of mortars: 250mm, 170mm, and 76mm.[30]

Other armies quickly copied the *Minenwerfer,* although initially the French had to make do with antique mortars from the 1830s. By August 1916 one mortar company supported each French infantry division, and other mortar companies helped suppress German defenses at the start of major offensives. Most significant, in March 1915 the English inventor Wilfred Stokes developed the grandfather of all current infantry mortars, the three-inch muzzle-loading Stokes mortar.[31] Unlike many previous mortars, the Stokes design was extremely simple. The crews could drop shells down the mortar tube as quickly as possible; as each shell made contact with a fixed firing pin at the bottom of the tube, the propellant charge detonated automatically. Such mortars were much easier to manufacture and operate than artillery and were employed extensively in all armies during the war. However, most armies segregated the larger mortars, like heavy machine guns, into specialized units, separated from the infantry.

As early as 1915 the French began to issue other weapons to the infantry, notably the Chachot automatic rifle and the rifle grenade launcher. Such weapons had slower rates of fire and shorter ranges than their counterparts, the heavy machine guns and mortars. The automatic rifle and grenade launcher were much more portable, however, giving the French infantry more mobile firepower and short-range (up to 150 meters) indirect-fire capability.

Such weapons were part of a major change in both organization and small-unit tactics. On 27 September 1916 France reorganized the infantry company to consist of a headquarters, which included communications and combat engineer personnel, plus four platoons of two sections each. Within these twelve-man sections, hand grenadiers, rifle grenadiers, and riflemen were organized around the automatic rifleman as the base of fire. The automatic weapon's fire would pin the enemy down while the rest of the section maneuvered to close with and eliminate him, using grenades and rifles. Such sections no longer aligned themselves in the long, thin lines of 1914. Three of these French infantry companies, plus a company of eight heavy machine guns and a 37mm gun, made up an infantry battalion that modern infantry-men can recognize as such.[32]

The resulting changes in infantry tactics were slow to take root. In May 1915 an obscure French captain, André Laffargue, privately published a pamphlet that suggested a variety of innovations, including not only trench mortars but also so-called skirmisher or sharpshooter groups. These groups, armed with light machine guns, rifle grenades, and hand grenades, would precede the main assault wave by fifty meters. Their mission was to provide covering fire for the main attack and, if possible, to infiltrate through the forward German positions to suppress and outflank German machine-gun posts. The French government distributed but did not endorse this pam-phlet; the British largely ignored it. Not until 1916 did the French officially reduce the density of their skirmish lines to one man every two, and later every five, paces, as opposed to every pace, and integrate the new weapons fully into infantry organization. Meanwhile, the Germans captured a copy of Laffargue's pamphlet during the summer of 1916 and may have added these ideas to their own developing tactical doctrine.[33]

Other armies adopted similar armament and organizations, although there was considerable institutional opposition. For a long time, the British in-fantry battalion remained almost exclusively a series of thin skirmish forma-tions armed with rifles. In their trench raids, conducted for intelligence and morale purposes, the British were highly innovative, but their official tac-tics remained unchanged. Finally, in February 1917 the British Army issued new tactical instructions that rejected the idea of a line of riflemen in favor of platoons organized into two groups. One group, consisting of riflemen and hand grenadiers, would maneuver to outflank and surround an enemy position; the second group, consisting of rifle grenadiers and a light machine gun, provided a firepower base. Thus the individual infantry platoon, at least in theory, combined automatic weapons, indirect fire, and flexible maneu-ver. In practice, the exhausted British troops had little opportunity to learn the new tactics, so the effectiveness of this idea varied.[34]

The most systematic change occurred, predictably, in the German infan-try. As was the case for the development of the elastic defense, the German Army as an institution developed and debated new infantry weapons and small-unit tactics. Beginning in 1915 the German High Command formed experimental companies to explore the tactical employment of light mor-tars, flamethrowers, and light, mobile-infantry support guns. By trial and error, two officers in these units—Captains Rohr and Reddemann—developed the assault tactics (*Stosstrupptaktik*) that became the German standard during both world wars and that influenced most other major armies.[35] In essence, the strengths of one weapon were used to compensate for the weaknesses of another. Infantry support guns, for example, were the most accurate means of destroying enemy strong points, but such guns were so difficult to move on the battlefield that, until a few lightweight 77mm guns were produced in 1918, they would generally take up overwatching positions to cover the assault forces. These forces took the more portable mortars with them to eliminate any obstacles that survived artillery fire. Similarly, heavy machine guns provided great volumes of fire, but the assault forces supplemented it with light machine guns, automatic rifles, and grenade launchers. In fact, the high explosive grenade replaced the rifle as the primary weapon for close combat. The Germans also produced a limited number of submachine guns and small carbines that would provide more portable firepower than the conventional rifle.

Different weapons also meant different formations. The German infantry gave up linear formations in favor of flexible groups—infantry guns and mortars to destroy strong points, preceded by teams of grenadiers and light machine gunners to win the firefight with the enemy.

An inevitable corollary of the German elastic defense and assault tactics was a greatly increased role for junior officers and sergeants. Senior commanders had to provide an unprecedented amount of training and trust to the young men who actually conducted the battle. The prewar training standards and social prestige of German junior leaders made such decentralization of authority much more possible than in many of the Allied armies, but the transition was not easy.

Armed with new tactics and weapons, and freed from the idea of advances in rigid lines, attacking German troops began to bypass some enemy strong points and to advance rapidly into rear areas. Although instances of such infiltration occurred as early as Verdun in 1916, the concept actually evolved in 1917 on the Russian and Italian fronts, leading to the German victories of Riga and Caporetto. These tactics are sometimes called, probably erroneously, "Hutier tactics." Gen. Oskar von Hutier directed such attacks against both the Russians and the Italians before commanding one of the field armies in the German spring offensives of 1918, but he did not invent the concept. Some German officers later denied the very existence of the "infiltration" or "soft-spot" tactics, and in fact the victories of 1918 were largely an intelligent application of the tactics evolved over the preceding three years.

Just as in the case of elastic defense, however, the German chief of staff, General von Ludendorff, issued a new set of tactical instructions for all infantry units. During early 1918 as many as seventy divisions rotated through a special training course on the new offensive tactics.[36] The result was the astonishing German success of March and April 1918, the so-called Ludendorff Offensives. The tactics involved were the culmination of German developments in combined arms during World War I. The spirit behind these tactics, when combined with armored equipment, had much to do with the later German blitzkrieg.

The Return of Mobility, 1918

The German tactics of 1918 can be summarized under four headings: Bruckmüller artillery preparation, the combined arms assault or storm battalion, rejection of linear advance in favor of bypassing enemy centers of resistance, and attacks to disorganize the enemy rear area.

Col. Georg Bruckmüller, an obscure officer retired for nervous problems in 1913 but recalled to duty for the war, developed German artillery techniques to a fine art. This development was by no means single-handed, and Bruckmüller, like Hutier, tended to exaggerate his contributions at the expense of other artillery experts. Still, Bruckmüller persuaded senior German commanders of the need for corps and higher level artillery headquarters to synchronize fire support for offensives. He also developed an unusual mix of weapons, ammunition, and timetables that had remarkable effects on the enemy defenders.[37]

The essence of a Bruckmüller artillery preparation was a carefully orchestrated, short but intense bombardment designed to isolate, demoralize, and disorganize enemy defenders. Before each of the great German offensives, Bruckmüller and his assistants held classes for junior leaders of both artillery and infantry, explaining what would take place. The result was not only unprecedented understanding and coordination but also a much more confident infantry. Indeed, Bruckmüller capitalized on the German collegiality, the willingness to cooperate to achieve victory rather than insisting on a fixed organizational hierarchy.

Next, Bruckmüller allocated different weapons against specific targets. For example, each trench mortar was given only twenty-five to thirty meters of enemy wire and trench to engage, and each artillery battery was assigned to suppress a specific enemy battery or to attack 100 to 150 meters of enemy positions. Bruckmüller also assigned the 77mm field guns of each division to counterbattery targets, while corps-level howitzers provided the rolling barrage to support the infantry of those divisions. He explained this apparent violation of organizational logic by noting that the field guns could fire rapidly to disrupt and disperse enemy gun crews, but they lacked the accuracy that the howitzers provided for firing close to friendly troops.[38] Bruckmüller avoided area targets, concentrating on such key points as artillery observation posts, command posts, radio and telephone centers, rearward troop concentrations, bridges, and major approach routes. He carefully pinpointed these targets on aerial photographs. The result was to cut enemy communications and to isolate forward units. The effect was increased by surprise. Using the survey techniques developed in all armies during 1916–1917, Bruckmüller was able to position and range his batteries in secret from points immediately behind the forward infantry trenches. Moreover, once the battle began, artillery commanders and forward observers oversaw the battlefront in person, adjusting artillery fire as necessary instead of sticking to a rigid timetable.

At the start of the German offensive on 21 March 1918, Bruckmüller began his bombardment with ten minutes of gas shells (to force the British to put on their protective masks), followed by four hours and twenty-five minutes of mixed gas and high explosives. The preparatory fires stopped and started repeatedly so that the British did not know when the artillery was actually lifting for the infantry advance. Meanwhile, automatic rifle teams moved as close as possible to the British positions during the bombardment.[39] When the Germans did advance, they moved behind a rolling barrage, further enhanced by intense fog. The combination of surprise, brevity, intensity, and carefully selected targets was unique.

Although all German infantrymen were eventually organized and trained to conduct assault tactics, the heart of the new tactics was special assault or storm battalions. These units developed the tactics, trained conventional infantry units, and provided detachments to lead the assault against particularly strong enemy positions. These assault troops, together with conventional infantry, were organized into an assault battalion task force in which all available weapons could be focused by a battalion commander. A typical assault battalion task force in 1918 consisted of three to four infantry companies and one trench mortar company, one accompanying artillery battery or half-battery of 77mm guns, one flamethrower section, one signal (communications) detachment, and one pioneer (combat engineer) section. The regimental commander might also attach additional machine-gun units and bicyclists.

The accompanying artillery pieces did not participate in the overall artillery preparation but waited behind the infantry, ready to move immediately. One of the principal tasks of the pioneers was to assist in the movement of the guns across obstacles and shell holes. Upon encountering a center of resistance, the automatic weapons and grenades provided suppressive fire while the guns, mortars, and flamethrowers attempted to eliminate it. Despite a specially constructed low carriage on some of the 77mm guns, they were such obvious targets that their crews suffered heavy casualties during the attack.[40]

The essence of the German tactics was for the first echelon of assault units to bypass centers of resistance, seeking to penetrate into the enemy positions in columns or squad groups, down defiles, or between outposts. Some skirmishers had to precede these dispersed columns, but skirmish lines and linear tactics were avoided. The local commander had authority to continue the advance through gaps in the enemy defenses without regard for events on his flanks. A second echelon, again equipped with light artillery and pioneers, was responsible for eliminating the bypassed enemy positions. This system of decentralized soft-spot advances was second nature to the Germans because of their flexible defensive experience. At the battle of Caporetto in 1917, the young Erwin Rommel, future hero of North Africa, used such tactics to bypass forward defenses and capture an Italian infantry regiment with only a few German companies.[41]

The final aspect of the German infiltration tactics was the effort to disorganize the enemy rear. The artillery preparation began by destroying communications and command centers; the infiltrating infantry also attacked such centers as well as artillery firing positions. The British defenders who opposed the German offensive of 1918, having recently undergone a major reorganization, quickly lost their unit organization and communications, retreating thirty-eight kilometers in four days. Col. J. F. C. Fuller, one of the foremost British tank tacticians, observed that the British seemed to collapse and retreat from the rear forward. Major British headquarters learned of multiple German attacks on forward units just before they lost telephone contact with some of them. The senior British commanders then ordered their remaining forces, which were often successfully defending their bypassed positions, to withdraw in order to restore a conventional linear front.[42] Finally, in July, the French thwarted the Germans by using their own techniques against them. The French constructed a true defense-in-depth, with very few troops in the forward area where the German artillery preparation fell. Moreover, from German prisoners the French learned the exact date and time of a German attack near Reims. Shortly before the scheduled start of the German bombardment, the French fired a massive counterpreparation, disorganizing and demoralizing the Germans.[43]

The German spring offensives ultimately failed for a variety of reasons, including lack of mobility to exploit initial successes and lack of clear operational-level objectives. As a result, Ludendorff dissipated his forces in a series of attacks that achieved tactical successes, including several true breakthroughs, but no operational or strategic decision.

Indeed, the Germans used tactics and organization that could be described as a blitzkrieg without tanks. Their success was due not so much to actually destroying their opponents but to disorganizing and demoralizing the defenders. This was especially easy to achieve when fighting a World War I army, where the static nature of deployments and telephone communications had combined with the elaborate planning necessary for a set-piece battle, producing a defender who had great difficulty reacting to sudden changes. Once the Germans achieved a breakthrough, both sides found that their soldiers no longer knew how to fight in multiple directions across open terrain. Feeling vulnerable out in the open, the German infantrymen dug in immediately whenever they broke through the enemy defensive system.

The German spring offensives of 1918 were the most obvious example of mobility returning to the battlefield, but in fact all armies in 1918 were technically more capable of attack and maneuver than they had been in the preceding three years. Beginning on 15 July 1918, the British, French, and Americans launched a sustained series of attacks that combined the Allied innovations made during the war. Infantry used renewed flexibility and firepower, with tanks to suppress enemy strong points. Airpower provided limited ground-attack support and reconnaissance both before and after the battle. This air reconnaissance focused on antitank threats to the advancing ground elements. Artillery had become much more sophisticated and effective since 1914. Most important, the different weapons and arms had learned to cooperate closely, at least in carefully planned set-piece operations. Commanders could no longer rely on using one or even two combat arms but had to coordinate every available means to overcome the stalemate of the trenches. Nevertheless, the individual 1918 attacks in France did not achieve a decisive result on the battlefield, and the Germans were defeated more by sustained attrition, food shortages, and demoralization than by any decisive breakthrough and exploitation.[44]

One of the few cases in which a 1918 army penetrated a prepared defense and then exploited with decisive results occurred not in France but in Palestine, where the British defeated Germany's ally, Turkey. This victory is known as the second battle of Armageddon, or Megiddo (see Map 2), because it was fought in the same area as the original battle of 1479 B.C.[45] The British commander, Sir Edmund Allenby, had steadily advanced from Egypt through Palestine against a Turkish army with a German commander, Liman von Sanders, and a few German units. The Turkish government had diverted its resources elsewhere; thus in 1918 the British Commonwealth forces outnumbered the Turks two to one. Allenby further increased his advantage by a detailed deception plan that convinced the Turks that the British would attack at the eastern end of the front, in the Jordan Valley.

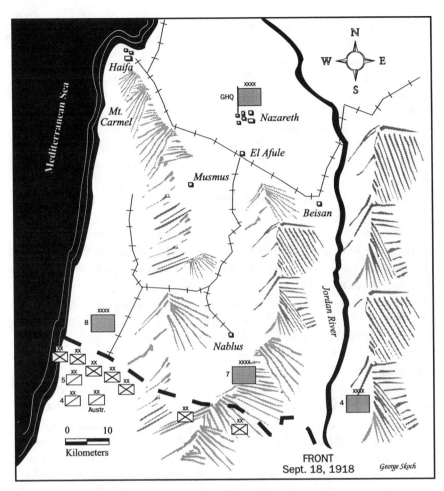

Map 2. Second Battle of Armageddon, 19–24 September 1918

The actual attack was then conducted in the west, near the seacoast. The fact that the British possessed a tremendous numerical advantage does not detract from the significance of the second battle of Armageddon in terms of its tactical methods and operational/strategic objectives.

Allenby used all available elements, beginning with irregular troops in the enemy rear areas. On 17 September 1918, two days before the planned offensive, the famous T. E. Lawrence and Prince Feisal of Arabia conducted a wave of guerrilla attacks on Turkish rail lines in order to divert attention and isolate the battlefront. The Royal Air Force also harassed Turkish lines of communications for days. At 0430 on 19 September the British infantry began to move forward behind a fifteen-minute artillery barrage. This short preparation achieved surprise and avoided tearing up the ground. Moreover, the long delays in assembling troops and supplies prior to the offensive had enabled the British and Commonwealth infantry to train to high standards of junior initiative and flexibility. Unlike the campaigns in France, exploitation forces did not have to wait for authority to pass through a penetration and advance. Instead, one Australian and two British cavalry divisions began the battle in assembly areas immediately behind the assaulting infantry, with exploitation objectives already designated. Because of such decentralized control, the 4th Cavalry Division had completed its passage of lines and begun the exploitation of the Turkish rear area within four hours of the initial assault.

The primary objectives of the campaign were the railroad junctions at El Afule and Beisan, forty miles behind the front; a secondary objective was Nazareth, the German-Turkish headquarters. Seizure of these points would cut off the forward Turkish units from their supplies, commanders, and route of retreat. The key was to move cavalry through the Mount Carmel heights so rapidly that the Turks could not react to block the passes. This was accomplished on the evening of the first day. The next morning, a brigade of the 4th Cavalry Division encountered a reinforced Turkish infantry battalion marching forward in a belated effort to block the pass at Musmus. A combination of armored-car machine-gun fire and horse-cavalry lances captured this battalion before it ever deployed from the march. Twenty-five hours after the offensive began, another British cavalry brigade surrounded Nazareth, which had been isolated and harassed by air attacks. The aerial destruction of the El Afule telephone exchange deprived Sanders of his control over the Turkish forces. Although the German commander escaped in the confusion, the British captured the documents in the enemy headquarters. The Turkish Seventh and Eighth Armies, except for a few hundred stragglers, surrendered en masse, and only the November armistice ended the British pursuit.

In the midst of this pursuit, on 21 September the RAF located a Turkish artillery column moving along a valley road at Wadi Fara'a. The British aircraft immobilized the first and last vehicles in the column, then repeatedly bombed and strafed the halted column until it lost all organization. Coincidentally, on the same day a similar fate befell a Bulgarian brigade at Kosturino Pass in Macedonia. After the war, airpower enthusiasts often pointed to these deep interdiction attacks as a more efficient alternative to the risks of attacking a dispersed enemy in the forward battle area.[46]

More generally, the significance of Second Armageddon was threefold. First, it represented a rare ability to make the transition from penetration to breakout and pursuit before the defender could react. The key to this success, apart from numerical superiority, was the fact that the exploitation force did not wait for permission from higher headquarters but was committed on the decision of division commanders and in execution of a previously arranged plan. Second, Allenby used all his weapons and units in a flexible and integrated manner that was matched in World War I only by the Germans. Finally, Second Armageddon influenced an entire generation of British cavalry officers, who considered it the model of a mobile, deep battle. After the frustrations of trench stalemate in France, the exploitation in Palestine seemed a dream come true. When these cavalry officers became armor commanders, they stressed the need for mobile, lightly armored vehicles. As a result, one-half of the British armored force in 1939 was equipped with inadequate guns and armor and was not prepared to cooperate with the other combat arms.

Intelligence

The failure of Sanders to identify the focus of Allenby's attack illustrates another problem area for commanders in World War I—intelligence. Techniques for collecting information multiplied, but interpretation remained highly subjective.

The empty, dispersed battlefield made intelligence collection more difficult and urgent. Traditional means, such as interrogation of prisoners of war, provided detailed information about the forward area but not necessarily about activities in the rear. Aerial reconnaissance and photography, however, collected unprecedented detail concerning enemy deployments. In addition, the development of radio and telephone communications brought with it the growth of signals intelligence (SIGINT). Because primitive telephone and telegraph systems worked by using the earth itself as one-half of the complete circuit, both sides found that they could intercept their adversary's conversations by connecting sensitive receivers to the ground near the forward trenches. Early radio operators soon became aware of transmissions by their counterparts and became de facto intercept operators to monitor them. Cryptographers had considerable success at deciphering enemy codes. As early as December 1914, the British Army began radio direction-finding efforts against enemy transmitters. A directional antenna would measure the compass angle from its location to an enemy radio signal; plotting out the angles from two or more such antennae would then give the location for the enemy transmitter.[47]

Unfortunately, this wealth of information often went unused. A recurring problem in every war is that wishful thinking clouds judgment. Commanders and sometimes even intelligence officers misinterpreted raw information about the enemy to suit their preconceived notions.[48]

Artillery target acquisition was less subjective, and the development of indirect fire stimulated efforts to locate enemy guns. Along with aerial observation, World War I saw a considerable expansion in flash and sound ranging. Each of these techniques operated on the same principle as radio direction-finding: when an adversary's artillery fired, observers at different points along the front attempted to determine the direction from their location toward the sound or flash of the gun; plotting out multiple angles on a map would identify the general area from which the shelling occurred. These techniques were only partially successful and were often defeated by the simple procedure of having multiple batteries fire at the same time.[49]

Organizational Results

In addition to the changes in infantry company and battalion structure, the rapid development of weapons and tactics during World War I significantly changed tactical organizations. By the end of that war, virtually all armies included mortars at regimental or brigade level and sometimes at battalion level. By 1918 the number of automatic weapons in an infantry division had risen considerably from a norm of twenty-four heavy machine guns in 1914 to the following totals four years later:[50]

Nation	Automatic Rifles/ Light Machine Guns	Machine Guns
Germany	144	54–108
France	216	72–108
Britain	192	64
Italy	288	72
United States	768	260

Artillery developed almost as dramatically, although most of the additional guns were concentrated in nondivisional units whose numbers varied, depending on the mission of the division being supported. As Gen. Wilhelm von Balck, a noted German tactician both before and after the war, remarked, "The question as to the proportion of the artillery is no longer: 'How many guns for each thousand men should be provided?' but far rather: 'How much infantry will be required to utilize the success of the fire of the artillery?' . . . [T]here are no longer principal arms. Each arm has its use, all are necessary."[51]

More complex problems drove other organizational changes. For example, both the French and the Germans found that the square division structure, with two brigades each of two regiments, was unsuited to positional warfare. Given the broad frontages involved in this type of war, no European power had enough manpower and units to deploy divisions with two regiments in the first line and two in the second. If, on the other hand, three regiments were in the first line and the fourth regiment served as a general reserve, one of the two infantry brigade commanders became superfluous, and the width of the division sector made it difficult to control and defend. So the Germans left one brigade commander in control of all infantry, and by 1916 the French as well as the Germans had reduced the number of infantry regiments in a division from four to three (Fig. 4). The British had entered the war with a three-brigade structure, which they retained; but in early 1918, when manpower shortages became acute, each of these brigades was reduced from four infantry battalions to three. This shift to a triangular structure had the added advantage of increasing the proportions of artillery

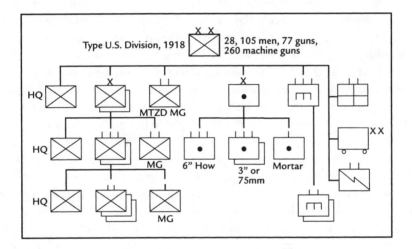

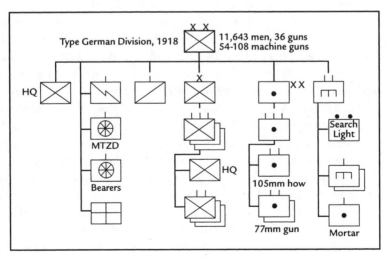

Figure 4. Type U.S. and German Divisions, 1918

and other branches to infantry, although the Germans moved part of their artillery into nondivisional units, which could be reassigned as necessary. Thus, a 1914 French infantry division consisted of 87 percent infantry, 10 percent artillery, and 3 percent support elements; the 1918 version had a proportion of 65 percent infantry, 27 percent artillery, and 8 percent support.[52]

The one exception to this trend was the United States Army, which not only insisted on a four-regiment structure but also actually increased the size of rifle companies during 1917 (see fig. 4). The result was a division that varied in size from 24,000 to over 28,000 men, a giant considering that the average strength of a European division was down to 8,000 men or fewer by 1918. In one instance, the 42d U.S. Infantry Division assumed the defense of a sector previously occupied by an entire French corps of three divisions.[53]

Why did the United States deviate from European experience? Gen. John J. Pershing and his subordinates had a number of concerns. First, they wanted a division that could remain in combat on a sustained basis, absorbing casualties and continuing to advance indefinitely. Each infantry brigade commander could attack with his two infantry regiments one behind the other—when one became exhausted or depleted by combat, the other could assume the attack, without a long time delay while the division or corps commander approved the change. The only reserve available to the division commander was the two-battalion combat engineer regiment, which was frequently pressed into service as infantry. Second, the square division made the best use of the relatively few professional soldiers in the U.S. Army. Pershing was critically short of experienced leaders, and particularly of general staff officers who could plan and coordinate the complex logistics and operations required for modern battle. The downside was that every commander and staff ran the risk of being overwhelmed by an excessive span of control—a typical company commander, who in many instances had less than one year's experience and training, had to control 250 infantrymen.

Third, and more subtly, the square division also represented Pershing's insistence on self-reliant infantry armed largely with rifles, as in previous wars.[54] He apparently regarded many of the new weapons, such as automatic rifles and grenade launchers, as suitable only for trench warfare. Because America had no defense industry, its troops had been armed with poorly designed weapons, such as the French Chachot automatic rifle. The Chachot jammed constantly, and after each jam the weapon had to be disassembled in the muddy trenches to eliminate the problem. Such unreliable weapons, combined with undertrained American infantry, made it ridiculous for U.S. commanders to aspire to the complex structure and tactics of European armies. Instead, the Americans concentrated on infantry armed primarily with rifles and bayonets. Heavy machine guns were regarded as specialized weapons, grouped into three machine-gun battalions per division instead of being distributed permanently to the infantry battalions and regiments (see Fig. 4).

Even though the Americans differed from their Allies about many details, all participants came away from World War I with certain common impressions: the tremendous problems of logistics and manpower; the necessity for detailed planning and coordination; the difficulty of advancing even when all arms worked closely together; the supreme problem of converting a penetration into a breakthrough. Under carefully planned and controlled circumstances, the Allies had been able to combine all weapons systems to maximize the effects of each one. Among the belligerents' doctrines for achieving this combination, Germany's solution proved to be most adaptable to new weapons and tactics.

French infantry leaving trenches after artillery barrage, c. 1916. (U.S. Army Military History Institute)

Soldiers of the 369th Infantry Regiment, 93d U.S. Division, with French equipment, May 1918. (U.S. Army Military History Institute)

Tank–infantry training for the 27th U.S. Division, Beauquesnes, France,
September 1918. (U.S. Army Military History Institute)

CHAPTER 3

• • • • • • • • • • • • • •

The Interwar Period

Common Issues

The conventional image of military affairs and doctrine between the two world wars depicts most armies as rigidly committed to a repetition of the positional warfare of 1914–1918. According to this view, only Hitler's Germany listened to the advocates of mechanized warfare, with the result that between 1939 and 1941 the German blitzkrieg achieved almost bloodless victories over the outdated Polish, French, and British armies.

The reality was much more varied and complex. No major army entered World War II with the same doctrine and weapons that it had used twenty years earlier, although some changed more than others. During the interwar period, the majority of professional soldiers recognized that some change was necessary if they were to perform better the battlefield functions of penetration and exploitation that had proven so difficult during World War I. Yet armies differed markedly in their solutions to these problems. Instead of a simple choice between trench warfare and blitzkrieg, each army was faced with a variety of possible changes, a series of degrees of modernization between the two extremes. In many cases, the choice was determined by social, economic, and political factors more than by the tactical conceptions of senior officers. Even in Germany, the advocates of mechanized warfare did not have a free hand. In a real sense, the German forces and doctrine of

1939 were not so much the perfect solution as they were simply a solution that was closer to the problems of the moment than were the structures and doctrine of Germany's early opponents.

Because of this tactical variety between the world wars, the doctrine and organization of each of the major powers must be considered up to the point at which that nation entered World War II. Before reviewing these armies, however, it is appropriate to note some common factors that hampered military change in all nations.

The first of these factors was a general revulsion toward warfare and all things military. After decades of peacetime preparation and years of incredible carnage, few people in Europe or America were interested in further military expenditures or experiments with new weapons and tactics. Particularly in France, firepower seemed so great that few soldiers foresaw any type of offensive success against prepared enemy positions without the combination of a mass infantry army with tanks, artillery, and attrition tactics, the means that had succeeded in 1918. Even after most armies concluded that trench warfare was a special kind of combat that would not necessarily recur, the general public and political leadership were unwilling to risk another war. In 1928, fifteen nations (later increased to more than sixty) signed the Kellogg-Briand Pact, renouncing the use of war except in national self-defense. During the 1920s and early 1930s, a series of international conferences attempted to limit military and naval armaments. Although these conferences ultimately failed, it was difficult for professional soldiers to justify the purchase of new weapons such as tanks and aircraft in a social and political environment that might outlaw such weapons at any time.

During the first fifteen years of peace, extremely tight defense budgets, a second factor, reflected the public distaste for warfare. The victorious armies were saddled with huge stockpiles of 1918-model weapons and ammunition and had to use up these stockpiles at peacetime training rates before major new expenditures could be justified. Thus, during the early 1930s the U.S. Army spent more money researching means to preserve aging ammunition than to develop new weapons.[1] Just as the stockpiles were finally consumed or worn out, the Great Depression caused even tighter defense budgets, which hampered development and procurement of tanks, aircraft, and other new weapons. When the armies of the Western democracies failed to field new weapons, the cause was more likely to be budget woes than any institutional resistance to new technology.

The Germans, by contrast, had been deprived of their weapons at the Versailles Peace Treaty of 1919 and could therefore start fresh. To some extent, the German tactical successes of 1939–1942 were due not to any superiority in equipment quality or quantity but to the fact that the German tanks and other vehicles were produced early enough to allow extensive experimentation and training before the war. In contrast, the British and French had few modern weapons with which to train until the very eve of World War II, when they mass-produced them on a crash basis. Nations with a smaller industrial base, such as Japan and Italy, could not compete fully in the arms race. The Japanese selectively built a few types of warship and aircraft of high quality. In land warfare, they relied on training and morale to compensate for weapons that they could not afford to mass-produce. Italy lacked not only production facilities but also equipment design capability and even public understanding of automobiles and other machinery. As a result, the Italians failed to produce any modern, well-designed weapons or mechanized units.[2]

A third factor was technology, which affected military change in two ways. On the one hand, rapid changes in technology made governments even more reluctant to invest in existing designs that would soon be outmoded. In 1938, for example, the inspector-general of the French Air Force had to advise the French and British governments to avoid a showdown at Munich because he believed that the majority of French combat aircraft was suddenly obsolescent; new developments such as flush-riveted metal construction gave the German Luftwaffe (Air Force) the appearance of technical superiority over its adversaries.[3] On the other hand, it was often difficult to determine how changes in technology would affect the tactics of 1918. Equipment designed to fulfill the requirements of those tactics might be unsuitable for different functions and concepts, yet new designs were sometimes fielded without a clear understanding of how that equipment would be employed tactically.

Fourth, there was considerable confusion in terminology. Both advocates and opponents of mechanism used the term "tank" loosely to mean not only an armored, tracked, turreted, gun-carrying fighting vehicle but also any form of armored vehicle or mechanized unit. Such vague terminology made it difficult for contemporaries or historians to determine whether a particular speaker was discussing pure tank forces, mechanized combined arms forces, or mechanization of infantry forces. Similar confusion existed about the term "mechanization." Strictly speaking, any use of the gasoline engine in warfare could be termed mechanization. However, the term is usually employed to describe the use of armored, tracked combat vehicles. By contrast, "motorization" describes the use of wheeled motor vehicles, usually trucks, that are not intended to go into combat but that improve logistical support and mobility off the battlefield. No nation in the world could afford to mechanize fully in the strict sense of using only armored vehicles, but between the world wars all armies made some motions in the direction of motorization.

Indeed, there was almost no choice about the matter. Before World War I, all nations relied on a pool of civilian horses as transportation in case of war. With the rise of motor vehicles during the 1920s, this supply of civilian animals declined to the point where many armies had to base their transportation planning on motor vehicles.[4] Thus, motorization was often seen as an easier, cheaper, less revolutionary change than mechanization.

Fifth, advocates of change did not always speak persuasively or with one voice even when their terminology was understood. Reformers with a clear vision of mechanized, combined arms war were often so extreme in their statements that they alienated the men they needed to convert, the commanders and politicians who set military policy. In the French and Soviet cases, political issues also retarded the development of new mechanized formations. At the same time, proponents of strategic airpower such as William Mitchell in the United States and Guilio Douhet in Italy made exaggerated claims that retarded the development of the tactical combined arms team.

This point leads to the sixth and final common factor of the interwar period, the opposition of the more traditional combat arms to new changes. Many commentators have blamed such opposition for thwarting or retarding the development of mechanized warfare. There is some truth to that allegation. Even when horse cavalry advocates accepted the need for mechanization, the resulting organizations and tactics tended to reflect cavalry ideas of reconnaissance, screening, and mounted combat instead of full-scale, deliberate battle offered by a combined arms, mechanized team.[5]

It is also worth noting that the tank and the aircraft were not the only weapons systems that underwent major development between the world wars. The older branches had genuine needs that competed with new weapons for funding and for roles in the combined arms team. The infantry had legitimate requirements for increased organic firepower, for antitank and antiaircraft defenses, for greater mobility, and for some type of armored support to assist it in the deliberate attack. The field artillery needed the same mobility as the armored forces in order to support them in the breakthrough. Fast-moving mechanized formations required more flexible communications, logistics, and fire support, yet these areas remained weak in many interwar armies. Combat engineers, who had become preoccupied with maintaining supply roads and rails during the positional warfare of 1914–1918, were more important than ever when mechanized units increased the problems of mobility and countermobility on the battlefield. For example, unlike infantry units, it would be very difficult for armored forces to cross canals and rivers without engineer bridging units.

In short, the development of mechanized formations and tactics must be viewed in the context of a more traditional mass army. Any nation that created a mechanized elite ran the risk of dividing its army and neglecting most of its forces, with catastrophic problems of coordination and morale.

Airpower and Air Support

One final common issue deserves general discussion before examining the different armed forces involved.[6] The period between the two world wars witnessed a decline in air–ground cooperation in most nations of the world. Intent on achieving independence from army control, the airpower advocates in most nations vigorously opposed tactical air support and air–ground cooperation. To some extent, this was a natural consequence of the fact that all the armed forces worked under severe budgetary restraints. Of equal importance, however, was the widespread belief that close air support was an inefficient and dangerous way of using airpower. The different viewpoints of air and ground officers are thus summarized here, although this debate was rarely expressed clearly between the wars. I have included issues that arose during this time and that have endured even into the 1990s.

The basic premise of most professional aviators was that airpower should operate as an independent armed force, capable of winning wars on its own. Unfortunately, like those of some of the early tank enthusiasts, their theories far exceeded the technology of the day. William Mitchell, Guilio Douhet, and a host of other theorists argued that long-range bombardment aircraft could strike directly at the industrial base and the civilian morale of an enemy nation, paralyzing an entire nation in a matter of days. Such theorists believed that, even on the battlefield, aviation should be independent of the ground forces. Their reasoning may be condensed into four premises.

First, aviators argued that the best battlefield use of airpower was interdiction, striking large units of the enemy deep in rear areas. In the final battles of 1918, there had been several occasions when primitive aircraft had caught and destroyed enemy units on the march. This, said the airpower enthusiasts, was the most efficient use of aircraft.

Second, aviators contended that all aircraft must be under centralized control rather than parceled out to different ground commanders. The inherent range and speed of aircraft gave them a flexibility and reach that exceeded the concerns of individual ground commanders, who might well waste scarce aircraft. Only central control would allow the true force of the air element to be used in concentrated form at critical points.

Third, regardless of how aircraft were controlled, the aviators believed that close air support was the most difficult, inefficient, and wasteful use of airpower. To be successful, close-support pilots had to train carefully in order to distinguish between enemy and friendly units. Primitive aircraft had great difficulty in striking small individual targets, at least until the advent of dive-bombing. As the RAF discovered in 1917–1918, close air support is extremely hazardous, wasting scarce pilots and aircraft to strike only a small number of targets. Even if the pilot were successful, at most he could destroy only a small number of enemy targets, because those targets were usually dispersed, concealed, and placed in defensive positions.

Finally, aviators argued that close air support targets could be engaged just as effectively by artillery but that many deeper targets, the appropriate objectives of any battlefield air operations, were beyond the range of field guns. In short, dedicating aircraft to ground operations, and particularly to close air support, appeared to the aviators as a ridiculous, dangerous, and expensive misuse of a precious asset. There were exceptions among professional pilots, but those exceptions were rare.

By contrast, ground soldiers had a completely different set of priorities. Some commanders undoubtedly remained rooted in the past, where aircraft were used primarily for reconnaissance and artillery spotting; but as time passed, more and more professionals saw airpower as a valuable part of the combined arms team.

First, the flexibility of aircraft appeared to be ideal for massing combat power quickly at critical points. Aircraft could be the ground commander's best means of quickly influencing key battles. As a result, that commander became increasingly interested in ensuring that he had control over how the available aircraft were employed. Moreover, as a practical matter the communications available to armed forces during the 1930s were far from efficient. Most armed forces, including the Germans, began World War II with communications that were so clumsy that centrally directed aircraft rarely received their orders in time to strike a desired battlefield target. Therefore, decentralization and subordination to ground commanders were practical necessities.

Second, ground officers contended that the experience of World War I showed the necessity of concentrating as much force as possible to achieve a breakthrough. The infantry and, later, mechanized forces deserved every possible support in this risky endeavor. Even granted that artillery might have the same destructive power as close air attacks, there was rarely enough artillery to achieve the desired results. Moreover, a limited air attack could achieve the same psychological impact as a much larger concentration of artillery. Attackers would be encouraged by the sense that aviation was supporting them, and defenders, especially poorly trained troops, would be daunted by the threat of death from above. As an added bonus, the fewer the artillery shells fired, the more trafficable the ground would be for the attacker, allowing an attack supported by air to move forward more easily than an attack supported only by conventional artillery.

Finally, ground officers recognized the difficulty of conducting close air support without endangering friendly troops. For that very reason, however, one could argue that this most difficult of tasks should be practiced most often—an aviator who could strike dispersed targets in close proximity to friendly troops would surely have little difficulty in the interdiction environment, where the dangers of friendly fire were much less.

Such, in essence, was the difference in attitudes that framed the debate on air–ground cooperation from World War I until the end of the century.

Great Britain: "Hasten Slowly"

In 1918 Great Britain led the world in both armored equipment and armored doctrine.[7] At a time when most soldiers regarded the tank as a specialized infantry-support weapon for crossing trenches, a significant number of officers in the Royal Tank Corps had gone on to envision much broader roles for mechanized formations, including breakthrough and exploitation. In May 1918, Col. J. F. C. Fuller, a staff officer in the Tank Corps headquarters, had used the example of German infiltration tactics to refine what he called Plan 1919. This was an elaborate concept for a large-scale armored offensive in 1919, one that would not only produce multiple penetrations of the German forward defenses but also totally disrupt the German command structure and rear organization.

Fuller's expressed goal was to defeat the enemy by a "pistol shot to the brain" of enemy headquarters and communications instead of by destroying the combat elements through systematic attrition. In order to attack German headquarters before they could pack up and move, Fuller relied on the Medium D tank, which had only just begun development at the time he wrote. Potentially, the Medium D could drive twenty miles per hour, a speed that would allow it to exploit the rupture of trenches caused by slower heavy tanks. In fact, the Medium D suffered the usual developmental problems of any radically new piece of equipment and might not have been available even if the war had continued into 1919. Moreover, then as later, Fuller was noteworthy for his neglect of infantry in the mechanized team. He did conceive of trucked infantry following the tanks under certain circumstances but did not recognize the need for close coordination between armor and infantry except at the point of a deliberate attack.[8]

Like so many of Fuller's writings, Plan 1919 was overly optimistic and even simplistic. It was, in fact, one of a number of similarly grandiose plans produced by British tank officers at the end of the war.[9] Given the later experience of blitzkrieg, however, Fuller's vision of future warfare appears prophetic, despite its fallacies.

Notwithstanding the efforts of numerous innovators, the British Army gradually lost its lead, not only in armor but also in most areas of tactical progress. Several special obstacles to continued British innovation existed. The most commonly cited was traditionalism within the British Army. This institutional resistance has often been exaggerated, but certainly the strong unit identity of the British regimental system discouraged radical changes within the traditional arms and services.

A related problem was that Great Britain was the first nation to create an independent air force. The Royal Air Force was intent on developing its own identity as a separate service and resisted any close relationship with the army. In 1922, for example, the army requested that eight Army Co-Operation Squadrons be permanently assigned for liaison and reconnaissance duties with ground troops. The RAF would provide only three squadrons, equipped with light aircraft that could not conduct combat operations. In fairness to the aviators, it should be noted that all the British services were extremely restricted in size and budget. During mechanized exercises in 1928, a number of RAF pilots voluntarily practiced close air support for armored units, but after this display the Air Ministry formally requested that the army refrain from encouraging pilots to violate RAF doctrine.[10] This limitation was clearly reflected in British Army regulations from 1924, where the RAF was described as providing only liaison and reconnaissance in the immediate proximity of ground attacks. Fighter aircraft might conduct strafing and other ground attacks "in exceptional circumstances," but only at the expense of their air superiority mission. Despite the efforts of many British armored theorists, close air support doctrine was not really developed in Britain until 1942.[11]

The problem of imperial defense also limited change. Since 1868 most British troop units stationed at home had exchanged places with units overseas on a regular basis. In particular, a large portion of the British Army was always stationed in the Middle East and India. These overseas garrisons required large numbers of infantrymen to control civil disorders and made logistical support of elaborate equipment and weapons difficult. Consequently, a unit in the British Isles could not be motorized or mechanized without considering the effects of this change on its performance in low intensity, imperial police operations. This did more than delay mechanization. It also meant that in designing armored fighting vehicles the British were often thinking about the requirements of warfare against relatively unsophisticated opponents and not against well-armed European forces.[12]

Despite these limitations on innovation, British doctrine did not stand still during the 1920s. A repetition of World War I seemed unthinkable, so positional warfare rapidly declined in British doctrine to the status of a special case. Instead, the British Army returned to the concepts of open, maneuver warfare that had been common before 1914, updating those concepts only to allow for the effects of firepower and motor vehicles. The 1924 edition of *Field Service Regulations* considered infantry support to be the chief mission of tanks, but it also recognized the possibility of tanks attacking enemy flanks and rear to disorganize the opponent, as Fuller envisioned. These regulations showed a serious and practical concern for the problems of antitank and antiaircraft defense of all arms, although actual weapons to address them were slow to appear. By 1929 British regulations had abandoned the old belief in the primacy of infantry, which instead became "the arm which confirms the victory and holds the ground won" by a close cooperation of all arms. Still, this cooperation was to be achieved by the same detailed, meticulous centralized planning and control that the British had used in 1918. Coordination in unanticipated encounter battles was much more difficult.[13]

Yet the British, despite significant budgetary restrictions, were able to motorize part of their artillery and supply units and to continue development of the small Royal Tank Corps. In 1927 and 1928, an Experimental Mechanized Force conducted brigade-level exercises in Britain. It included a light tank battalion for reconnaissance, a medium tank battalion for assault, a machine-gun battalion for security and limited infantry operations, five motorized or mechanized artillery batteries, and a motorized engineer company. Unfortunately, the equipment varied widely in its cross-country mobility and mechanical reliability. It was a mixture of tracked and wheeled, experimental and well-developed vehicles that could not move together except at very slow speeds. As a result, some officers of the Royal Tank Corps decided that the other arms were incompatible with armored operations and focused their attention on almost pure tank formations. In the typically acerbic words of Fuller, "To combine tanks and infantry is tantamount to yoking a tractor to a draft horse. To expect them to operate together under fire is equally absurd."[14]

The British War Office dissolved the Experimental Mechanized Force in 1928 for a variety of reasons, including budgetary restrictions and the opposition of some military conservatives. This force did, however, provide the basis for Col. Charles Broad to produce a new regulation, *Mechanized and Armoured Formations,* in 1929. This regulation was a considerable advance in describing the roles and missions of separate armored formations, but it also reflected the pure-tank attitude that was coming to dominate the Royal

Tank Corps. Even when Broad proposed a Royal Armoured Corps that included tanks, mechanized cavalry, and mechanized infantry, he explicitly excluded artillery and engineers.[15] Still, Broad recognized different models of armored vehicles and different roles for them. In particular, the standard "mixed" tank battalion of an independent tank brigade was a combination of three different types of vehicles. Within each company, seven light tanks would reconnoiter the enemy positions and then provide fire support for five medium tanks that actually conducted the assault. In addition, two "close support tanks"—really self-propelled howitzers or mortars—would provide smoke and suppressive fire for the assault.[16] Since in practice the light tanks were often small armored personnel carriers, the parallel with more recent American armored cavalry should be obvious.

British armored theorists did not always agree with each other. Basil Liddell Hart, a noted publicist of armor, wanted a true combined arms force with a major role for mechanized infantry. Fuller, Broad, and other officers were more interested in a pure-tank role, in part because the other arms of the British Army appeared to lack the equipment and the desire to cooperate closely with tank units. G. L. Martel, one of the most innovative theorists and tank designers of the period, was fascinated with the idea of using small armored personnel carriers, capable of transporting one to three men and a machine gun, to assist the infantry in its attacks. Unfortunately, the machine-gun carriers designed at Martel's instigation participated in experiments both as reconnaissance vehicles and infantry carriers and proved inadequate for either function.[17] Not until the eve of World War II did the British develop a reliable machine-gun carrier, and even then it was dispersed in small numbers within infantry battalions that otherwise attacked on foot.

Despite these differences of opinion, the next step in developing the role of armor was to form an independent mechanized force of division size. This was undertaken as an experiment in 1934, using Col. Percy Hobart's 1st Tank Brigade, a newly formed unit of the type envisioned by Broad, and Maj. Gen. George Lindsay's partially mechanized 7th Infantry Brigade. Unfortunately for the British, personality differences, lack of training, and artificial restrictions from the umpires turned the ensuing armored exercise into a farce. General Lindsay, one of the few senior officers who was genuinely committed to the development of a combined arms mechanized division, was so discredited by this fiasco that he ceased to have any influence over policy.[18]

Instead, the moderate chief of the Imperial General Staff, Gen. Sir Archibald Montgomery-Massingberd, chose to create a Mobile Division by mechanizing large parts of the British cavalry. The Mobile Division authorized in December 1937 consisted of two armored cavalry brigades, each composed almost entirely of light tanks and armored cars, plus the one exist-ing tank brigade, two mechanized infantry battalions, and limited numbers of artillery, engineers, and support units.

The result was a huge division authorized (but never fully equipped) with 500 tanks, most of them very lightly armed and armored. In the absence of clear guidance, the cavalry officers appointed to lead this Mobile Division naturally concluded that it should perform the same functions of reconnais-sance, screening, and security that horse cavalry formations had done in previous eras. Indeed, the division, though rather unwieldy, would undoubt-edly have been useful for imperial policing or cavalry reconnaissance. It was not, however, capable of integrating the different combat arms at the small-unit level in order to conduct serious armored breakthrough and exploita-tion operations against a sophisticated enemy. In general, reconnaissance, medium armor, infantry, and artillery were isolated under separate brigade-level commands. In 1939 this Mobile Division became the 1st Armoured Division, a smaller force that was still unbalanced and segregated by arm. It consisted of one brigade each of mechanized cavalry and armor, a single motorized infantry battalion, and towed artillery and engineers. A second mobile division formed in Egypt, providing the basis for later British opera-tions there.[19]

There were also problems in equipment. The Royal Tank Corps had to make do with the same basic tanks from 1922 until 1938, despite frequent changes in design and technology. Almost the only improvement came from 1930 to 1932, when radio communications advanced markedly. Before that time, each vehicle crew had to tune its radio by hand to a common frequency, and the motion of a moving tank could easily throw the radio off. Colonel Broad instigated a series of developments that eventually provided crystal-controlled, preset frequencies. Only such radios could allow a commander to control his rapidly moving units while observing and leading from the front. The complexity and expense of such equipment, however, made dis-tribution of radios down to individual tanks very slow. Worse still, the Brit-ish Army neglected the development of high-powered, long-range radios for higher level command and control. British signals units entered World War II with equipment whose essential design dated from World War I, equipment that could not respond quickly to the fluidity of battle.[20]

During the 1930s, disagreements about tank roles combined with frequent changes in the defense bureaucratic structure to thwart good armored-vehicle design. In the confusion, the sixteen-ton Mark III medium tank was terminated. The Royal Tank Corps had counted on this design to perform a multitude of functions, and its demise resulted in a hodgepodge of de-signs when rearmament finally began in 1936. Generally speaking, British armored vehicles tended to maximize either mobility or protection. Both the cavalry and the Royal Tank Corps enthusiasts wanted fast, lightly armored, mobile vehicles for reconnaissance and raiding. Thus, for example, the only vehicle ready for immediate mass production was the Mark VI light tank, armed only with machine guns. This seemed appropriate for converted cavalry units, but the Royal Tanks had to make do with a variety of half-baked, poorly designed medium or "cruiser" tanks. At the other extreme, the army tank battalions assigned to provide traditional infantry support had extremely heavy armored protection to advance against prepared enemy defenses that might include antitank guns.[21]

In short, Britain dissipated its design and production resources to provide vehicles for a variety of disconnected roles—machine-gun carriers for the infantry, tiny light tanks for the reconnaissance elements of infantry divisions and for the cavalry in mobile divisions, medium or cruiser tanks for the pure-tank theorists of the Royal Tank Corps, and heavy infantry-support tanks. All these designs neglected firepower. As late as 1937, the thin armor on most tanks in the world made armor-piercing machine guns, or at most a 20mm cannon, seem entirely adequate for antitank defense and therefore for the armament of tanks themselves. In fact, many soldiers believed that the tank was more vulnerable than ever because infantry had acquired some antitank training and equipment. Anticipating some improvements in tank armor, the British Army concluded that it could safely standardize on a two-pounder (40mm) antitank gun. This was also the standard weapon mounted in most British medium tanks well into World War II. Yet such a weapon could penetrate the German medium tanks of 1939–1942 vintage only at ranges of 500 meters or less; moreover, the two-pounder gun was designed to fire only armor-piercing ammunition instead of the high explosive it would need to suppress enemy infantry or towed antitank gun crews. Although Percy Hobart called for a six-pounder (57mm) tank gun in 1938, this was not included as a formal requirement for tank design until after the fall of France in 1940.[22] Even then, most of the turrets that had been designed for the two-pounder were too small to be upgunned.

As Britain drifted in the area of mechanization, developments in the more traditional arms were equally mixed. Cavalry in essence merged into the mechanization process, although too late to learn all the mechanical and tactical differences between horses and light armor. Infantry was saddled with inappropriate weapons throughout the 1920s. It had no useful antitank capability, and the Lewis machine gun was too heavy to maneuver as a squad weapon.

Between 1936 and 1939, new equipment and organization finally restored the firepower and mobility of British infantry, but at a price. The excellent

Bren light machine gun, with its accompanying small armored carrier, was a significant advance. Each squad of a rifle platoon had a dismounted Bren gun, and the platoon had a two-inch smoke mortar and a caliber .55 Boyes antitank rifle. The Boyes was extremely heavy to carry, but it could penetrate most armor plate, and it gave the infantry a new sense of confidence. An infantry battalion consisted of four rifle companies and a headquarters with platoons of Bren gun carriers, two-pounder antitank guns, three-inch mortars, and antiaircraft machine guns. Heavy machine guns and 4.1-inch mortars were centralized into separate support battalions. Consequently, the British infantry battalion was much lighter and more mobile than it had been, but it had a somewhat reduced firepower and only limited antitank capability. On the eve of World War II, increasing thicknesses of armor made the Boyes rifle inadequate. This in turn forced the artillery to assume primary responsibility for antitank defense.[23] British field artillery had indeed developed excellent pieces that had an additional antitank capacity. In the process, however, the British had neglected the scientific procedures of indirect fire developed during World War I. Only the School of Artillery continued to teach these techniques, so that a few officers were familiar with them. In 1939 the prejudice of many artillerymen against artillery survey techniques, techniques that were necessary to site the guns accurately, led to a reorganization that briefly eliminated survey parties from artillery headquarters.[24]

Thus, by 1939 the British Army had lost much of its pioneering advantage in both equipment and technology. Outside the infantry battalion, cooperation among different weapons systems and combat arms was little better than it had been in 1914.

Germany: "Strike Concentrated, Not Dispersed"

France, Britain, and the United States, the victors of 1918, had a natural tendency to employ at least some of the materiel and doctrine of 1918 during the immediate postwar years. Moreover, these same nations tended to rely on large armies composed of a mixture of regular and reserve component forces. A defeated Germany, by contrast, had every reason to embrace new weapons, tactics, and training concepts.[25]

Even if it had wished to, Germany could not reproduce the mass armies and static defenses of 1914–1918. In 1919 the Treaty of Versailles limited the German Army to 100,000 long-term professional soldiers, without reservists except for the paramilitary police forces. The same treaty forbade Germany the possession of tanks, poison gas, combat aircraft, and heavy artillery. Paradoxically, for the Germans this prohibition may have been a blessing in disguise. The German defense budget and tactical thought were less restricted to, or dependent on, 1918-era technology than were other armies.

Gen. Hans von Seeckt, the man who rebuilt the German Army after its defeat, directed a complete review of the German method of warfare. Based on his experience fighting the Russians during World War I, Seeckt believed that a highly trained, mobile army could outmaneuver a much larger but immobile mass army.[26] Under his direction, German planners studied concepts and then developed the organization and equipment to realize them. Doctrine led technological development, in contrast to the situation in other armies. In those instances where field trials had to be conducted, the Germans used mock-ups, or tested forbidden equipment and concepts in secret within the Soviet Union.*

German planners did not start from scratch, of course. No army can completely escape its past, and Germany retained many of its World War I tactics. Nevertheless, Seeckt had an enormous advantage over the victorious Allies in freedom to innovate.

Since the 1860s, the German tradition of tactics and operations had favored outflanking and encircling the enemy or, if that failed, breaking through to disrupt his command and control structure from within. This was in contrast to the frontal battles of attrition that most of Germany's opponents had fought in World War I. This German tradition had two meanings. First, unlike the French and British, who had learned to attack on a broad front in order to protect their flanks, the Germans believed in concentrating all their resources on a relatively narrow front for breakthrough.[27] Second, this concentration of forces required the careful integration of all weapons and arms at battalion level and below to overcome the enemy's defenses. The infiltration tactics of 1917–1918 reflected this viewpoint and were retained after the armistice. Despite the restrictions of the Versailles Treaty, the 1921 German regulation, *Command and Combat of the Combined Arms,* included not only the infantry assault battalion and the careful artillery preparations of 1918 but also close air support, gas warfare, and tanks in an infantry-support role.[28] Again, the Germans were free to develop doctrine on the basis of their experience without being restricted to the available technology. Despite later manuals, Seeckt's sophisticated regulation remained the basis of German doctrine between the wars.

Another aspect of the German military tradition was decentralized execution. German commanders moved forward to observe and make tactical

---

*As the two outcast nations of Europe during the 1920s, Germany and the Soviet Union had much in common. Their secret exchange of military knowledge continued until Adolf Hitler came to power in 1933.

decisions for themselves. This ability enabled them to communicate their decisions to subordinates much more rapidly than was possible from a command post in the rear. This decentralization was facilitated by a mutual understanding among German leaders based on common doctrine such as the *Command and Combat of the Combined Arms.* Aware of both a commander's intentions and the common doctrine, subordinate leaders could execute those intentions in accordance with that doctrine and, thereby, reduce the need for detailed instructions from higher echelons. This decentralization and rapidity of decision making were ideally suited to any form of fluid combat, including mechanized operations—they were completely different from the lockstep form of command and control found in the democratic armies during World War I. The small career army permitted under the Versailles Treaty allowed German commanders to train their troops to much higher standards than seemed possible in the large, citizen forces of their former adversaries.

To facilitate such complicated maneuver warfare required more than just commanders operating close to the battle. The postwar German Army placed great stress on flexible radio communications. From the mid-1920s, each new vehicle, including the prototypes for future German tanks, always contained a radio mount. By the time that Hitler began to expand the army in the mid-1930s, the Germans had developed an entire family of radio systems, from short-range radios between combat vehicles to higher powered, lower frequency radios that could connect divisions to higher headquarters.[29] And the Germans needed to purchase only small numbers of radios, equipment that was far too expensive for the conscript French Army or even for the smaller British and U.S. forces.

As in so many other aspects of modern warfare, Hans von Seeckt stressed the importance of tanks. Although the German Army could not officially possess tanks under the Versailles Treaty, it could and did include armored warfare in its theoretical studies, manuals, and exercises. Seeckt's emphasis on mobility made the Inspectorate of Motor Troops a focus of doctrinal development. Throughout the 1920s, a series of bright officers worked on armored doctrine in this office and then went on to high command in other portions of the German Army. Although they studied foreign armored theorists such as Fuller, the Germans generally reached their own conclusions about armored warfare. Meanwhile, a small number of tanks were fielded, either secretly in the Soviet Union or openly under the fiction of armored cars for the German police.[30]

Given this official sponsorship of tanks, it might seem inevitable that the German Army would develop the style of warfare later known as blitzkrieg, or lightning war. The German experience of the psychological effects of tanks during World War I, their infiltration tactics, the belief in penetration on a narrow frontage, and the concept of decentralized execution of the commander's intent were components of blitzkrieg. In fact, however, the German Army as a whole did not completely accept the concept of mechanized warfare until the defeat of France in 1940. Before that, the majority of senior German commanders, like their counterparts in other armies, were open to other ideas but wanted tanks primarily to support infantry attacks. Adolf Hitler's patronage of the new armored forces for propaganda reasons undoubtedly added to the discontent and suspicion of the senior German officers.

Hitler's patronage did, however, enable the armored theorists of the German Army to implement at least some of the ideas nurtured during the 1920s. Heinz Guderian, a brilliant if self-aggrandizing product of the Inspectorate of Motor Troops, was the most famous and perhaps most influential of the German proponents of mechanization. Like Percy Hobart in Great Britain, Guderian had considerable experience with the early military use of radio communications. His 1914 service with radiotelegraphs supporting cavalry units taught him the difficulties of integrating new doctrine and equipment and then overcoming institutional resistance to them.[31] By 1929, when many British students of armor tended toward a pure armor formation, Guderian had become convinced that it was useless to develop just tanks, or even to mechanize parts of the traditional arms. What was needed was an entirely new mechanized formation of all arms that would maximize the effects of the tank. Only such a formation could sustain mobile warfare, whether offensive or defensive.[32]

Guderian and his contemporaries were not deterred by the general belief among military theoreticians that antitank defenses were becoming stronger. Unlike most advocates of armor, he considered antitank weapons to be an essential part of the mechanized combined arms team, not the defender of the traditional arms against the new weapons. Most early tanks were too small and unstable to carry accurate, high-velocity antitank guns. By contrast, the towed antitank gun was specially designed for maximum effectiveness against armor, and its small silhouette made it difficult to detect and hit on the battlefield. The German armored units trained to avoid fighting other tanks or antitank guns and instead to exploit in areas of little or no enemy resistance. In the event of tank-versus-tank combat, the German tanks would withdraw temporarily, luring an unwary enemy into a hidden screen of antitank weapons that had deployed behind the German spearhead. To do this, tanks needed reconnaissance units to lead the way and to screen the flanks of the advance, with combat engineers to sustain the mobility of the mechanized force. Motorized or mechanized infantry and artillery were necessary to reduce bypassed centers of resistance, to support tanks in the attack, and to hold areas seized by such attacks. The entire force required support units that could keep up with a rapid advance.

In 1931 Guderian became commander of the 3d Motor Transport Battalion. Using dummy equipment because of the limitations of the Versailles Treaty, this battalion was actually an experimental mechanized force consisting of one company each of motorcycles, armored cars, tanks, and antitank guns. A similar small-scale demonstration, using some of the first light tanks openly produced in Germany, impressed Hitler in 1934.[33] That same year, experimental maneuvers for a full panzer (armored) division took place, and in 1935 Hitler formed the first three such divisions on a permanent basis (Fig. 5). As in other armies at the time, Germany's first effort at armored organization included a tremendous number of tanks (561 per division).[34] Otherwise, this organization showed considerable balance in numbers and types of weapons. Thus from the beginning, the panzer divisions reflected the concept of balanced combined arms that had developed in the German Army of the 1920s. Moreover, during the later 1930s the various brigade and regimental headquarters began practicing to control cross-attached units and weapons systems, forming combined arms teams.

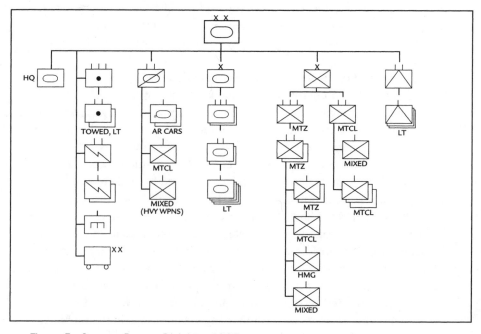

Figure 5. German Panzer Division, 1935

Such a system required considerable training and put great stress on the maintenance and logistical support of the cross-attached elements; for example, mechanics trained to repair one type of vehicle might have to do the same for other types borrowed from other units. Still, this system enabled the panzer division to combine different weapons systems as needed. In practice, it must be admitted, the German infantry and armor did not yet cooperate with one another at the level of company or platoon, but the entire German approach came closer to a mechanized combined arms team than most opponents.

Guderian did not accomplish this feat alone, nor did he succeed without opposition and difficulties. The other branches of the German Army resisted the creation of this new arm and naturally demanded a share of mechanization and motorization for themselves. During the later 1930s, the chief of the German General Staff directed the motorization of all antitank units and one engineer company in each infantry division, in order to give those divisions a flexible defense against enemy tanks. The same directive motorized four selected infantry divisions at a time when the panzer divisions were still short of transportation. In 1937–1938, two separate tank brigades were formed for infantry support, isolated from the other arms. At the same time, four light divisions, based on cavalry units in most cases, absorbed still more motorized and mechanized equipment. The actual composition of these units varied, but the most common pattern was an armored reconnaissance regiment, two motorized infantry regiments, one light tank battalion, and two towed howitzer battalions.[35] A frustrated Guderian found himself shunted aside as "chief of Mobile Troops," with little control over the motorized infantry and light divisions.

Another issue, however, was that early German tanks did not live up to the concepts of theorists like Guderian. Despite Hitler's support for panzer units, they had to compete for funds and production capacity not only with the rest of the expanding German Army but also with the German Air Force. Hitler placed first priority on the Luftwaffe because of the intimidation value that airpower gave him when dealing with the rest of Europe. Hermann Goering's influence as head of both the air force and the German industrialization plan also distorted production in favor of the Luftwaffe. Under the circumstances, Guderian had to settle for tanks that were not completely battleworthy. The *Panzerkampfwagen* I (tank) was really a machine-gun–armed tankette, derived from the British Carden-Loyd personnel carrier. The Panzer II did have a 20mm cannon, but little armor protection. These two vehicles made up the bulk of panzer units until late 1940.[36] Their value lay not so much in their armor and armament as in the fact that they were available early, in considerable numbers, and with radio communications. This allowed the new panzer force to conduct extensive training, establish battle procedures, identify and solve problems, and develop changes in organization and equipment. By 1939 the panzer divisions were still not completely ready for war, but they had gone through their first, most necessary stages of organization and training. Such an advantage was denied to most of Germany's opponents.

Another German advantage was in the field of close air support for ground operations. During the 1920s, Seeckt had kept air doctrine alive and had ensured that German civil aviation was led by former military aviators. When the Luftwaffe was established in 1933, most of the higher commanders were World War I airmen. By doctrine, the Luftwaffe, like other air services, favored missions such as strategic bombing and air superiority rather than supporting ground forces. However, the experience of the Spanish Civil War (1936–1939) gave the Germans experience in ground support as well as in strategic bombing. The small force of German "volunteers" sent to aid Gen. Francisco Franco against the Spanish Republican government used a limited number of obsolete fighters in a ground-attack role, with considerable effect. These experiences provided the impetus for Germany to create five ground-attack aviation groups in the fall of 1938. To some extent, this was an inevitable outgrowth of the fact that Germany was a continental power—unlike airpower advocates in the United States and Britain, proponents of the Luftwaffe could not ignore the need for support to ground operations. Indeed, German Air Force Manual 16, published in 1938, called for all aircraft to be concentrated at the decisive point. This did not mean, however, that the Luftwaffe would permit its aircraft to be controlled by ground commanders. Quite the contrary: German air doctrine was firmly wedded to the concept of airpower and strategic bombing; the only difference lay in the practical German willingness to commit some air force resources to ground support.[37]

As early as 1934, the Germans had become interested in dive-bombing after joint maneuvers with their Swedish counterparts. Unlike bombing from high altitude, dive-bombing allowed the pilot literally to aim the bomb at the target before releasing his weapon, thereby achieving much greater accuracy. At the same time, it was more difficult for antiaircraft gunners to follow the descent of an aircraft whose speed and altitude changed so precipitously. As an added advantage, the Germans soon learned that dive-bombing had a great psychological effect on the targeted troops, as the aircraft appeared to be aiming personally at the individual soldier. Ernst Udet, the chief of the Luftwaffe's development branch after 1936, persuaded his superiors to procure a limited number of close-support dive-bombers patterned after the U.S. Navy's Curtiss Helldiver. The resulting JU-87 Stuka dive-bombers equipped four of the five group-attack groups, each of twenty-seven aircraft, during 1939, although in fact there were more Stukas assigned to other units for nonclose-support missions.[38]

Further, in both Spain and Poland a very small number of air liaison detachments were attached to the infantry corps and armored division headquarters making the main attacks. Organized by the brilliant Wolfram von Richthofen, these detachments could pass air support requests from ground commanders directly to the Luftwaffe and could monitor in-flight reconnaissance reports. They could not, however, guide the aircraft onto specific targets, nor were they trained for such a role. Moreover, the handful of dive-bomber groups and air liaison detachments was available only to the army units at the point of main effort; all other army headquarters had to submit preplanned requests that might or might not be honored in a timely manner. As late as 1940, the commander of the Luftwaffe, Hermann Goering, insisted on absolute centralized control of his aircraft; on-call air support against targets of opportunity was still well in the future for the German armed forces.[39]

Despite these problems, the tradition of combined arms integration was generally continued and updated in the German Army between the world wars. Guderian was tactically incorrect when he denied the need to provide armor and motorized equipment for the other elements of an army that remained essentially foot-mobile and horse-drawn. This diversion of mechanized equipment to nonpanzer units was not too unusual. Still, despite or perhaps because of Guderian's complaints, the German Army kept its available armor more tightly concentrated than did most opposing armies. In September 1939, twenty-four of thirty-three tank battalions and 1,944 of 3,195 tanks were concentrated in the six panzer divisions.[40] The contrast with other armies, where large numbers of tanks were dedicated to infantry support and cavalry roles, is striking.

France: The Methodical Battle

The existence of a 100,000-man professional German Army during the 1920s forced the French to develop plans to counter a sudden invasion. The postwar French Army was huge on paper but ill-prepared to stop a surprise attack by even the small German force. If pre-1914 French officers had wrongly suspected the abilities of their reservists, their suspicions were more than confirmed between the wars. Except in its colonies, the French Army was essentially a cadre for reservists, who required weeks or even months to mobilize. After 1918 French war weariness eliminated the highly developed mobilization system of 1914. By 1928 conscripted service had been reduced to less than twelve months of training.

· ·

To protect itself from a sudden attack by the Germans, France chose to construct a sophisticated version of the defenses that had apparently worked so well in the attrition battle of Verdun. The Maginot Line (Map 3) was a string of self-contained concrete forts with gun turrets. These forts were built between 1930 and 1936 in northeastern France; their function was to protect the land regained in 1918 and to force any German invasion to pass through Belgian territory before reaching France. This extra distance would give France time to mobilize.

After the fact, many observers derided the Maginot Line because it appeared child's play for the Germans to outflank these fortifications. In practice, French industry was too close to the Belgian border to be defended by an extended Maginot Line, and France felt a moral obligation to support the Belgians, the gallant allies who had been victimized by German attack in 1914. Thus from 1920, French military planners expected to fight any future war inside Belgium instead of on French soil. The Maginot Line was a secure anchor, a base around which the mobile field forces of the French Army would maneuver.[41] In the later 1930s, both France and Britain ex-

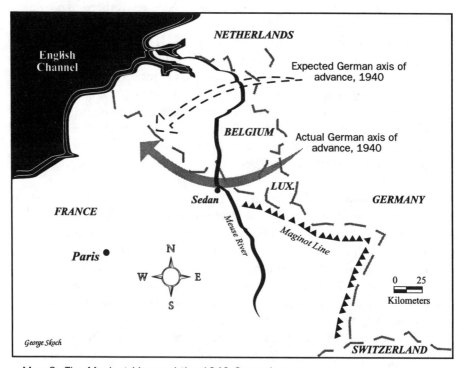

Map 3. The Maginot Line and the 1940 Campaign

pected the approaching war with Germany to be a repetition of 1914, with Germany advancing through Belgium and possibly through the Netherlands as well. Because Belgium had reasserted its neutrality in 1936, France and Britain could not enter that country to help defend it until the Germans had already invaded. Thus, the majority of French and British mobile forces planned to make a headlong rush into Belgium and the western Netherlands. The surprise to the Allies in 1940 was that the main German effort was shifted southward, passing through Luxembourg and southern Belgium toward Sedan. Almost by accident, this German advance chopped the weak hinge that connected the mobile forces in the north with the Maginot Line to the south.

Moreover, despite the intended purpose of the Maginot Line (delaying a German invasion), the practical effects were much less positive for French national security. The tremendous expense of fortress construction during the depths of the Great Depression restricted the depth of the forts, their size, and the number of weapons in them. Only a few positions included the lavishly equipped works shown in contemporary propaganda. In case of war, the line had to be supplemented by troops in trenches and hasty bunkers between the fixed forts. More important, the Maginot Line had a negative psychological effect on both politicians and commanders. Its apparently invincible defensive strength reinforced the general left-wing political belief that France should avoid any aggressive actions and be content to defend its frontiers. Thus a sensible precaution against surprise attack helped encourage French appeasement policies before 1939 and discouraged any real offensive once the war began.

This defensive orientation influenced not only national budgets but also French military doctrine. More than any other participant in World War I, in postwar regulations France retained its belief in carefully controlled battles of positions. Under the influence of Marshal Philippe Pétain, the master of defensive battle, the French Army produced *Provisional Instructions for the Tactical Employment of Larger Units (1921)*. This regulation was not entirely oriented on the defensive, but to minimize casualties it did insist on elaborate, methodical preparations before attacking. Within the carefully coordinated circumstances of a set-piece offensive, battle would employ all the combat arms to assist the infantry: "The infantry is charged with the principal mission in combat. Preceded, protected, and accompanied by artillery fire, aided where possible by tanks and aviation, it conquers, occupies, organizes, and holds the terrain."[42]

This conception had two flaws. First, such a meticulously planned, centrally controlled operation was unable to react to sudden changes. The German offensives of 1918 had already demonstrated that any enemy action that disrupted the defender's linear deployments and lockstep planning would catch the French headquarters off guard, unable to reorganize a defense against a highly mobile, flexible attacker. To cite but one example, even in 1940 most French tactical communications relied on field telephones rather than on the radio networks developed by their German opponents.

More generally, the French doctrine viewed combined arms as a process by which all other weapons systems assisted the infantry in its forward progress. Tanks were considered to be "a sort of armored infantry," subordinated to the infantry branch.[43] This approach at least had the advantage that armor was not restricted purely to tanks. During the 1920s, the French cavalry experimented extensively with armored cars and ultimately with half-tracks, vehicles with wheels in front and caterpillar tracks in the rear. These half-tracks sometimes formed combat teams with armored cars, towed artillery, motorcycles, and light tanks carried on trucks until contact was made.[44] In fact, the French half-tracks may have been the models for later German and American infantry carriers. Still, the subordination of tanks to infantry impeded the development of roles for armor other than close infantry support. Moreover, though half-tracks might be useful in colonial wars or for reconnaissance tasks, infantry still walked in the deliberate assault. Armor was tied to the rate of advance of foot-mobile infantry and creeping artillery barrages. The alternative of finding ways to increase the mobility and protection of the infantry in order to keep pace with the tanks was rarely considered. The slow speed of the World War I–vintage FT light tank, which equipped most French armored units throughout the 1920s, reinforced this attitude.

Not all Frenchmen held this view, of course. Gen. Jean-Baptiste Estienne, the commander of the World War I French tank corps before it was disbanded, was quite farsighted in his concept of mechanized warfare. In 1919 Estienne submitted his "Study of the Mission of Tanks in the Field" to Pétain's headquarters. This remarkable document explained the need to provide armored, tracked vehicles not only for tanks but also for reconnaissance, infantry, artillery, and even battlefield repair and maintenance teams. Estienne's vision of this force, supported by air bombardment and attacking in depth against a narrow segment of the enemy front, closely resembled the best mechanized practice of World War II. In 1920 Estienne proposed a 100,000-man armored army with 4,000 tanks and 8,000 other vehicles. Instead of rejecting the use of infantry, he argued that infantry mounted in armored vehicles would be able to attack using its organic weapons.[45] Estienne's concept was not only radical militarily but also seemed too offensive in orientation, too aggressive to be acceptable to French politicians after the enormous casualties of the war. Nevertheless, Estienne remained Inspector of Tanks until his retirement in 1927.

Despite the restrictions imposed by the Great Depression and by the enormous cost of the Maginot Line, Chief of Staff Maxime Weygand took significant steps toward motorization and mechanization during the early 1930s. Ultimately, seven infantry divisions became motorized, and one brigade in each of four light cavalry divisions was equipped with half-tracks and armored cars. The resulting cavalry division was a strange hybrid, consisting of 10,000 men, 2,200 horses, and 50 armored vehicles, but only 24 field guns and 20 antitank guns. In 1934 Weygand continued the trend toward armored cavalry by forming the first "light mechanized division," Division Légère Mécanique (DLM), (see Fig. 6). This division, with its combination of reconnaissance, light tanks, truck-mounted infantry, and towed artillery, was remarkably similar to the German panzer division being developed at the same time. Yet Weygand was a cavalryman, and it was politically easier to justify a defensive screening force than an "offensive" armored assault unit. As a result, the four DLMs ultimately formed received standard cavalry missions of reconnaissance and security instead of mechanized main battle tasks.[46]

Just as the French Army was cautiously moving forward in the area of mechanization, this development was almost aborted by the writings of Charles de Gaulle. In 1934 Lieutenant Colonel de Gaulle published *Towards the Professional Army*. This call for a 100,000-man armored army was based heavily on Estienne's work, although de Gaulle is usually regarded as an original theorist. De Gaulle's book was hardly innovative in terms of doctrine and organization in that it envisioned a pure tank brigade operating in linear formation, followed by a motorized infantry force for mopping-up operations. The real problem was political. In a nation that had become extremely pacifistic and was dedicated to the principle of the citizen-soldier, de Gaulle was advocating an aggressive, professional standing army of technicians. His "instrument of repressive and preventive maneuver"[47] might well be used to start an offensive war with Germany or to support a rightwing coup d'etat in republican France.

De Gaulle's sensational book not only jeopardized the more gradual efforts of Weygand but also set extremely high standards for what constituted an armored division. In 1936 France belatedly decided to produce armor and other equipment in larger quantities, including 385 B-1bis (or B-1b) tanks. The B-1bis, developed by Estienne in the early 1920s, was still one of the best tank designs in the world fifteen years later. It had sixty millimeters of frontal armor in a cast hull, hydromatic transmission, and other advanced features. It was limited by the small size of its turret, where one man

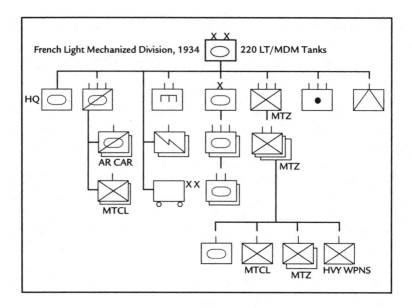

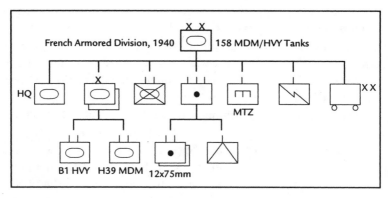

Figure 6. French Light Mechanized Division, 1934, and Armored Division, 1940

had to be both tank commander and gunner for a 47mm gun, but a lower velocity 75mm gun was mounted in the hull. The B-1bis also had an extremely limited fuel capacity. Despite these difficulties, it was an excellent weapon that caused the Germans much difficulty in 1940. Yet, given the fine craftsmanship involved in B-1bis production and the weakened state of French industry, it took years to produce sufficient tanks to organize an armored division on the pattern advocated by Estienne and de Gaulle. Even after the war started, France could never produce more than fifty of these tanks per month, and the rate prior to 1939 was much lower.[48]

As a result, France did not form its first two armored divisions (Division Cuirassée [DCR], Fig. 6) until after the war began, and, even then, it had to substitute medium tanks for half of the authorized B-1bis in each division. The resulting unit was primarily a collection of tanks for an armored breakthrough; it lacked sufficient reconnaissance, antitank, infantry, artillery, and engineer support. Similar problems plagued the production of other tanks and military equipment; thus, French troops rarely had the time for realistic training and experimentation that the Germans had achieved before 1939. The French regulation for large armored-unit tactics was not even issued until March 1940, a few weeks before the German invasion of France.[49]

Despite such limitations, France slowly modernized during the 1930s. The 1921 *Provisional Instructions* gave way to a much more sophisticated regulation in 1936. These new *Instructions* recognized the major changes in warfare, including fortified fronts such as the Maginot Line, motorized and mechanized units, antitank weapons, increased air and antiaircraft involvement in combat, and improved communications. The regulation no longer classified tanks by size but designated the particular mission they would perform at any given time. Tanks could either accompany infantry, precede infantry by bounds to the next terrain feature, or operate independently, especially after the enemy's defenses had already been penetrated. The 1936 regulation, however, still insisted on the primacy of infantry, the careful organization of artillery, and the methodical advance of all elements in accordance with an elaborate plan. The regulation repeatedly emphasized the need for "defense without thought of retreat," which tended to mean rigid orientation toward the terrain and the enemy to one's front rather than toward maneuvering to deal with a threat to the flanks or rear. References to antitank defense-in-depth also appeared frequently in this regulation, but France lacked the troops and especially the antitank mines to establish such a defense in 1940. Finally, because of the possibility of enemy signals intelligence, radios were to be used only when no other means of communications were available. In any event, most French tank radios were capable of only short-range communications with dismounted infantry in a deliberate attack; they were consequently useless in mobile operations. Thus, most of the French command and control still moved at the pace of communications in World War I.[50]

As in Britain, French air support to ground forces consisted primarily of aerial reconnaissance in the battle area, with bombing only beyond the range of artillery. Even when the air force was called upon to fight on the battlefield, the French air planners emphasized low-level strafing rather than dive-bombing, a decision that made the attackers easy prey for air defenses. For much of the interwar period, the French Air Force tried to design aircraft capable of multiple roles, jacks-of-all-trades that were usually ineffective. Responding to the frequent changes in technology, the French tended to produce only a few prototype aircraft of any kind, delaying production until it was too late to refine the aircraft designs or their tactics. By doctrine, available air support was centrally controlled at the level of an army group, three echelons above a division; given the telephone communications and staff procedures of the era, aircraft would never reach the needed area in time.[51]

France entered World War II with two armies: a mass militia that required months to organize and train, and a smaller number of new mechanized formations. Typical of the former army was the 55th Infantry Division, the low-priority unit that ultimately faced the German offensive at Sedan in May 1940. Ninety-six percent of the 55th Division were reservists, men who had undergone basic training as long as twenty years before the battle. The division was not even organized until after the war began in 1939 and then devoted most of its effort to building defensive fortifications instead of training as coherent units. The entire division sector had only one antiaircraft battery to respond to more than 800 German dive-bombers in a single day.[52] As for the French mechanized formations, like their counterparts in Britain most armored units were specialized either for cavalry missions or for deliberate breakthrough attacks; they were not balanced for all types of mobile operations. Given these limitations, the French concept of slow, methodical action based on firepower appeared as the only course that would allow them to attack at all. Unfortunately, the Germans did not wait for the French to plan and execute such attacks.

The Soviet Union: Deep Battle

The Soviet Union's military development after World War I differed from that of the rest of Europe for two reasons. First, the Red Army was created in 1918 after the Bolshevik Revolution and therefore lacked the traditions and training of other major armies. Many of the new Red commanders had been noncommissioned or junior commissioned officers during World War I, but few trained senior officers of the tsarist army remained with the new regime, and those who did were often suspected of antibolshevik sympathies. As a result, the Red Army was open to change, unhampered by excessive traditions or past habits. It was also subject to the blunders of ignorance. Second, the Russian Civil War (1918–1921) was markedly different from most of the European campaigns of World War I. Because of the vast distances and understrength armies involved in the civil war, penetration and encirclement were no longer difficult, and fluid maneuver was the rule. The elite of the Red Army by the end of the civil war was Marshal S. M. Budenny's 1st Cavalry Army, which had patterned its encirclements and pursuits after the best tsarist cavalrymen. The veterans of this army received the patronage of Joseph Stalin, dictator of the Soviet Union, who had been the political commissar of the higher headquarters that controlled Budenny's cavalry. As a result, many officers from this army rose to senior positions before World War II.[53]

Like Hitler's Germany, but unlike France and Britain, the Soviet Union was openly interested in offensive warfare as a means of spreading its political doctrine. As a practical matter, Stalin chose to concentrate on developing the Soviet Union before trying to revolutionize Europe. Still, the Red Army could expect that any future war would be offensive, using weapons that democratic societies abhorred as too aggressive. This offensive orientation was reinforced by the close relationship that existed between the Red Army and the German Army from 1923 to 1932. Soviet officers studied in Germany, and the Germans secretly manufactured and tested tanks, aircraft, and poison gas in European Russia. Soviet military doctrine, however, appeared to be largely independent of similar developments in Germany; Soviet concepts were official policy long before Guderian gained even partial approval from his government.

During the course of the 1920s and early 1930s, a group of Soviet officers led by Marshal Mikhail Tukhachevsky developed a concept of Deep Battle to employ conventional infantry and cavalry divisions, mechanized formations, and aviation in concert.[54] These efforts culminated in the 1936 *Field Regulations*. Instead of regarding the infantry as the premier combat arm, Tukhachevsky envisioned all available arms and weapons systems working together in a two-part battle. First, a massed, echeloned attack on a narrow front would rupture the defender's conventional defenses of infantry, artillery, and antitank weapons. The attacker's artillery and mortars would suppress defending artillery and especially defending antitank guns. Moving behind the artillery barrage and a few meters in front of the infantry, the tanks could safely crush wire, overrun machine-gun posts, and reduce other centers of resistance.

Up to this point, the battle envisioned was similar to those of 1918. However, once the enemy's forward defenses were disrupted, tanks would not be tied strictly to the infantry rate of advance but could take advantage of local opportunities to penetrate and attack enemy reserves, artillery, headquarters, and supply dumps. This action would duplicate on a small scale the second part of the battle, which was to disrupt and destroy the enemy by deep attacks. "Mobile groups," composed of cavalry, mechanized formations, or both, would exploit their advantage in speed to outflank the enemy or to develop a penetration in order to reach the enemy rear areas. The object was to attack the entire depth of the enemy defenses simultaneously, with conventional frontal attacks, long-range artillery fires, deep penetrations by mobile forces, and bombing and parachute attacks on key points. Smoke and deception operations would distract the enemy from the attacker's real intentions.[55]

This remarkably sophisticated doctrine was backed up by a force structure that, by 1937, was well on its way to implementing Tukhachevsky's concepts. Using the expanded production facilities of the Soviet government's first Five-Year Plan (1929–1934), and employing some design features taken from the American tank inventor Walter Christie, the Soviets produced 5,000 armored vehicles by 1934.[56] This wealth of equipment enabled the Red Army to create tank organizations for both infantry support and combined arms mechanized operations. Virtually every rifle (infantry) division had a tank company or battalion attached to it, with an entire regiment of 190 or more tanks for each of the horse cavalry divisions. Beginning in 1930, the Red Army experimented with integrating all arms into functional mechanized groups at battalion, brigade, and higher levels. Although unit structures changed frequently as equipment and tactical techniques evolved, the 1935 mechanized "corps" was typical of these developments (Fig. 7). The four corps organized under this concept were really small armored divisions, because the Soviets often used the terms "corps" and "brigade" to designate experimental, combined arms organizations of division and regimental size, respectively. The 1935 mechanized corps were still top heavy with tanks, like most contemporary organizations in the West, but they nevertheless integrated the essential combat arms at a relatively low level. The trend during the later 1930s was for these corps, redesignated "tank corps" in 1938, to become increasingly large and armor-heavy.

This Soviet force structure had its problems, of course. To begin with, despite the massive industrial support of the Soviet Union, the armored force was so ambitious that not all units could be fully equipped. Given this shortage, Soviet historians later criticized the separation of available equipment into infantry-support and independent formations, a common problem in most armies.[57] More specifically, the average Soviet citizen had little experience with motor vehicles, so that maintenance was often a problem, particularly as the vehicles wore out during the later 1930s. Soviet radios were notoriously unreliable, making command and control of this mass of moving vehicles difficult. Despite frequent major exercises during the mid-1930s, the Soviet armored force needed several additional years of experimentation and training before it could reach its full potential.

It never had that time. On 12 June 1937, the Soviet government executed Tukhachevsky and eight other high-ranking officers, as Stalin shifted his purge of Soviet society against the Red Army, the last power group that had

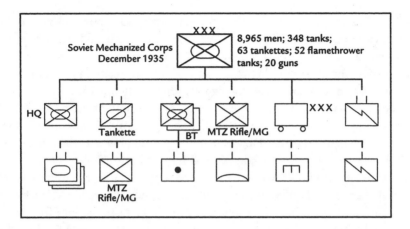

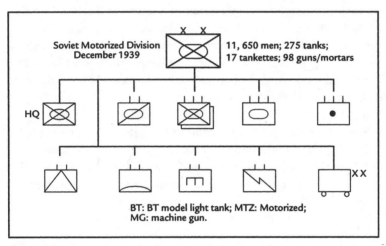

Figure 7. Soviet Mechanized Corps, December 1935, and Motorized Division, December 1939

even the potential to threaten him. In the ensuing four years, the Soviet government imprisoned or executed at least 40 percent of the officer corps, including a majority of all commanders of units of regimental size or larger. Of an estimated 75,000 to 80,000 officers in the armed forces, at least 30,000 were imprisoned or executed, and another 10,000 were dismissed in disgrace. Thus, at the same time the Red Army was expanding because of the threat from Nazi Germany and Imperial Japan, it was losing its most experienced planners and leaders. The politically reliable survivors were promoted into positions far above their previous training and experience, with disastrous effects on unit training and tactics.[58]

At the same time that Tukhachevsky's thought was under suspicion, the Soviet experience in the Spanish Civil War caused the Red Army to reassess mechanization. Dmitri Pavlov, chief of tank troops and one of the senior Soviet officers to serve in Spain, came back with an extremely pessimistic attitude. The Soviet tanks were too lightly armored, their Russian crews could not communicate with the Spanish troops, and in combat the tanks tended to run away from the supporting artillery and infantry. Pavlov argued that the new mechanized formations were too large and unwieldy to control, too vulnerable to antitank fire, and would have great difficulty penetrating enemy defenses in order to conduct a Deep Battle. The fact that Pavlov had been able to use only fifty tanks without any chance of surprise at the battle of Esquivas (29 October 1936) apparently did not dissuade him from generalizations.[59] In any event, many observers from other armies reached the same conclusions based on the limited experience in Spain.

In July 1939, one of Stalin's cronies, Marshal G. I. Kulik, chaired a special commission to review the question of tank force organization. With most of Tukhachevsky's followers dead or imprisoned, there were few advocates for large combined arms mechanized units. The commission therefore directed the partial dismantling of such units and reemphasized the infantry support role for tanks. A few months later, however, the commission created a new, more balanced organization, the motorized division of December 1939 (Fig. 7). This continued support for the 1936 doctrine and force structure may have been in response to German armored success in Poland in September 1939 and to the Soviet success that year against Japan. Four of the planned fifteen motorized divisions were organized in early 1940, representing a better all-purpose organization than the tank corps they replaced.[60]

This reorganization only contributed to the confusion of the Red Army, which was unable to occupy eastern Poland effectively in 1939 and had to make a major effort to defeat Finland in 1940. These battlefield failures prompted a series of reforms in organization, leadership, and tactics that slowly began to improve Soviet military capabilities. The only successful campaign of this period was in the undeclared war against Japan. Stalin was apparently so concerned about Japanese expansion in northeastern Asia that he gave one of Tukhachevsky's most able students, Gen. Georgi Zhukov, a free hand in commanding the Soviet forces there. The Red Army in Siberia was among the last units to be affected by Stalin's purge. Thus, with the exception of some reserve units mobilized in the crisis, the training and command structures of the Siberian forces were still intact when hostilities with the Japanese Army erupted in the summer of 1939 on the Khalkin-Gol River of Manchuria (Map 4). The Japanese chose to fight the Soviets in this

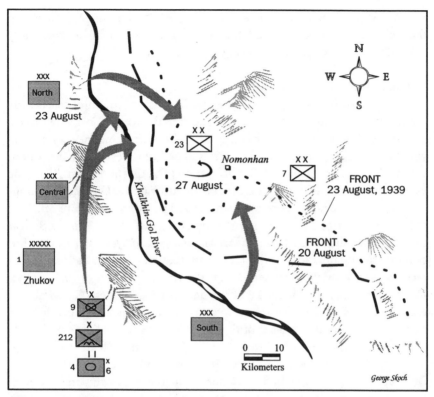

Map 4. Khalkin-Gol, 20–31 August 1939

remote area on the border between Japanese-occupied Manchuria and
Soviet-dominated Outer Mongolia, believing that the Soviets would be unable to assemble, much less supply, a large force in such a barren region.
To the surprise of the Japanese, the Soviets massed 469 light tanks, 426 other
armored vehicles, 679 guns and mortars, and over 500 aircraft, all supplied
by thousands of trucks. Zhukov organized a classic double envelopment,
surrounding the Japanese from both north and south simultaneously. First,
a series of Soviet probing attacks in the center pinned down the Japanese
defenders. Then the two Soviet flank attacks pressed forward, encircling the
Japanese 23d Infantry Division and part of the 7th Infantry Division. The
Soviet attacks used tank and machine-gun direct fire, as well as coordinated
artillery fire, to protect their advancing infantry. In some cases, the infantry
rode on the outside of armored cars, reducing the time needed to close with
the enemy but exposing both vehicles and riders to concentrated enemy fire.
Some Soviet commanders were unimaginative in executing Zhukov's plan,
however, making repeated frontal attacks instead of bypassing Japanese resis-

tance.[61] Still, Khalkin-Gol provided an excellent trial of Soviet doctrine on the very eve of World War II. Zhukov and his subordinates in this battle naturally rose to prominence during that war.

United States: Streamlining

The U.S. Army, despite its unique square division structure, was heavily under the influence of French tactical and staff doctrine in 1918. Of necessity, American officers had learned to do business in a manner compatible with the French units they dealt with daily. The French even gave the Americans their system of general staff organization (G1, G2, and so on). To some extent, therefore, the immediate postwar doctrine of the U.S. Army continued to parallel that of the French Army. Initial postwar regulations reflected the French view of combined arms so faithfully that in 1923 the U.S. War Department issued a draft, *Provisional Manual of Tactics for Large Units,* without even mentioning that it was a direct translation of the 1921 French *Provisional Instructions.*[62] That same year, the revised version of *Field Service Regulations, U.S. Army* insisted that "no one arm wins battles. The combined employment of all arms is essential to success." In the next paragraph, however, it stated that the mission of the entire force "is that of the infantry."[63]

Still, this rigid view of tactics did not affect all American soldiers, nor did it last for a long period of time. As early as 1920, staff officers such as Brig. Gen. Fox Conner had concluded that the requirements of trench warfare were inappropriate for operations on the American continent, the only politically acceptable arena for planning future American wars. Conner asked Gen. Pershing, the U.S. wartime commander in France, to discard the square division structure because it was too immobile and unwieldy for future operations. Pershing recommended that the U.S. infantry division be reorganized along the lines of European triangular divisions and that units needed only for specialized operations be "pooled," or controlled centrally by higher headquarters such as a corps or a field army.[64]

This plan for streamlining and pooling was only one product of an exhaustive series of postwar studies of tactics and organization, studies carried out at Pershing's direction by investigative boards of the American Expeditionary Force (AEF) in Europe. Unfortunately, most of these proposals fell afoul of bitter political wrangling between Pershing and his counterpart in the War Department in Washington. As a result, the square division's authorized structure changed only slightly during the 1920s, even though in fact the active, reserve, and National Guard divisions were skeletonized for lack of troops.

Still, by 1925 American officer education was again focused on mobile warfare, with trench warfare relegated to the status of a special situation.

Financial restrictions and the general neglect of the U.S. Army prevented major changes in equipment and organization until the mid-1930s. Then the army was able to use public works funds allocated to restart the depression economy as a means of achieving limited improvements in equipment. These included partial motorization of active army and National Guard divisions and production of different gun carriages with pneumatic tires for existing artillery pieces. Although still armed with the aging, French-designed 75mm gun, the U.S. artillerymen could tow their guns behind motor vehicles and elevate the gun tubes high enough to use them in a limited antiaircraft role.

In 1935 Gen. Malin Craig became Chief of Staff of the U.S. Army. Craig had apparently been influenced by Conner and the other reformers of 1920, and he instigated a review of all combat organizations and tactics.[65] Craig specifically suggested development of a smaller, more mobile division using mechanical power to replace human power wherever possible. A General Staff board drew up a proposed division structure that totalled only 13,552 men and closely paralleled European divisions of the same period. From 1936 through 1939, the 2d U.S. Infantry Division conducted extensive tests of this concept, reviewing such matters as the amount of firepower and frontage that could be allocated per man and per unit, the proportion of artillery and transportation that should support each infantry unit, and the echelon (platoon, company, battalion, or regiment) at which different infantry weapons should be pooled.

One of the driving forces behind these tests was then–Brig. Gen. Leslie J. McNair, who later designed and trained all U.S. Army ground troops during World War II. As a junior officer two decades before, McNair had participated in the post–World War I studies conducted by the AEF, and many of the concepts of those studies were revived in the 2d Division tests. Unfortunately, McNair also absorbed the Pershing attitude that infantry platoons should still be armed essentially with rifles; as a result, he tended to pool more complicated weapons at higher headquarters.

By 1939 the resulting organization was remarkably close to the AEF ideas of 1920. In essence, the machine gun and other specialized heavy weapons were integrated into the infantry rifle organization at most levels of command. To avoid an excessive span of control, a commander at any level had a headquarters, three subordinate rifle units, and a specialized weapons unit—three rifle platoons and a heavy weapons platoon in each company, with three such companies and a heavy weapons company in each battalion. In practice, regimental commanders might shift companies from one battalion to another, and divisional commanders might move entire battalions between regiments, allowing those commanders to concentrate most of their force at the critical point. Doctrinally, however, all units operated with three subordinate maneuver units.

Each echelon also had a combination of weapons that fired in a flat trajectory and in a curved trajectory or high angle of fire. Although the infantry received greater firepower in terms of automatic weapons and mortars, it was echeloned in an effort to keep each level of infantry relatively lightly equipped and mobile. Thus, for example, the infantry platoon had nothing heavier than the Browning Automatic Rifle (BAR), and the company had nothing heavier than the 60mm mortar.[66] This dedication to mobility, when combined with a continued faith in the individual rifleman, meant that an American army platoon had less firepower than its European counterparts— the BAR had a much slower rate of fire than most light machine guns found in European platoons. This deficiency was only partially corrected by the rapid-fire ability of the M1 Garand rifle developed in the United States between the wars. Since American tactics were based on the premise of using one element to put down suppressive fire while another element maneuvered toward the enemy, this organization left U.S. infantry at a disadvantage.

The same principle of weapons pooling was continued throughout the new triangular division. Light antitank guns, heavy mortars, and machine guns were relegated to the heavy weapons company of each battalion. Specialized arms such as tanks, antiaircraft, and most antitank weapons were not even authorized within the division, because McNair believed that such weapons should be held in a central mass and used only against a major enemy force. Similarly, the division received only one company-sized reconnaissance troop, with long-range reconnaissance being assigned to higher headquarters. This arrangement limited the division commander's view of what was in front of him on the battlefield. The general result was an infantry force that was at once more mobile and more heavily armed than its predecessors yet deficient compared to foreign armies. Its principal drawback, besides a shortage of automatic weapons, was its limited capacity for antiaircraft and antitank defense. During the later 1930s heavy machine guns still seemed adequate against aircraft and armored vehicles. Thus, American planners believed that these machine guns, with a limited number of 37mm antitank guns, would suffice for the triangular division. Once the German blitzkrieg demonstrated its psychological and physical effect on infantry, the U.S. Army realized that it had to add more antitank defenses.

The controversies about the triangular division included the proportion of engineers and artillery for the infantry component. The army was conditioned to regard engineers only in their World War I role of road construction and limited fortification support. At one point, General Craig suggested eliminating all engineers from the division structure. In 1938 General McNair recommended a single engineer company of 175 men, or 1.7 percent of the division, because he believed that only hasty road repair and limited road-block construction would occur in the next war. The engineers had to campaign vigorously for their very existence in the division, arguing that an increasingly motorized and mechanized army had greater need for engineers to construct and eliminate antitank defenses and other obstacles. Only the German use of combat engineers for such tasks in 1939–1940 finally convinced the United States to retain an engineer battalion in each division.[67] Even this was a mixed blessing for the engineers, because they were frequently used as the division's infantry reserve force.

The 1935 division proposal had conceived a division artillery consisting of three combined 75mm-gun/81mm-mortar battalions for direct support of the three infantry regiments, with a 105mm howitzer battalion for general support of the entire division. All other artillery was to be nondivisional, pooled at higher headquarters and allocated to support specific operations. In actual testing, however, the artillery found that the 81mm mortar was essentially an infantry weapon; its short range meant that it was best combined with the infantry it supported instead of operating under artillery control from the rear. In any event, as an artilleryman McNair objected to this emphasis on dedicated support to the infantry, arguing that longer range weapons with greater centralized control would lead to more flexible massed fires. No unit, he contended, needed weapons whose range exceeded the parent unit's assigned area of operations. Ultimately, the army decided to have three battalions of 75mm guns, to be replaced by 105mm howitzers when more of the new weapons were produced. The general support artillery battalion would therefore consist of twelve 155mm howitzers and an additional battery of heavier antitank guns. The June 1941 organization represented the final step before American entry into the war (see Fig. 8).

The debate over artillery in the division organization occurred at the same time that the U.S. Army Field Artillery School was developing the next major step in infantry–artillery fire coordination, the ability to focus or mass artillery fires from many units on targets of opportunity. During World War I, massed fires were normally the result of carefully planned artillery concentrations, in which known targets were predesignated well ahead of time, their positions plotted on maps or overlays. If the infantry needed artillery fire on an unexpected target of opportunity, however, it was difficult to bring more than one battery to bear on it. Each artillery battery had its own forward observer, who had to be able both to see the target and to communicate with his battery. In practical terms, the observer had to keep in field telephone contact with the battery. Such reliance on landline communications greatly restricted his ability to accompany the infantry in the advance,

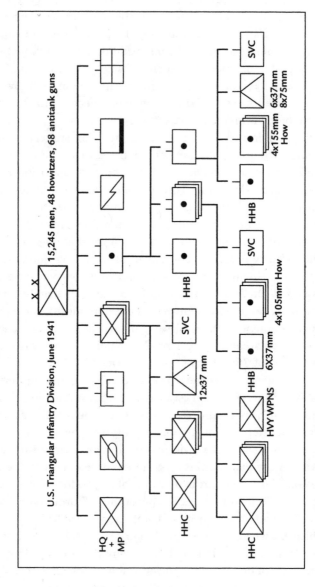

Figure 8. Type U.S. Triangular Infantry Division, June 1941

although some forward observers managed this feat. Even if the forward observer could adjust his own battery's fires to strike the target, he had no accurate way of guiding other batteries, unless the target's map location was known precisely. Thus concentrating or massing the fires of different batteries usually meant that each battery had to have its own forward observer in a position to see the target.

Between 1929 and 1941, a series of instructors at the Field Artillery School gradually developed a means of concentrating any amount of available artillery fire on a target of opportunity.[68] One obvious step in this process was to have observers use new, more reliable radios instead of field telephones to communicate. More important, the gunnery instructors developed forward observer procedures and a firing chart that together would allow an artillery headquarters to record adjustments to the impact of artillery shells as viewed from the observer's location instead of from the battery's location. Graphical firing tables compensated for differences in the locations of different batteries. Ultimately, the precise location of one artillery piece in each battalion was surveyed in relation to a common reference point for all artillery units in that division area. The resulting fire-direction centers (FDCs) could provide infantry units with an entire battalion, or even multiple battalions, of guns firing on a target that only one observer could see. The different batteries could fire together even when they were dispersed over a wide area, reducing the target they presented to enemy counterbattery fire.

By contrast, throughout World War II German artillerymen had to use well-known terrain features to adjust on a target of opportunity; massed fires of multiple batteries remained extremely difficult and time-consuming. Fire-direction centers gave the U.S. Army a new and unprecedented degree of infantry–artillery integration. It also encouraged the U.S. Army to maintain large amounts of nondivisional artillery to reinforce divisions as needed.

The United States was not nearly so advanced in the development of armored and mechanized forces, however.[69] As in France, the supply of slow World War I tanks and the subordination of most tanks to the infantry branch impeded the development of any role other than direct infantry support for a breakthrough attack. Yet the British experiments of the later 1920s and the persistent efforts of a cavalry officer named Adna Chaffee Jr. led to a series of limited steps in mechanization.

In 1928 and again in 1929, an ad hoc Experimental Armored Force (EAF) was organized at the Tank School in Fort Meade, Maryland. Two battalions of obsolescent tanks; a battalion of infantry in trucks; an armored car troop (i.e., company); a field artillery battalion; and small elements of engineer, communication, medical, ammunition, chemical warfare, and maintenance troops formed the EAF. Despite frequent mechanical breakdowns in the equipment, these experiments aroused sufficient interest for a more permanent force to be established at Fort Eustis, Virginia, in 1930. Unfortunately, the austere budgets of the Great Depression forced the army to disband this unit a year later for lack of funds. The Infantry School at Ft. Benning, Georgia, absorbed the Tank School and the remaining infantry tank units.

As chief of staff from 1930 to 1935, Douglas MacArthur wanted to advance motorization and mechanization throughout the army rather than confining them to one combat arm or unit. Restricted budgets made this impossible, but Chaffee did persuade MacArthur to conduct limited mechanized experiments with cavalry units, because the horse-mounted cavalry's existence was threatened by its apparent obsolescence on the battlefield. By law, tanks belong to the infantry branch, so the cavalry gradually bought a group of "combat cars," lightly armored and armed vehicles that were often indistinguishable from the newer infantry tanks. In 1932, a single squadron (i.e., battalion) of mechanized cavalry moved to Camp Knox, Kentucky, to be followed by another squadron in late 1936. These units were the nucleus of the 7th Cavalry Brigade (Mechanized). A series of early armor advocates commanded this brigade, including Adna Chaffee himself in 1938–1940. However, this force was plagued by the same difficulties as mechanized cavalry were in Britain and France. Its weapons were too small, its armor was too thin, and it was generally viewed as a raiding or pursuit force in the cavalry tradition, not as a combined arms combat force in its own right. Despite Chaffee's efforts, the other combat arms trained with the brigade only on intermittent exercises. Not until January 1940 was a mechanized engineer troop (i.e., company) approved for the 7th Brigade.[70] About the same time, the 6th Infantry Regiment joined the 7th Brigade as truck-mounted infantry, and a Provisional Tank Brigade grew out of the infantry tank units at Ft. Benning.

The German armored attack on France in May 1940 gave further impetus to mechanized experiments already conducted in U.S. Army maneuvers. To avoid branch prejudices, Chaffee convinced the War Department to create an "Armored Force" outside the traditional combat arms. Consequently, in July 1940 the 7th Cavalry Brigade and the Provisional Tank Brigade became the nuclei for the first two armored divisions. These divisions, like the first organizations of the European powers, were unbalanced with too many tanks. Each was authorized six battalions of light tanks and two battalions of medium tanks, for an approximate total of 400 armored vehicles. Yet each division originally had only two battalions of armored infantry (mounted, as they became available, in half-tracks) and three battalions of artillery. The predominance of light tanks reflected both the ease of manufacturing and the cavalry heritage of the new division design. Such a structure left inadequate infantry to support the tanks and too many lightly armored vehicles to fight the increasingly better-armored German tanks. Considerably more production and development were needed before the lopsided American armored units became a cohesive mechanized force.

Close air support was also lacking in the American combat team. The postwar U.S. Army Air Service, which became the Army Air Corps in 1926, established two squadrons (out of thirty-two in the air corps) of specialized aircraft known as attack aviation, for battlefield operations. Paradoxically, the fact that these squadrons were dedicated to ground support meant that the other air corps units were even less willing to spend scarce training resources practicing for such a mission. Moreover, under the influence of the strategic bombing theorists, air corps doctrine for attack aviation moved steadily away from close air support toward interdiction of enemy supply lines and other targets that were deeper behind enemy lines. The air corps discussed but did little about the psychological effects of air attack on ground forces. By 1934 destruction of enemy ground forces had fallen to fourth priority as missions for attack aviation, behind attacks on enemy aircraft, bases, and logistics. Contrary to later accusations of narrow-mindedness, the army's senior leaders permitted the air corps to practice such theories during peacetime exercises. By 1939 the doctrine and aircraft devoted to attack aviation had been redesignated as "light bombardment," with targets that, like those of heavy bombers, were oriented toward weakening enemy air forces instead of meeting the needs of the ground forces. The results proved to be ineffective even at their given interdiction missions, let alone for close air support.[71]

Like Guderian, Adna Chaffee hoped to use air support to avoid the delays and logistical buildup necessary for a deliberate breakthrough attack. Yet he received little support until the Stuka dive-bomber captured attention in 1939–1940. Thereafter, army air forces chief Henry Arnold hurriedly ordered the creation of two groups of dive-bombers and made some effort to improve cooperation with ground troops.

PART I CONCLUSION

The preceding discussion of five different armies appears to go in many different directions, and yet certain common threads are evident. First, antiwar sentiment, limited defense budgets, and similar restrictions hampered the development of new weapons and doctrine in the armies of the Western democracies. Even Germany and the Soviet Union encountered limitations, so that no nation was fully equipped with modern weapons when it entered World War II. The Germans had, however, achieved an advantage of several years of practical experience with the new weapons.

Second, even within the peacetime armies, the World War I traditions of infantry-artillery dominance delayed new developments designed to broaden the nature of the combined arms, although the Red Army was an exception until the 1937 purges. In the British, French, and American armies, mechanization developed in two divergent directions. Heavy, almost pure armor formations supported conventional infantry attacks, and highly mobile but poorly armed and armored light forces performed cavalry functions. Even the Germans and Soviets diverted some armor to specialized cavalry and infantry-support roles.

During the 1930s, professional soldiers in most armies gradually broke free of traditional 1918 views on the role of various combat arms. The Germans had the advantage in these new developments, especially after the purges had shattered the Red High Command. Thanks to Hitler, Guderian, and others, the Germans funneled more of their assets into fewer Panzer units than did their opponents, who tended to modernize slightly a much larger part of their armies and who therefore had almost no forces really trained and equipped for mechanized combat in 1939–1941. Worse still, the British and French commanders retained the methodical staff and communication procedures of trench warfare and were unable to respond to rapidly changing situations.

Finally, the airpower advocates of all nations retarded the development of close air support for ground operations. Even the Germans had only the embryo of an air–ground command and control system when the war began.

Had World War II come in 1936 or 1937, Tukhachevsky's developments in the Red Army probably would have triumphed, despite problems of materiel and training. Had the war begun in 1942 or later, the British, French, and Americans would have had time to experiment with and adjust their mechanized organizations, weapons, and doctrine. Germany's military success from 1939 to 1941 was therefore the product of a transitory set of circumstances. The Germans had produced equipment and fielded mechanized units in the mid-1930s so that this equipment was still usable and the units were well organized and trained when war began in 1939. Moreover, Germany had two advantages that the other powers lacked: a primitive but developing air support system, and a command and control network that allowed for much more rapid maneuver than any opponent could achieve.

• • • • • • • • • • • • • •

TOTAL WAR, 1939–1945

The Defense of St. Vith, 1944

On 16 December 1944, four German field armies attacked the overextended U.S. VIII Corps in the hilly Ardennes region of France and Belgium. Adolf Hitler hoped to repeat his 1940 success in the same region, but the Germans found themselves stymied by stubborn American delaying actions that held the key road junctions. The U.S. 101st Airborne Division gained most of the credit for the Allied victory by its defense of the town of Bastogne, located on the southern shoulder of the German penetration, a penetration immortalized by the name Battle of the Bulge. Almost equally important, if less famous, was the defense of the village of St. Vith on the northern shoulder of the German bulge.

When the German offensive began, St. Vith seemed destined to be overrun in a matter of days or even hours. The half-trained U.S. 106th Infantry Division, which had first entered combat only four days earlier in a supposedly quiet sector of the Ardennes, almost dissolved in the face of the German onslaught. Lacking air and armor support, antitank weapons, and above all experience, the American troops seemed doomed from the beginning. Because the division commander refused to withdraw, two of his three infantry regiments were surrounded and ultimately surrendered. Two German columns advanced on the northern and southern flanks of this encirclement, closing in on the division headquarters at the road junction of St. Vith.*

Recognizing the danger, the Allied commander, Dwight Eisenhower, directed the movement of the U.S. 7th Armored Division from northern Belgium to support the beleaguered VIII Corps. Unfortunately, the same poor road network and snowy weather that impeded the German advance also slowed the movement of American forces, a situation complicated by large numbers of rear echelon American troops fleeing the enemy advance. In the meantime, Brig. Gen. Bruce C. Clarke of the 7th Armored arrived in St. Vith on 17 December. Clarke and other officers cobbled together a scratch defense, organized around combat engineers, lightly armored cavalry recon-naissance troops, and even the 106th Division's band. On the afternoon of 17 December, these defenders halted a premature German advance north of St. Vith. American antitank shells penetrated the thin bellies of the first German Panther tanks as they came over a ridge, and the startled German brigade commander pulled back to rethink the situation. Meanwhile, the German 18th Volksgrenadier Division, a scratch formation that was as unskilled as the 106th, made the mistake of assem-bling its troops on open ground east and northeast of the town, in plain sight of the American defenders. The self-propelled 105mm howitzers of the 275th Armored Field Artillery Regiment gave a stunning display of the superiority of American artillery, shattering the Volksgrenadiers before their attack could even begin.

By that evening, Clarke's Combat Command B, consisting of two armored infantry battalions, a tank battalion, an armored engineer battalion, and reconnaissance and artillery units, finally reached the vicinity of St. Vith. Instead of standing on the defensive, however, Clarke repeatedly launched local counterattacks, usually conducted by a battalion-sized task force of tank, armored infantry, and tank de-stroyer (antitank) companies. Supported by artillery fire, these task forces swept around open German flanks, disrupting the enemy assault units and then pulling back before German commanders could react. These counterattacks were so fre-quent and aggressive that Hasso von Manteuffel, the commander of the German 5th Panzer Army, was convinced that he was fighting an entire armored corps of several divisions instead of one-half of the 7th Armored Division (Morelock, 306).

Ultimately, German successes both north and south of the town forced the Ameri-cans to evacuate it on 23 December. Yet the defense of St. Vith, Bastogne, and other road junctions had derailed the German plan to achieve a rapid victory. The defenders of St. Vith not only redeemed the disaster that had befallen the 106th Division but also demonstrated the degree of combined arms sophistication that American commanders had reached after three years of mechanized warfare.

*The best recent study of the 106th's defeat and of the ensuing defense of St. Vith is in Jerry D. Morelock, *Generals of the Ardennes: American Leadership in the Battle of the Bulge* (Washington, DC, 1993) 275–344. See also Hugh M. Cole, *The Ardennes: Battle of the Bulge,* U.S. Army in World War II series (Washington, DC, 1965), and William D. Ellis and Thomas J. Cunningham, *Clarke of St. Vith: The Sergeants' General* (Cleveland, OH, 1974).

CHAPTER 4

• • • • • • • • • • • • • • •

The Axis Advance, 1939–1942

World War II did more than force armies to integrate all available weapons and arms into a mobile, flexible team. It also demanded that they adjust to a variety of threats, climates, and terrain.

Despite the vast scope of this struggle, several major trends are evident. First, the mechanized combined arms force came of age in this war. In 1939 most armies still thought of the armored division as a mass of tanks with relatively limited support from the other arms. By 1943 those same armies had evolved armored divisions that were a balance of different combat arms and support services, with each such service trying to be as mobile and as protected as the tanks they accompanied. The Soviet, German, and American armies cannibalized infantry-support tank units to form more armored divisions.

Second, this concentration of mechanized forces in a small number of mobile divisions left the ordinary infantry units deficient in both antitank weapons for the defense and armored vehicles to accompany attacks against prepared enemy positions. Several armies therefore developed a number of tank surrogates such as tank destroyers (which sought to kill enemy tanks) and assault guns (which were intended to support the infantry's advance).

Third, one of the driving forces behind these two trends was the gradual development of the means and the will to counter the blitzkrieg. During the

107

period 1939 to 1941, conventional infantry units were unprepared psychologically and technologically to defeat a rapidly moving armored foe, especially when it broke into rear areas and disrupted the often rigid, World War I–style communications and organization of the defender. By 1943 those same infantry units had lost their paralyzing fear of armored penetration and had acquired a much greater antitank capability. Successful armored penetrations were still possible, as the Soviets demonstrated, but they were increasingly difficult.

Fourth, the many different environments and fluid tactics of the war encouraged the development of all manner of specialized combat units. Some of these forces, such as amphibious and airborne divisions, used unusual methods to get to the battlefield but then fought as conventional troops. Other forces used infiltration and surprise to accomplish extremely risky operations against superior enemy forces. These latter units, generally called special operations forces, have probably always existed in history, but the global nature of World War II gave such units unprecedented opportunities and challenges.

Fifth, World War II represented the end of pure ground operations. Although no nation's air force really fulfilled the promises of interwar airpower theorists, the airmen of 1939–1945 made themselves an indispensable ingredient for victory. Seaborne support was equally essential. Mechanized attack required air superiority and close air support, airborne landings required close coordination between air transport and ground forces, and amphibious landings developed as the most sophisticated and complicated form of combined arms and joint operations. Such joint service interaction was not achieved without grievous operational errors and doctrinal arguments, but by the end of the war ground commanders had reached a temporary working compromise with the other services on most questions.

Finally, coordinating such a complicated, multidimensional conflict required major changes in command, control, communications, and intelligence. The French and British commanders learned the costly lesson that World War I staff procedures and communications could not react promptly to mechanized assaults. A successful leader had to balance his requirements for flexible communications, for an effective staff operation, and for personal knowledge of the battlefield. Every commander profited from his intelligence and reconnaissance assets, but none was satisfied with the scope and timeliness of the information he received in this manner. Yet when he went onto the battlefield to observe for himself, the commander was handicapped by leaving much of his staff and communications capability behind at the main headquarters.

The best way to examine these developments is to consider the actions and reactions of the opposing armies during the course of the war. Our starting point will be the reasons for the German success of 1939 and 1940, followed by a discussion of British reactions and adjustments to it. As we turn to the next cycle of developments, the German victories in Russia during 1941 and 1942 must be compared with Soviet efforts to adjust their tactics and organization both before and after the German invasion. After reviewing American developments in divisional structure, we shall consider the many technological advances of the war (chapter 5) and then survey the development of mechanized forces, close air support, and specialized operations during the conflict's final years (chapter 6).

Poland, 1939

During the first seventeen days of September 1939, Germany overwhelmed Poland and occupied more than half its territory. The Western Allies, who were still mobilizing and training their reservists, were unable (and unwilling) to make more than a symbolic advance along the Franco-German border during this period. Yet the speed of the German conquest obscured a number of problems that their army encountered, problems the Germans attempted to solve during the winter of 1939–1940. As a result of their efforts, they widened the gap of experience and experimentation that separated them from their future opponents, Great Britain and France.[1]

To begin with, the German higher commanders intended to conduct a war of maneuver and penetration, but they had not accepted Heinz Guderian's theories in this regard. The panzer and light motorized divisions were parceled out among the various armies rather than being concentrated at a few points. The only exception was the German Tenth Army, which had two panzer, two motorized, and three light divisions in addition to its six conventional infantry divisions. In general, the mechanized and motorized forces were employed as the cutting edges of a more conventional advance on a broad front, with relatively shallow penetrations of the Polish defenses. Armored forces exploited into the rear only after organized Polish resistance had collapsed.[2]

Although German tanks and motorized infantry had developed techniques for close interaction, the same was not true of relationships between these ground elements and their fire support. Within hours of the first attack, General Guderian was bracketed by near misses from his own artillery, which had violated orders by firing blindly into the morning fog. The Luftwaffe concentrated, appropriately, on achieving air superiority and interdicting Polish lines of communication, but this meant that little air support was

· ·

initially available for direct support of the ground troops. The complexity of close air support operations, the problems of coordinating and communicating between air and ground units, and the lack of training in such methods made it difficult for the Luftwaffe and the army to work together.

Many German tactical commanders were too cautious, allowing themselves to be halted by even minor Polish resistance. This was a natural response for an army that had not seen combat for decades, but it was not appropriate to the situation. The Poles were probably doomed from the outset, because they had dispersed their forces along the entire Polish-German border in an effort to thwart any limited German grab for territory. Under the circumstances, German forces needed to punch through the thin Polish frontier defenses rapidly instead of stopping to fight a pitched battle whenever they encountered Polish troops.

The German system of division and higher level commanders going forward to make on-the-spot decisions greatly increased the tempo of operations. However, that same system had several drawbacks that were evident even in this first campaign. The presence of a higher commander on the scene tended to inhibit the initiative of the battalion or regimental commander who should have directed operations. This inhibition may have been partially responsible for the caution displayed by German units in Poland. Moreover, the senior commanders were extremely vulnerable to enemy attack while moving about in a fluid battle. Guderian, as a corps commander, was pinned down for hours by a few bypassed Polish troops and was therefore unable to control his forces. This was a recurring problem for leaders in several armies during World War II, especially for the more daring German commanders in North Africa. Ultimately some, like Erwin Rommel, organized ad hoc security task forces to travel with them. Yet such a force reduced the combat power of subordinate units and at the same time increased the temptation for a senior commander to become involved personally in the small-unit actions he saw while visiting the front. If he lost radio contact with his headquarters, the senior commander became isolated and largely ineffective.

Although no German unit advanced more than 250 kilometers into Poland, significant problems of supply and maintenance developed. All major tank repairs required evacuation of the damaged vehicle back to Germany, and forward maintenance units were unprepared for the new demands of active campaigning. By the end of the Polish campaign, the German mechanized force was almost immobilized for maintenance reasons. The Soviet mechanized troops experienced similar logistical problems when they occupied eastern Poland in that same September 1939, often outrunning their fuel supplies.

A related problem was the unsuitability of early German equipment. The Germans had intended the Panzer I tank for training rather than for combat, and the Panzer II was scarcely better. The use of such vehicles in Poland reflected two problems: Germany had begun the war before its mechanized forces had developed completely, and those forces still did not have priority for industrial production. During September 1939, for example, the Germans lost 218 tanks in battle, approximately 10 percent of their entire forces while manufacturing only 57 new ones. Even at the time of the invasion of France eight months later, the second-generation Panzer III and IV medium tanks constituted less than one-fourth of the German tanks in field units.[3] The Polish campaign did accelerate the retirement of Panzer Is by revealing their deficiencies and may have hastened the movement of Panzer IIs out of line tank companies and into reconnaissance, engineer, and headquarters units. As a result, the relatively few Panzer III and IV tanks bore the brunt of the effort in 1940.

By contrast, other German equipment had exhibited unexpected uses. The half-tracked vehicles originally intended to tow artillery guns proved to be so mobile that infantry units in panzer divisions sought to acquire them as armored personnel carriers. The vast majority of panzer grenadiers (armored infantry), however, continued to travel in trucks and motorcycles throughout the war; there were never sufficient half-tracks available. The 88mm antiaircraft gun proved to be extremely useful in a ground-support role, foreshadowing its later use as the premier antitank weapon of the German ground forces.

A basic result of the German invasion of Poland was the beginning of the slow evolution of the German panzer division structure toward greater balance among the combat arms. At the time of the Polish campaign, each of the six panzer divisions had between 276 and 302 tanks, organized into a panzer brigade of four battalions. These same divisions each had only three battalions of infantry and two of artillery. This tank-heavy force proved too unwieldy for some commanders, and in any event Hitler was interested in creating more panzer divisions. At the same time, the German light divisions, built around two motorized infantry regiments and one tank battalion, proved to be too weak for sustained operations, lacking the combat power of either a panzer division or a conventional infantry division. Given the limited number of tanks in the German inventory, the solution was obvious—tanks moved from the existing panzer divisions to the light divisions, three of which officially became panzer divisions during the winter of 1939–1940. In addition, during the Polish campaign an ad hoc panzer division had formed around one of the infantry-support tank brigades created in 1938; this formation became the 10th Panzer Division. Thus, by the time of the French campaign in 1940, even more of the available German tanks were concentrated into panzer divisions, some of which were reduced from a four-battalion tank brigade to a three-battalion tank regiment, with a total of 160 to 200 tanks per division. This put the tank element more in balance with the rest of the division, which normally consisted of three infantry battalions and two or three towed artillery battalions, an armored reconnaissance battalion, antitank battalion, engineer battalion, and communications troops.[4] This trend toward a more balanced division with fewer tanks continued throughout the war.

Regardless of exact organization, the panzer divisions were slowly developing the habit of task organizing for combat. This process was by no means complete even in 1940, when one of Guderian's divisions had its tank brigade operating separately from its infantry. Increasingly, however, the brigade, regimental, and battalion headquarters practiced attaching and detaching elements of other arms in order to have a combination of tanks, infantry, artillery, engineers, and, on occasion, air defense. The balance between these arms varied with the mission, terrain, and enemy forces involved.

Beyond these organizational changes, German tactical concepts and structures seemed essentially sound. With the exception of a few technical problems with a particular machine-gun design, the infantry divisions had functioned well. The only other lesson of the Polish campaign was the predictable discovery that armored forces were at a disadvantage when fighting on urban terrain, where the tanks were vulnerable to short-range attacks from nearby buildings in the narrow streets. The Germans lost fifty-seven tanks in one day while attempting to seize Warsaw.[5] This experience only reinforced the need for a higher proportion of infantry to tanks in armored units, in order to provide close-in security for the tanks in built-up areas.

The German Advance, 1940

Between the fall of Poland in September 1939 and the beginning of the Belgian-French campaign in May 1940, another German operation unsettled Allied morale and foreshadowed the future complexity of joint operations. On 9 April 1940, an improvised German force used motorized troops, small-scale parachute drops, and seaborne landings to occupy Denmark and Norway by surprise. Only one of the six German divisions sent to Norway was a fully trained, established organization; and at one point Hitler feared that the invasion would collapse, yet all German units performed remarkably well. Despite the shoestring nature of the German operation, this "warfare in three dimensions" (land, air, and sea) caused a shift of Allied resources and planning away from the battlefields. From the German point of view, the most serious consequence of this adventure was the loss of thirteen warships of the small German surface navy. This loss had significant implications for the later German attempt to invade England.[6]

By contrast, the British and French performance in Norway was deplorable, especially considering the fact that the Allies had begun planning for their own invasion even before the German attack. The first British troops in central Norway lacked antiaircraft guns, aircraft, and artillery; indeed, the entire operation was conducted outside the range of Allied air cover.[7] More ominously, command and coordination between the two Allies was weak at best, a harbinger of similar problems in France itself.

The stunning operations in Denmark and Norway preceded another surprise when the main battle in France and Belgium was joined. The invaders used small airborne forces in an attempt to seize key points at various locations along the front. Some of these airborne invasions were at best only partial successes, as when the German paratroops failed to seize the governmental center of the Netherlands, and infantrymen carried by tiny liaison aircraft became lost over the Ardennes Forest, failing to seize key roadblocks ahead of the advancing armored units.[8] By contrast, German airborne forces had a stunning success at the fortress of Eben Emael, the key to Belgium's defensive position. On 10 May 1940, a small party of German glider troops landed on top of Eben Emael. Using shaped-charge explosives* and the element of surprise, they blinded and neutralized the huge fortress until ground troops arrived, thereby eliminating one of Belgium's main defenses.[9] This surprise, coming on the heels of the Norwegian invasion, caused many Allied military and civilian leaders to become excessively concerned about the rear area threat posed by airborne and unconventional warfare forces. Such concern was the first step to creating the psychological uncertainty on which the blitzkrieg thrived.

Conquering Belgium and France required more than propaganda and a few paratroopers to create psychological paralysis. Contrary to frequent stereotypes, the Western armies were generally well armed in 1940, having greatly increased their production during the later 1930s. By one calculation, Britain and France had a combined total of 4,340 tanks on the Continent during the 1940 campaign, as compared to only 3,863 for Germany.

---

*A shaped charge allowed the user to focus the blast of a small amount of explosive in order to achieve a much greater effect than the same explosive would produce if detonated normally. The essence of this shaping was to mold the explosive with a cone-shaped hollow on one end, so that the blast effect that centered within that hollow would produce a shock wave in one direction, toward the target at the wide end of the cone.

Despite weaknesses such as a shortage of radio communications and crowded turrets, most of the Allied tanks were actually better armed and armored than their German counterparts. Only the light British cruiser tanks were more vulnerable. One obsolete French FCM tank absorbed forty-two hits from German 37mm antitank guns without being knocked out of action. To deal with the more heavily armored French B-1bis and British infantry support tanks, the Germans had to bring up 88mm antiaircraft guns. Indeed, they were disturbed by the general ineffectiveness of their antitank weapons in 1940. By contrast, the outnumbered French 25mm and 47mm antitank guns had much higher muzzle velocities and therefore greater armor penetration capacity than the equivalent German and British guns.[10] The French Army, however, did have serious equipment shortages in antiaircraft guns and antitank mines.

Yet the Germans defeated the Allies so rapidly that they seemed to validate the concept of blitzkrieg in Germany and abroad, even when its details were not well understood. The reasons for this success have been reviewed in the text thus far. First, in contrast to their own performance in Poland and to the French dispositions in 1940, the Germans concentrated their available mechanized forces into a few large masses at critical points. Seven of the ten panzer divisions, with five motorized divisions following closely to mop up and protect the flanks, advanced through the Ardennes Forest on a seventy-kilometer front. Thus, even when one of the three mobile corps was held up by French defenders, the other two continued the advance into the French rear.

By contrast, the French Army dispersed thirty-six tank battalions evenly along its borders in support of infantry armies, even in the Maginot Line region. In most cases these battalions had never trained to cooperate with infantry and artillery in conducting a deliberate attack or counterattack. Much of the remaining British and French armor was in the extreme north, moving into Belgium in a direction away from the main German advance on Sedan (see Map 3, chapter 3). Four French armored divisions were still forming, but these were scattered at wide distances behind the front and were sometimes broken up when committed to battle.[11]

Moreover, the Western Allies had organized a linear defense, spreading their forces thinly across a wide front. The French command structure in particular was geared to methodical, set-piece battles but lacked the forces to create a true defense-in-depth on the World War I model. By rushing through the Ardennes Forest, the main German attack shattered this linear defense at one of its weakest points. By the fifth day of the campaign (14 May 1940), the German mobile forces were conducting the type of deep exploitation envisioned by many theorists during the 1930s. Such penetrations were psychologically unnerving to the defenders, who were suddenly facing major enemy forces in the rear and who lacked a procedure to redeploy units rapidly to meet and contain that threat. The French command and control structure, which was still tied to telephones and written reports, could not react quickly to the German advance. The British commander, Field Marshal John, Viscount Gort, separated himself from his main headquarters staff, and even that staff did not receive information in time to act on it.[12] As Gerhard Weinberg has observed, "The basic factor was surely that a poorly led and badly coordinated Allied force was pierced at a critical point by concentrated German armor and was never able to regain even its balance, to say nothing of the initiative."[13]

At the unit level, some of the French and British weaknesses were intangible. During the winter of 1939–1940 some individual commanders had succeeded in training tough, effective divisions, especially in the higher priority mechanized and motorized units filled with young, physically fit soldiers. Other commanders, however, had failed at the nearly impossible task of welding older reservists into effective teams without experienced junior leaders, all the while building the elaborate field fortifications required by the high command. In any event, units composed of undertrained, overaged reservists had difficulty absorbing the new equipment that the Western democracies belatedly produced.[14]

Because there was so little resistance in the rear areas, the German commanders did not always use tanks to lead their advance. Instead, the armored reconnaissance battalions, accompanied in some cases by engineers to eliminate obstacles, scouted up to a day's march ahead of the main body, the slower elements strung out in column behind. Commanders used armored vehicles or light aircraft for control during the pursuit. Of course, this advance in column made the Germans vulnerable if the defenders were able to mount a counterattack, as Erwin Rommel discovered when the British struck the flank of his panzer division at Arras on 21 May. Only the improvised use of 88mm antiaircraft guns and 105mm howitzers in an antitank role halted the heavy infantry-support tanks of the British 1st Army Tank Brigade. (The British did not realize that the 88mm gun was responsible for their defeat until they encountered the weapon again in North Africa.) Even this poorly coordinated and unsuccessful British counterattack at Arras put some of the fear of tanks into German higher commanders, causing German armor leaders to accelerate development of larger antitank weapons and higher velocity tank guns.[15]

At the tactical level, both the British and the French were at a distinct disadvantage in force structure and experience. German armored divisions were clearly better organized than those of the French. The French Division Cuirassée was too tank-heavy, with four tank battalions but only one infantry and two artillery battalions. When ordinary infantry or artillery units were attached to this division to correct the imbalance, they had had no training for cooperation with tanks. French logistical support was too dependent on roads and rails to follow the all-terrain maneuver elements of these divisions, and in fact a rail movement once separated all the tracked elements from all the wheeled vehicles of a division. Finally, the inexperienced French commander of an armored division had to control most of his subordinate units directly, giving him an impossibly broad span of control. The "demibrigade" headquarters that controlled his tank battalions were neither intended nor trained to integrate the other combat arms. By contrast, the panzer division commanders had a number of subordinate headquarters, each of which was capable of controlling and coordinating the different weapons systems.

German training in combined arms was especially evident during the penetration of the Ardennes. The rapid German advance over a poor road network was made possible only by road repairs conducted by combat engineers. At the critical crossing of the Meuse River on 13 May, the German infantry and some engineers crossed under the covering fire of tanks, artillery, and tactical aircraft. Indeed, the Germans had relied on air support to reduce the need for artillery units and ammunition resupply while moving through the Ardennes.

The celebrated image of air–ground cooperation at the crossing of the Meuse illustrated the weakness as well as the strength of German joint operations in 1940, however. In war games before the campaign, Guderian and his air force counterparts had agreed that the best way for the Luftwaffe to support the river crossing was by a continuous stream of Stukas and other ground attack aircraft, who would be able to keep the French defenders ducking without endangering the German troops making the crossing. At the last minute, however, Guderian's immediate superior insisted on a full-scale, medium bomber attack to destroy the French defenders. Such an attack would be worse than useless, because the Germans might be hit by stray bombs dropped from high altitude. Thus the river crossing would have to wait until the bombers had departed, giving the French time to recover. Fortunately, by the time the Luftwaffe agreed with the army's request, it was too late to implement the massive attack. Instead, the supporting Luftwaffe commander simply followed the schedule of air strikes that he had devised with Guderian during the war games. The result, from the French point of view, was decisive. Not only were French telephone lines disrupted by German bombs, but the French artillery also became demoralized. Although few guns were hit, the French gunners ceased firing and took cover whenever a German air attack appeared. Again, the psychological effect of airpower was out of all proportion to its actual destructive power.[16]

Even after the breakthrough at Sedan, the advancing German ground forces found it difficult to communicate with their air elements. A series of bomb lines were drawn on the map, with the ground troops expected to stay on one side while the Luftwaffe attacked on the far side. In practice, the rapid advance made such control measures clumsy at best. Midranking Luftwaffe and army officers established informal coordination procedures while their superiors insisted on a time-consuming system of passing air support requests up the ground chain of command and down the air hierarchy. Meanwhile, on the rare occasions when British or French fighters intercepted the Stukas, the Germans suffered heavy losses.

Yet the German air campaign was far better conducted than that of its opponents. The French command in particular showed little interest in aerial reconnaissance or in coordinating the efforts of its air elements. British and French pilots attacked German columns and bridges with suicidal gallantry, flying in level bombing runs that made them easy prey for German antiaircraft gunners and fighter pilots.[17] By contrast, Allied antiaircraft gunners found it difficult to adjust for sudden changes in both range and altitude as a Stuka plummeted from the sky to dive-bomb the Allied troops.

The fall of France demonstrated more than just the importance of combined arms mechanized formations, penetration attacks, and maneuver for exploitation into the rear. The German advantage over the British and French was equally apparent in combined arms training and procedures of all types. Yet the images of parachutists, tanks, and screaming Stukas tended to obscure the combined arms nature of the blitzkrieg from many contemporary observers and subsequent popular chroniclers.

The British Response, 1940–1942

The sudden collapse of France in 1940 caused professional soldiers in many armies to reassess their organizations, training, and doctrine. As the only major belligerent still at war with Hitler, Great Britain had the most urgent need to reorganize its forces and restructure its doctrine in the months after Dunkirk. Unfortunately for the British, the period 1940–1942 seems in retrospect to have witnessed the development of two diverging British armies—the army at home, which gradually rebuilt and developed new doctrines and organizations, and the field army in the Middle East, which after initial success against the Italians found itself repeatedly outmaneuvered by the small forces of the German Afrika Korps. Once the experienced

British desert units were sacrificed to defend Greece in 1941, subsequent forces in North Africa never had the opportunity to reorganize and retrain as had the army at home. Yet these two armies were connected in doctrine if not in practice, and the British victories from 1942 to 1945 owed a great deal to the quiet process of rebuilding forces at home.

Faced with the possibility of German invasion after the French surrender, the British at first believed that there was no time for major changes in organization, doctrine, or equipment. In a desperate effort to rearm the troops evacuated from Dunkirk, British industry continued to produce weapons whose designs were obsolescent if not obsolete. Lightly armored cruiser tanks and armored cars, together with two-pounder antitank guns, appeared by the hundred because there was no time to redesign and build better weapons.[18] Some British commanders became preoccupied with the material difficulties of obtaining trucks to motorize infantry elements within the newly formed armored divisions, thereby obscuring the more fundamental need for doctrine and techniques of close infantry–tank cooperation. The British did develop some new weapons during this period, most notably the six-pound (57mm) gun for use both as an antitank weapon and as the main armament on new tanks. Yet it did not appear in the field until 1942 and even then was too large to be retrofitted into the turrets of older model tanks.[19]

As the threat of invasion lessened, the British Army could reemphasize training and reconsider its prewar doctrine in light of the experiences of 1940.[20] The General Staff published a series of notes from various theaters of war, identifying such points as the need for combined arms organization below division level and the German use of antitank weapons instead of tanks to defeat enemy tanks. Under the direction of Gen. Alan Brooke, Commander in Chief, Home Forces, and later Chief of the Imperial General Staff, the units of the expanded active and reserve (Territorial) forces conducted more realistic training at all levels. Some of it was simply an improvement on prewar principles, such as the development of fire-and-movement battle drills, that is, standardized, automatic responses, for small infantry units. Indeed, in some cases these battle drills became just that—stylized, rigid drills that did not encourage initiative in battle. Meanwhile, Col. H. J. Parham experimented with a single radio network to mass artillery on the basis of an estimated map reference for the target. The results were rather inaccurate, but in the absence of the American fire-direction center, Parham's ideas allowed the Royal Artillery to provide at least some response to targets of opportunity. These innovations were limited by two other tendencies: the continued staff reliance on lockstep fire support planning and the habit of decentralizing artillery control, dedicating individual gun battalions to support maneuver brigades.[21]

The most unusual feature of the period 1940–1942 was the conduct of large-unit command post exercises and field maneuvers, with detailed study before and critiques after each step. Lt. Gen. Bernard L. Montgomery had pioneered such exercises as a division commander in France during 1939 and 1940, enabling his division to move more rapidly and flexibly than most other British units. After Dunkirk, Montgomery applied the same training techniques as commander of two different corps and finally of an army-level force. He also acted as chief umpire for exercises involving other units in Britain. Similar if less elaborate training took place in the newly formed armored divisions under Lt. Gen. Giffard Martel, the commander of the Royal Armoured Corps after December 1940.

Montgomery contended that few British officers had experience maneuvering any unit larger than a brigade, and certainly his exercises helped to produce commanders, staffs, and units that were capable of more rapid changes in deployment and mission than those of the World War I British Army. More important, Montgomery and others developed a common conception of the interaction of different weapons and of how to commit divisions and larger units to battle. For example, Montgomery believed that the decentralized nature of German mechanized pursuit and exploitation had caused many British commanders to lose sight of the necessity for centralized control in the deliberate attack and defense. In the best traditions of the carefully orchestrated, World War I–style British staff planning, Montgomery argued that reconnaissance, artillery, tanks, infantry, engineers, and airpower had to be "stage-managed" at the highest levels of command in order to concentrate combat power at critical points for attack and defense. Only in a fluid situation could commanders decentralize control of these arms and push them forward, so that subordinate leaders would have the different weapons readily available. Defense to Montgomery meant not a series of fixed lines on the terrain but blocking positions in depth plus massive counterattacks of the kind Germany had used so well in 1917 and 1918. All arms needed to employ night attacks to reduce the lethal effects of aimed enemy fire. Finally, Montgomery opposed the traditional British concept that tank units should maneuver independently, like cavalry or ships at sea. Instead, he saw the armored division as a combined arms force that would seize key terrain in order to use the advantages of being on the defensive when the enemy armor counterattacked. Infantry and antitank forces would follow the initial armored assault to mop up and hold terrain, releasing the tanks to refit or attack again.[22]

Inside the Royal Armoured Corps (RAC), Martel developed these same concepts in a series of exercises, until in June 1942 the senior armor commanders in Britain agreed to an RAC Creed. This creed—a product of exercises and of a critical analysis of events in North Africa—began with the statement that "an armoured division is a formation of all arms. Each arm or branch of the service is a member of the team, and has its vital part to play." Like the Germans before them, British commanders concluded that antitank guns were the best means to defeat enemy tanks, although tank-to-tank combat would still occur. Motorized infantry and antitank weapons together would hold key terrain, around which the armored forces would maneuver.[23]

Changes in organization accompanied changes in doctrine. Immediately after Dunkirk, the pure tank brigades of the army armored divisions had given way to brigades composed of three tank battalions and one motorized infantry battalion.* A 1940 British armored division therefore consisted of an armored car reconnaissance battalion, two armored brigades, and a support group, which in turn contained battalions of field, antitank, and light antiaircraft artillery as well as an additional infantry battalion, two engineer companies, and support troops. Martel and his subordinates deliberately retained this organization well into 1942 to avoid constant changes that would disrupt training.

By 1942, however, this structure was obviously too heavy with tanks, and so the War Office removed one of the two armored brigades from the division structure (see Fig. 9). The separate brigades that resulted from this amputation could reinforce any division as needed for a particular mission. Moreover, the term "support group" had apparently caused the nonarmored elements of the division to be regarded as an afterthought to the tanks. To correct this perception, an armored infantry brigade and a division artillery regiment replaced the support group. Division commanders could hold their artillery, antitank, antiaircraft, engineer, and support elements under centralized control or attach them to one of the two brigades as needed. At the same time, the British created two different types of infantry division. The "division" per se, apparently intended for Asian operations, retained the traditional configuration of three infantry brigades of three battalions each. By contrast, the "infantry division" lost one infantry brigade in favor of an infantry-support tank brigade. Martel and the new Commander in Chief, Home Forces, Bernard Paget, strongly advocated this latter change to improve training and cooperation between infantry and supporting tanks.[24]

---

*The British normally used the term "regiment" to designate an armored force equivalent to an American battalion. American terms and symbols are used here for consistency.

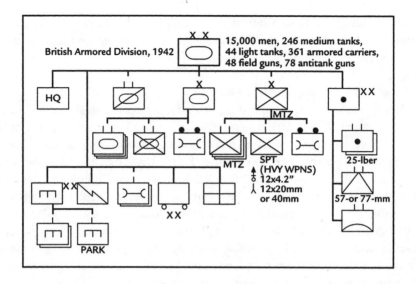

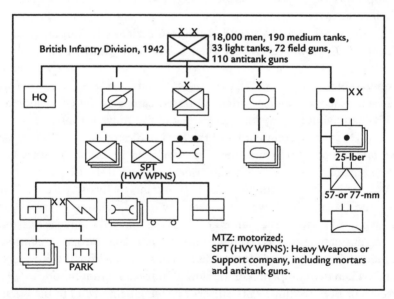

Figure 9. British Armored and Infantry Divisions, 1942

Unfortunately, the British returned to a division of three infantry brigades by 1944. As a result, the quality of tank–infantry cooperation in 1944 and 1945 varied widely among divisions.

War in the Desert, 1940–1942

The battles of North Africa did not always reflect the state of the British Army at home. In late 1940 the small force in the Middle East was the only British field army still trained to high prewar standards, although its equipment was little better than that found at home. Once Italy joined the war on Germany's side in mid-1940, Prime Minister Winston Churchill took a calculated risk and sent a portion of his scarce resources to defend Egypt against the threat from Libya, which was an Italian colony at the time. The shipment included a single battalion (7th Royal Tank Regiment) of forty-eight heavily armored Mark II (Matilda) infantry support tanks. This battalion, in combination with the two understrength but well-trained divisions already in Egypt, was the basis for a classic demonstration of prewar British tactical doctrine (see Map 5).

In September 1940, Marshal Rudolfo Graziani's Italian army of ten divisions had advanced from Italian Libya into the western desert of British Egypt. Graziani was cautious, however, and his force was largely composed of foot soldiers with poor logistical support. He therefore halted and established a chain of widely scattered camps in the general area of Sidi Barrani, about eighty kilometers east of the Libyan frontier. Lt. Gen. Richard O'Connor, commander of the British Western Desert Force, used the infantry support tanks in conjunction with the 4th Indian Infantry Division to reduce these camps in a surprise advance on 8–10 December 1940. The tactics involved exemplified the best of interwar British practice.[25] Because the Italian camps were protected by minefields and obstacles, the British passed between the camps and attacked them from the far (western) side, aiming at the unmined entrance road to each camp. Artillery and mortar fire pinned the defenders down and distracted attention from the unexpected assault. Then two companies of the slow infantry tanks moved forward, with platoons of Bren machine-gun carriers following behind and to the outside flanks, providing flank security and machine-gun fire to support the tanks. As soon as the British tanks broke into the enemy positions and came to close-quarters fighting, infantry in trucks drove up to (and in some cases into) the camp entrance, dismounted, and accompanied the tanks in mopping-up operations.

After the tank-artillery-infantry teams had reduced the enemy defensive system, the 7th British Armoured Division used its light, mobile armored vehicles to conduct a high-speed pursuit. The retreating Italians lacked effec-

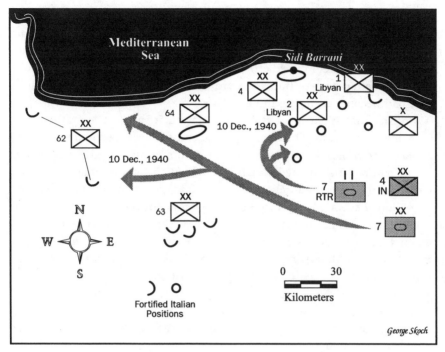

Map 5. Sidi Barrani, December 1940

tive tanks or antitank weapons and were tied to the single road that paralleled the Mediterranean Coast. The 7th Armoured therefore made a series of wide flanking movements south of the road, repeatedly turning back northward to the coast to intercept the Italian retreat. Thus a British force, which for much of the operation was spearheaded by only one armored car and three tank battalions, captured 130,000 prisoners.[26]

The roots of the British victory lay in advantages of superior training, mobility, and equipment. Yet these advantages were negated when the Italian disaster led to the introduction of German forces to North Africa. In early 1941 and again a year later, the British reduced their forces in Egypt in favor of needs in other theaters, sending forces first to Greece, and then, after Japan entered the war, to Southeast Asia. As a result, when the German Afrika Korps attacked in March 1941, it encountered only partly trained British troops equipped with worn-out and inferior weapons. Thereafter, German victories and London's repeated demands for British counteroffensives meant that the British desert forces had little time to analyze their mistakes and to train to correct them. With few exceptions, the senior British commanders did not stay in office long enough to learn and apply the lessons of desert war. The Germans had arrived in Africa with a system of combined arms battlegroups, flexible commanders, and variable tactics to mass combat power on the basis of battle drills. By contrast, the British units had rarely studied combined arms tactics. Units newly arrived from Britain might be better trained, but they were often squandered piecemeal before they had become acclimated to the desert.

British commanders were also betrayed by their habits of centralized staff planning and control. The time delays inherent in such a system were magnified by the weaknesses of British combat communications. Like the French, the interwar British Army had relied on telephones rather than on radios, regarding the latter as unreliable and prone to enemy intercept. Budgetary limitations had discouraged the development of tactical communications systems. As a result, until 1943 British division and corps headquarters had no encrypted, secure radio communications and few trained radio operators. Yet in the desert, radio was the only feasible method of long-range communication. In the crisis of battle, British generals seized on the few available voice radio sets as the only means of collecting information and disseminating orders to their subordinates. German intercept operators who monitored these unencrypted voice communications obtained a gold mine of information about British operations.[27]

The Germans also had a considerable technological advantage in weapons.[28] After their shocking encounter with heavy British and French tanks in France, the Germans had experimented with the 88mm antiaircraft gun to test its effectiveness as an antitank weapon against captured British equipment. The German divisions sent to Africa had a number of organizational modifications, such as fewer howitzers but more antitank artillery, including a small number of 88mm guns. Moreover, the German tanks in Africa were largely Panzer III and IV mediums, with Panzer IIs in reconnaissance and command elements. These medium tanks were considerably better armed and armored than the British cruiser and light tanks.

During the course of 1941, a 50mm medium-velocity main gun replaced the 37mm on most Panzer IIIs. Then in mid-1942 the Germans installed an even higher velocity 50mm (the Panzer IIIJ with a long-barreled gun) on some Panzer IIIs, giving them the same penetration power as the 50mm towed antitank gun that had already replaced the ineffective German 37mm. This new 50mm tank gun had improved sights and fired special "arrowhead" ammunition, an early form of armor-piercing discarding Sabot shells in which only a small, high-velocity penetrating rod actually reaches the enemy tank. The arrowhead was capable of penetrating even thickly armored infantry support tanks at short ranges. By contrast, the Germans had designed the Panzer IV to provide area fire support for other tanks, suppressing enemy antitank defenses while the Panzer IIIs closed in the attack. As such, the Panzer IV's original armament was a 75mm low-velocity gun capable of damaging British tracks and roadwheels at 1,000 meters, but not of penetrating thick armor. Again, during 1942, the continuing German quest for gunpower caused some Panzer IVs to receive a higher velocity 75mm gun. One should note, however, that even in May 1942 Rommel had only four of these Panzer IV Specials and 19 Panzer IIIJs with high-velocity guns.[29]

The two-pounder (40mm) gun installed in most British tank turrets and issued to the infantry as an antitank weapon was totally outclassed by German tanks, where the frontal armor was face-hardened for extra strength. British gunners often had to hit a German tank twice—once to shatter the face hardening and a second time to penetrate the armor. Moreover, the two-pounder could not fire smoke or high explosive rounds to discourage German antitank crews. As late as May 1942, the British forces in North Africa had only 100 six-pound antitank guns and were just receiving their first American-built Grant tanks. This much-maligned improvisation, in which a 75mm medium-velocity gun was mounted on the side of the hull because it would not fit into the turret, at least gave the British the capability of penetrating German armor to a distance of 650 to 850 yards. The 75mm gun also gave the British their first capability to fire high explosive rounds from a tank, and the Grant's armor was proof against the typical 50mm short German gun beyond 250 yards.[30]

Without armored vehicles or effective antitank guns, British and Commonwealth infantry were extremely vulnerable in the desert battles of 1941 and 1942. British tankers came to regard the infantry as a nuisance, something that had to be protected, rather than as an equal member of the combined arms team. It was not surprising, therefore, that many British armored units reverted to the pure-armor cavalry mindset of the interwar years. Despite the training efforts of Martel, British tank battalions in Britain and North Africa found it difficult to resist the temptation to close with the enemy. For some time, the British did not realize that the German antitank guns, and not the German tanks, were their true enemies; even after this discovery, British units charged without first locating the antitank gun line. The results were often disastrous. On 15 June 1941, for example, a few German tanks decoyed the 16th Royal Tank Regiment into a screen of 50mm antitank guns; the British lost seventeen tanks in a matter of minutes. That same day, a single German 88mm claimed to have destroyed eleven Matildas single-handedly.[31]

Such bitter lessons only convinced the British to value gunpower over all other elements. The armor's tendency to maneuver on its own often left the infantry exposed, and the resulting mistrust made any attempt at cooperation between these arms extremely difficult. In those cases where the British and Commonwealth infantry were able to entrench effectively, the commanders chose positions that were too dispersed to provide supporting fire to one another; thus the Germans could concentrate all available firepower against one British unit at a time.

Early in the desert war, British commanders apparently identified the German concept of combined arms task organization at the small-unit level but did not always use the correct tactics to complement that organization. As Montgomery was preaching in Great Britain, the tendency to form combined arms task forces of brigade and battalion size was not always appropriate or sufficient and caused the divisions to fight as uncoordinated and dispersed collections of small units. British artillery in particular could not mass its fires. The concentrated efforts of the German Afrika Korps often defeated these more numerous British task forces in detail.

The British tried to reverse this process. General Martel visited North Africa in early 1942, and the local armor commanders agreed to the newer concepts of a combined arms armored division. However, North African units did not implement these changes in organization and tactics before the next German offensive, so the British again lost armored "brigade groups" piecemeal, despite their intentions to employ their divisions as unified forces. After losing most of their tanks, the British resorted to small motorized columns built around the few remaining field and antitank units, with just enough motorized infantry to provide local security for them. "Excess" infantry went to the rear.[32]

This was the situation when Montgomery took command of the Eighth British Army in August 1942. Lt. Gen. Brian Horrocks, who had participated in Martel's training exercises as an armored division commander, arrived soon thereafter to command one of the corps in the Eighth. In effect, Montgomery began to retrain that army from scratch. It would have been difficult if not impossible for him to completely change the procedures of the British Army. Instead, he focused on ways to make the traditional British method of centralized command and control for a set-piece battle work in the desert.

The British gained time by halting the Germans at Alam Haifa (31 August–5 September 1942). Having predicted the key terrain that the Germans would have to seize, British and Commonwealth defenders dug in to deny it to the enemy. The Royal Air Force attacked German armor while it was immobilized in British minefields. The main British defenses included Grant tank fire at long range, towed antitank guns at closer range, and finally massed artillery protective fires at short range. These successive layers of defense, at a place that could not easily be bypassed, exhausted the German attackers.[33]

After Alam Haifa, Montgomery used an abbreviated form of his training program from Britain to prepare the Eighth Army for the deliberate attack known as the second battle of Alamein (October–November 1942). To ensure that the entire army attacked in a coordinated manner, he resorted to the elaborate planning and centralized direction characteristic of British attacks in World War I. Each corps directed its artillery, for example. Such procedures were more familiar to British staff officers than the fluid, improvisational tactics that they had attempted to copy from the Germans. Engineers, infantry, and artillery conducted a night penetration of the German-Italian defensive positions, seizing high ground on which to establish infantry–antitank defenses. Next, Montgomery planned to move armor forward under their protection, tempting the Germans to counterattack.

Montgomery's plans allowed the British Army to perform according to its own doctrine and training. At the same time, shortages in fuel and equipment almost immobilized the Germans and Italians, negating much of Rommel's advantage in flexible maneuver. The resulting battle was an attrition contest in which Montgomery had to adjust his plans frequently, largely because the armored units still had difficulty cooperating with the artillery and infantry. The ultimate British success clearly owed as much to Montgomery's methods of forcing combined arms cooperation on his subordinates as it did to the superiority of British materiel at the time. Historians have frequently criticized Montgomery for the cautious manner in which he conducted both deliberate attacks and more fluid exploitation and pursuit. Yet this caution enabled him to minimize or to avoid the errors of his predecessors, errors caused largely by an inability to coordinate the different arms without advanced planning.[34]

The German Advance in Russia, 1941

While Germany went from victory to victory during 1939 to 1941, the Red Army stood nearly impotent, thanks in part to Stalin's purge of its officer corps. The administrative occupation of eastern Poland in the fall of 1939 strained Soviet logistics to the breaking point, and the disastrous Russo-Finnish War (1939–1940) demonstrated the Soviet inability to coordinate units for a deliberate attack.[35] Indeed, the Soviets eventually learned from their mistakes, redoubled their efforts, and forced the Finns to negotiate an armistice in March 1940. Nevertheless, the Red Army was a shambles.

In light of these experiences, during 1940 and 1941 the Soviet government undertook major reforms in military organization, equipment, command structure, and deployment. The Soviets mismanaged many of these changes, and none was complete by the time Germany attacked in June 1941. The Germans caught the Red Army in transition and ripped it apart.[36]

The most noteworthy Soviet change before the German invasion, for our purposes, was the reintroduction of large combined arms mechanized formations. In reaction to the German victories of 1940, the Soviet government reversed its previous decision against such units. Thus, twenty-nine mechanized corps were created in 1940 and 1941, each consisting of two tank and one motorized rifle division, for use as the exploitation forces in each field army. On paper, each of these huge corps was authorized 1,031 tanks. Unfortunately, the Soviets had neither the manpower nor the equipment to implement their ambitious plan immediately. By removing all tanks from infantry and cavalry support units, the Red Army collected approximately 17,000 tanks, but the new organizations called for a total of 29,899. Worse still, the existing tanks were almost entirely the lightly armed and armored variety produced in the mid-1930s. By 1941 such equipment was tactically obsolete and mechanically worn out. In late 1939 the Red Army had approved designs for new, second-generation equipment, including the T-34 medium and KV-1 heavy tanks. Both were armed with an excellent 76.2mm main gun, and each had more armor protection than any existing German tank. Each had various weaknesses, such as the design of its turret; however, until the installation of higher velocity guns in 1942, none of the German panzers was equal to the T-34 and KV-1. Yet the Great Purges had also affected industrial management in the Soviet Union so that only 1,475 of these outstanding new weapons were delivered before the German attack.[37] Similar managerial and bureaucratic problems deprived the Soviets of enough trucks to move infantry and artillery, of mines to stop tanks, and of modern fighter planes to contest German air superiority. Typical Soviet units were a mixture of worn-out, obsolescent weapons and newer equipment whose crews were not yet fully trained.

In contrast to the Soviet disarray, the German Army that invaded on 22 June 1941 was at the top of its form. Hitler's continuing desire for more panzer divisions had unintentionally improved the balance of arms within those divisions. In order to assemble enough tanks to create additional units, the Germans had reduced all panzer divisions to an establishment of only two or three tank battalions of three companies each, for a total of 150 to 202 tanks per division. This action, and an increase in infantry to a total of one motorcycle and four truck-mounted battalions, meant that a 1941 division had six to nine tank companies but fifteen motorized infantry companies; the other arms remained unchanged. Considering the high casualties and many demands for motorized infantry, this ratio was probably the most effective for most forms of mechanized combat.

Armored enthusiasts have frequently criticized Hitler for this reduction in tank strength, arguing that the resulting panzer division lacked the combat power for sustained advances of the type necessary across the roadless expanse of European Russia.[38] It would be more accurate to argue that German planners had geared their entire armed forces for relatively limited distances and tied them to railroads and horse-drawn logistics. The problems in the German maintenance system had been evident as early as the Polish campaign of 1939, but the Russian campaign involved much greater distances and longer operations. Under these circumstances, the German system of centralizing spare parts and shipping damaged tanks back to the factory for major repairs was completely inadequate. In August 1941 the field commanders in Russia had to mount a major argument to convince Hitler to release 300 tank engines to replace those already worn out in the campaign. Prior to invading the Soviet Union, much of the German mechanized force had been involved in a Balkan campaign, wearing down equipment and using up spare parts. During the summer and fall of 1941, these same vehicles covered hundreds of additional miles over uneven, dusty, and sometimes muddy roads, causing many breakdowns. If each panzer division had retained another tank battalion, those additional tanks would have worn out at the same rate as the rest of the division, leaving at best only a handful of additional vehicles still operating by the time the division reached the gates of Moscow in December 1941. What the Germans needed more than additional tanks were additional trucks for resupply and a better field maintenance system to repair existing equipment. They eventually developed the latter, but not in 1941.[39]

These problems, however, were not immediately evident. Operationally, the 1941 campaign was the heyday of the German blitzkrieg and especially of the encirclement battle. The Soviet analysis and description of these encirclements offers the best summary (see Fig. 10).[40]

First, the attacker had to penetrate or outflank the enemy's defenders. This was relatively easy in 1941, when the Germans caught the Soviets in their peacetime garrison, unorganized for any coherent defense. Under these circumstances, the attacker would immediately launch armored units to exploit into the enemy rear areas. If a deliberate attack proved necessary to create the penetration, however, the Germans preferred to conduct it with a conventional infantry force, supported by engineers to clear obstacles and by artillery and preplanned air strikes to suppress enemy defensive fires. As the war lengthened, such penetrations became increasingly difficult for all armies.

Once penetrations or flanking maneuvers had succeeded, the German armored forces sought to encircle the enemy by using one or two pincers. A combined arms battlegroup of battalion or regimental size (600 to 2,000 men) usually led each pincer. After the jaws of the pincers closed, the at-

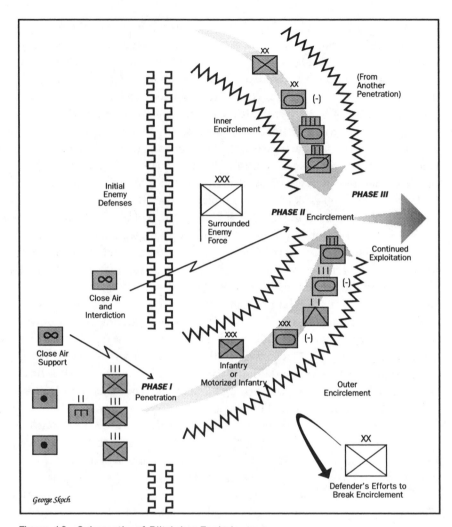

Figure 10. Schematic of Blitzkrieg Encirclement

tacker had to create two encirclements—one facing inward, to hold the surrounded force and gradually reduce it, and another facing outward, to ward off any efforts to relieve the encircled units. In order to establish these two encirclements, the Germans tried to give each panzer corps one or more motorized infantry divisions to follow and support two to three panzer divisions. In practice, the Germans never had enough force in a panzer corps to seal off an encirclement, so the process of holding and reducing encirclements had to wait until the arrival on foot of the conventional infantry divisions. During the interim, individual Soviet soldiers and sometimes entire

Red Army units were able to exfiltrate or break out of the loosely cordoned encirclement, escaping to join local partisans or to return to their own lines and fight again. This time delay also immobilized the panzer units, prevented further exploitation, and gave the defender time to reorganize his forces for a new effort farther to the rear. Only when the infantry and logistics troops had caught up with the panzer units could the latter resume the exploitation and pursuit.

The Soviet Response, 1941–1942

As the Germans advanced into European Russia, encircling one Soviet field army after another, the Soviet military took desperate measures to overcome their weaknesses.[41] Two basic problems were immediately apparent. First, the average Soviet commander or staff officer lacked the skills necessary to orchestrate the different arms and weapons for an effective defense or counterattack. The survivors of the purges had been promoted far beyond their level of experience and competence. The general staff finally had to reprimand these commanders for continually deploying their forces evenly across a defensive sector as if on a textbook exercise, without regard for the terrain or the high-speed avenues of approach that required antitank defenses-in-depth. Second, the Red Army was seriously short of the specialized units and weapons that its commanders found so difficult to employ effectively—engineers, tanks, antitank guns, and artillery.

In retrospect, the solution to both problems seems obvious, but it took considerable nerve to undertake major changes while struggling to meet the German onslaught. *Stavka* (Supreme Headquarters) Circular no. 1, dated 15 July 1941, ordered the simplification of the commander's span of control by centralizing specialized units in pools held at higher headquarters than heretofore. This allowed more experienced commanders and staffs to focus scarce resources at the critical points. Specifically, the circular disestablished the rifle corps headquarters as a level of command. For the next two years, a Soviet field army consisted of only four to six divisions or separate brigades, plus specialized units such as artillery, tanks, and antitank weapons. Similarly, the circular removed tank and antitank units and reduced the amount of artillery in each division. The Soviet rifle division, which until that time had closely resembled divisions in other European armies, was cut in its authorized strength from 14,483 men to only 10,859.[42] Much of the equipment existed only on paper in any case, and what was actually available could then be centralized at the level of field army or higher. The same order disestablished the oversized mechanized corps of 1940–1941. Some of the individual tank divisions within them survived as separate formations, but in general the first German onslaught had already shattered the mechanized corps.

The remainder of 1941 was a desperate struggle for the Red Army, a struggle in which its traditional doctrines of Deep Battle and large mechanized units were inappropriate because of the German advantage in equipment and initiative. The few tanks coming off Soviet assembly lines were formed into small brigades used solely for infantry support.

Once the Red Army halted and threw the invaders back from Moscow in December 1941, the Soviet commanders began to revive their organizations and doctrine.[43] The Soviet regime had succeeded in evacuating its factories across the Ural Mountains, where they produced a phenomenal stream of new weapons in the spring of 1942. Col. Gen. Yakov Fedorenko, chief of the Armored Forces Administration, used these weapons to construct new tank corps in April. By July 1942 these corps had settled on an organization of one rifle and three tank brigades, plus supporting arms—a relatively tank-heavy force that the Soviets intended to use as the mobile exploitation force of a field army (Fig. 11, top). In the fall of 1942 Fedorenko added mechanized corps, which had a larger proportion of motorized infantry and were therefore more expensive in manpower and trucks. Truck production was in fact a major problem throughout World War II, and the Soviets depended on imported American wheeled vehicles to move and supply their mobile formations.

Unlike the mobile corps of 1940, those of 1942 were actually of division size or smaller. To conduct the type of deep exploitations envisaged in the 1930s, ranging 150 kilometers or more into the enemy rear, the Red Army needed a larger mobile formation, on the order of a German panzer corps or panzer army. In May 1942 the Commissariat of Defense took this next logical step, uniting some of the existing tank corps into tank armies. These 1942 tank armies, however, were usually improvised combinations of armored, cavalry, and infantry units, combinations that lacked a common rate of mobility or doctrine for employment. Moreover, these first tank armies rushed into battle against the Germans during the summer of 1942 and were largely destroyed before they had even trained to operate as a team.

Not until January 1943 did the Soviets finally produce a coherent tank army (Fig. 11); the six tank armies formed that year became the spearheads for all Soviet offensives for the remainder of the war. Although the actual organization varied, a typical 1943 tank army consisted of two tank corps, one mechanized corps, and various specialized supporting formations. Each of these armies was actually a corps-sized formation in Western terminology and, like the tank "corps" on which it was built, was extremely tank-heavy. This was probably an appropriate organization, both because of the

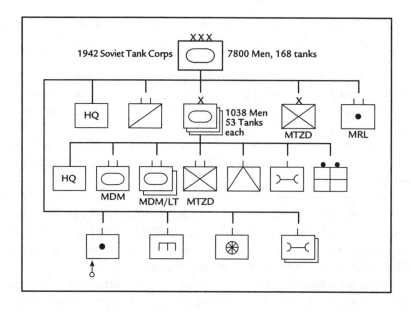

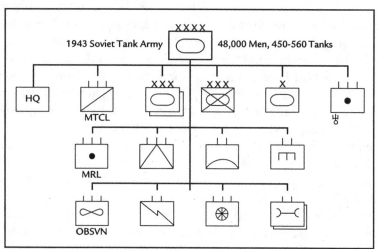

Figure 11. Soviet Tank Corps, 1942, and Tank Army, 1943

open tank country of European Russia and because of the high Soviet tank losses against the Germans. Given the inexperience of most tank crews and junior leaders in the Red Army in 1941–1943, it was inevitable that the better-trained German tank and antitank formations would inflict such disproportionate losses on the Reds. Thus, the Soviet Union's armored forces remained much more tank-heavy than those of other armies. Yet throughout the war, the Soviets also maintained corps-sized formations of horse cavalry, with limited tank and artillery support, for use in swamps, mountains, and other terrain that did not favor heavily mechanized forces.

The new mechanized formations must be understood in the context of their accompanying doctrine. During 1942 the Soviets digested the lessons of the first year of war and issued a series of orders to correct their errors, which greatly increased the effectiveness of their counteroffensive that encircled Stalingrad in November 1942. Senior Red commanders held conferences before Stalingrad to ensure that their subordinates understood the new doctrine.

From the Soviet point of view, the first problem was to penetrate the German defenses in order to conduct a counteroffensive. The initial Soviet counterattacks of December 1941–January 1942 had been made with widely dispersed forces that in some cases were actually outnumbered by the German defenders. To prevent such an error in future, *Stavka* Circular no. 3 was issued on 10 January 1942. It directed the formation of shock groups, that is, strong concentrations of combat power on a narrow attack frontage in order to break into the enemy defenses. Division and larger units were instructed to mass their available forces at a specific point or points in this manner. Stalin's Order no. 306, dated 8 October 1942, supplemented this directive by explicitly forbidding successive waves of infantry in the attack. Given the continuing shortages of equipment and firepower, the Soviets decided to maximize their available force by putting all the infantry in an attack into a single echelon. Thus, in a typical rifle division, as many as nineteen of the twenty-seven rifle companies would be on the front line for a deliberate attack.[44] The German defenses in 1942 were stretched so thinly that this massing of infantry was more important than successive echelons to sustain the attack. Later in the war, when both sides defended in greater depth, the Soviets learned to echelon their attack accordingly. Even in 1945, however, shallow German defenses prompted one-echelon Soviet attacks. Other orders in October 1942 governed the correct use of those tanks still assigned to assist the infantry assault. Because infantry commanders were still inexperienced, all such tank units were to be employed en masse under their own commanders.

Once the Soviets completed a penetration, their mobile groups would pass through for exploitation and encirclement operations. In effect, one such encirclement might include other, small encirclements within its pincers, like smaller Russian dolls nested inside larger ones. Each field army attempted to use its own mobile group, composed of a tank, cavalry, or mechanized corps, to exploit penetrations to a relatively shallow depth of perhaps fifty kilometers. This shallow penetration aimed to disrupt the German defenses, defeat enemy reserves, or link up with a similar penetration from a neighboring army. Simultaneously, the tank armies acted as mobile groups for larger commands, such as a front (army group). As such, the tank armies penetrated even deeper into the German rear areas, seeking to seize river crossings and other key terrain before the Germans could organize new defenses. This, at least, was the theory.

The first of these large, operational-level Soviet encirclements was in November 1942, when the German Sixth Army was surrounded at Stalingrad. Once the Germans were encircled, Lt. Gen. M. M. Popov led four understrength tank corps deep into the German rear in February 1943. Although most of Popov's force was defeated by a German counteroffensive, this experience became the model for the employment of the new tank armies.[45]

Thus, by late 1942 the German techniques for mechanized warfare had passed their peak and were no longer achieving the success of 1939–1941. On the contrary, Great Britain and the Soviet Union had reorganized and retrained their armies and were beginning to conduct their own successful mechanized offensives. Both German and British armored formations had evolved into balanced structures in which tanks no longer outnumbered the other arms. Moreover, the three armies were discovering the need for effective and mobile logistical support to sustain the mechanized offensives. The stage was set for a conflict in which logistics, technology, and defense-in-depth would determine as many battles as the panzer division had decided in the first half of the war.

A-20 aircraft of Army Air Forces Light Attack aviation using flour bombs on U.S. light tanks, California maneuvers, 1942. (U.S. Army Military History Institute)

German panzer column on the Russian steppes, c. 1942. (U.S. Army Military History Institute)

M10 tank destroyers acting as indirect artillery, Italy, 1944. (U.S. Army Military History Institute)

Soviet infantry and Sturmovik ground attack aircraft. (David Glantz)

Soviet forward detachment in Manchuria, August 1945. (David Glantz)

CHAPTER 5

• • • • • • • • • • • • • • •

Allied Response and Armored Clashes

By deferring any consideration of the war in the Pacific, I have reviewed the evolution of combined arms in World War II from the simple perspective of German advance and Allied response. The participation of the United States, Japan, and the Soviet Union made the war a much more complex affair, a war of production and technology as much as of battlefield maneuver. Thus those aspects of technology and tactics that affected the development of combined arms troops and tactics during the second half of World War II need to be identified. After examining the evolution of American force structure and doctrine, I shall consider the changes in tank design and employment that made the latter half of the war so different from the first half.

The American Response, 1941–1944

Prior to the Japanese attack on Pearl Harbor in December 1941, the United States was an interested observer of World War II. Most of the U.S. Army did not become involved in major ground operations for a year or even more after entering the war. During the period 1941–1942, however, the United States drew certain conclusions about the nature of weapons, organizations, and tactics and implemented these conclusions in a continuing evolution of the triangular infantry division and the 1940 armored division. Then, on the basis of large-scale maneuvers held in the United States and of the initial combat experiences overseas, additional changes in American doctrine and organization occurred in the middle of the war. The resulting tactical system dominated American military thought into the 1950s.

In March 1942 Lt. Gen. Lesley McNair, one of the designers of the triangular division, became head of Army Ground Forces, in charge of all troop unit training and organization. McNair continued to follow the concepts that had guided him in the 1930s, and thus the basic organization of the triangular division did not change significantly until after the war.[1]

First, McNair wanted each unit to have only the minimum essential forces necessary to conduct offensive operations in fluid, maneuver warfare against relatively limited resistance. In the case of the triangular infantry division, this meant that its standard base remained the three infantry regiments, four artillery battalions, reconnaissance troop, and engineer battalion developed during 1937 to 1941.

In McNair's opinion, a division did not need specialized units that were required only for specific situations or missions. This rule applied particularly to arms with an essentially defensive mission, such as antitank and antiaircraft artillery. These units that McNair had "streamlined" out of the infantry division became a "pool" of specialized nondivisional companies and battalions. Higher headquarters, particularly corps and field army, were supposed to control this pool; such headquarters could either attach units to a division for a particular mission or else employ the pool en masse at critical points on the battlefield. The actual combat power of a division might change from day to day, depending on requirements and missions. Thus, apparently on his own, McNair arrived at the same solution that the Red Army had adopted during the crisis of 1941.

In December 1942 McNair extended this trend to form ad hoc forces from nondivisional units by persuading the War Department to abolish most independent or nondivisional regiments (that is, units of one combat arm that had a fixed table of organization but that were not included in a division structure). He then created flexible groups of two or more separate battalions. Nondivisional armor, antiaircraft, field artillery, mechanized cavalry, and combat engineer battalions reported to group headquarters. These group headquarters, notably those of the mechanized cavalry, also acted as tactical headquarters to control their subordinate battalions when they were used together.[2]

Another of McNair's principles was that staff and support elements should be as small as possible, in order to maximize the proportion of forces actually available for combat and to reduce paperwork and other organizational obstacles to rapid decision making and communication. Logisticians should bypass divisional and corps headquarters on routine supply matters in order to keep those headquarters small, mobile, and oriented on the tactical situation. Wherever possible, a specialized unit or person should have weapons to perform a secondary role as infantry or rear area security forces. Of course, this model worked smoothly only if a full hierarchy of headquarters and support units, from field army downward, was present to support the combat units.

Finally, McNair sought to restrict the amount of motor transportation in a unit in order to facilitate strategic deployment. The fewer vehicles that were permanently assigned or were organic to a division, the less shipping space it would need when sent to Europe or the Pacific. For example, McNair sought to authorize only the minimum number of trucks needed to shuttle necessary supplies and ammunition to the infantry regiments during a twenty-four-hour period rather than the number that could transport all necessary materiel in one lift. Rifle units were not motorized but could become so temporarily by the attachment of six truck companies to the division. Alternatively, if the division had attached elements such as a tank battalion, the infantry could mount the tanks and the organic trucks borrowed from the artillery, allowing short-range motor movements with some loss in logistical support. Although this solution may appear extreme in retrospect, American infantry divisions still had far more motor vehicles than the foot-mobile and horse-drawn forces of their opponents.

When the U.S. Army finally employed these concepts overseas, they proved to be only partially successful. Regardless of the terrain or enemy involved, most infantry divisions in Europe and many in the Pacific believed that they needed tank, antiaircraft, "tank destroyer" (antitank), and nondivisional engineer support in virtually all circumstances. Corps and field army commanders who followed doctrine by shifting these nondivisional units from division to division according to the situation found that they could do so only at the cost of much confusion and inefficiency. Attachment to a different division meant dealing with a different set of personalities and procedures before the attached units could mesh smoothly with it.

Nowhere was this issue clearer than in the U.S. struggle to break through the hedgerow country of Normandy in June–July 1944. Nondivisional tank units had to learn to cooperate closely with engineers, infantry, and mortars to breach the hedgerow obstacles, resulting in a closely knit team. While mortars pinned down the enemy, engineer demolitions and tank-mounted plows tore holes in the thick hedges, after which the tanks and infantry protected each other in the advance across each enclosed field.[3] Once such a smooth relationship was established, the division was reluctant to release its attachments as ordered. In many instances, tactical commanders found it expedient to leave the same nondivisional elements attached to the same divisions and regiments on a habitual basis that might last for months. A typical U.S. infantry division in France during 1944 normally had attached battalions of tanks, tank destroyers, antiaircraft automatic weapons, and corps engineers. In some cases this division also had attached units of 4.2-inch chemical mortars, trucks, and logistical support from the pools at corps and field army level. Thus, the triangular division in combat was often much larger, more rigid, and more motorized than McNair had envisioned. An augmented infantry division of this kind might well have the mobility and firepower of a motorized division or even an understrength armored division, which goes far to explain the superior mobility of American infantry units when compared with standard German infantry forces.

Many of these attached forces were subdivided and further attached to infantry regiments, as were the division's organic assets such as engineers and medical support. Minor changes in the regiment's organization in 1942 and 1943 had added six 105mm howitzers, so that the regiment had its own artillery even without the direct-support field artillery battalion. In practice, a majority of infantry regiments normally operated as "regimental combat teams" (RCTs). At a minimum, this meant that they had their share of the division's medical, engineer, and field artillery attached or in direct support. Many RCTs also had companies of tank destroyers, tanks, and self-propelled antiaircraft guns. Thus, it was a combined arms force, a small division in itself.[4]

During the same period, the armored division underwent many more changes than the infantry division.[5] Of the six different changes in armored organization during the war, two were most significant. The 1940 U.S. armored division was composed largely of light tanks that greatly outnumbered the medium tanks, infantry, and artillery; this division also had several headquarters designed to control only one type of unit, such as the headquarters for armored and infantry regiments. When Maj. Gen. Jacob Devers became chief of the Armored Force in August 1941, he sought to establish a more flexible, functional organization within the armored division. His efforts culminated in the reorganization of 1 March 1942 (Fig. 12, top). It eliminated the armored brigade headquarters and substituted two Combat Commands, A and B (CCA and CCB), headquarters that might control any mixture of subordinate battalions given them for a particular mission. This was an American way to institutionalize the battle group *(kampfgruppe)* concept that the German panzer forces achieved by improvisation. The 1942 organization also reversed the ratio of medium and light tanks, leaving the armored division with two armored regiments, each consisting of one light and two medium tank battalions. The new structure still had six tank battalions but only three armored infantry and three armored field artillery

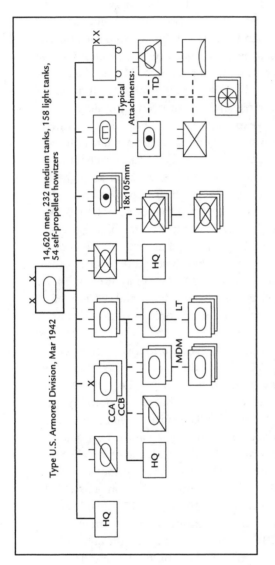
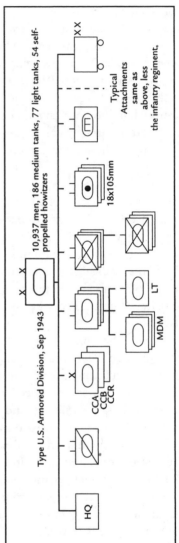

Figure 12. Type U.S. Armored Division, March 1942 and September 1943

battalions. This imbalance existed in part because the Armored Force planned to create a large number of armored corps that, like the German panzer corps, would have two armored and one motorized infantry division each.

By early 1943 intelligence studies of the more balanced German and British armored divisions had reinforced General McNair's desire for a less cumbersome division structure in his own army. The one U.S. armored division used in the North African campaign of 1942–1943 never operated as a coherent division, but its dispersal into three or four different battle groups only illustrated the difficulties of maneuvering such a large formation. At the same time, the U.S. Army had discarded the concepts of armored corps and motorized infantry divisions, making the imbalance of arms within the 1942 armored division structure even more significant. Technically, the U.S. light tanks had been no match for the increasingly well-armed and armored German vehicles, and therefore the United States, like Britain before it, lost enthusiasm for the concept of deep raids by lightly armored vehicles.

As a result, in September 1943 the War Department instituted a new, smaller armored division structure (Fig. 12, bottom). It eliminated the regimental headquarters that had theoretically controlled only one type of battalion and reduced the tank component to only three tank battalions of four companies each. Thus, the 1943 structure had three battalions each of tanks, armored infantry, and armored field artillery, although in practice there were still twelve tank companies to only nine infantry. A third, smaller combat headquarters, designated reserve (R), was added to control units under division control and not currently subordinated to the other two combat commands. Some division commanders used this CCR as a third tactical control element like CCA and CCB.

Two U.S. armored divisions, the 2d and 3d, continued under the heavier 1942 table of organization throughout the war. Corps or army headquarters frequently augmented each of these divisions with an infantry regiment borrowed from a normal division. As a result, the balance of tanks and infantry in American divisions, as in the German and British armored divisions, came to be approximately equal. Both types of U.S. armored division received attachments similar to those given to infantry divisions. In addition, virtually every American armored division habitually controlled two quartermaster truck companies capable of handling the great logistical requirements of a mobile division.[6]

The combat organization within each of these divisions varied greatly, but a typical combat command within a 1943 (light) armored division usually had two battalion-sized task forces. The combat command headquarters created these by trading a medium tank company from a tank battalion for an armored infantry company from an infantry battalion, producing one task force of three tank companies and one armored infantry company and one task force of two armored infantry companies and one tank company. In some units, the process of integrating the two arms continued down to the level of combining individual tanks with half-track mounted squads of armored infantry. These battalion task forces also had attached platoons of tank destroyers, armored engineers, and in some cases self-propelled anti-aircraft guns. An armored artillery battalion could be either in direct support of the combat command or attached to it if the division were widely dispersed.[7]

Antitank Technology

Effective force structure and tactics are intimately related to effective weapons design, and therefore any study of combined arms warfare must consider the major effects of technology. During World War II, one obvious intersection of technology and tactics was the critical question of tank and antitank warfare. Even if defending troops managed to overcome their fear of deep mechanized penetration, the blitzkrieg would still succeed unless the defense acquired effective antitank weapons and doctrine.

Antitank ditches and similar obstacles may slow the movement of armored units or channel them into antiarmor kill zones, but such obstacles usually take a great deal of time and effort to erect and do not actually destroy many tanks. Ultimately there are only two ways to defeat armored vehicles.[8] *Kinetic energy weapons* puncture armor plate by sheer momentum, as if they were punching through the metal; *chemical energy weapons* use explosive blasts to destroy the armor. Until the middle years of World War II, chemical energy weapons were generally ineffective against armor. Antitank design therefore concentrated on the kinetic energy systems. Mathematically, the energy of an object is equal to one-half the product of that object's mass multiplied by the square of its velocity ($1/2\ MV^2$); therefore, improving the armor penetration of a kinetic energy weapon requires increasing either its mass, its velocity, or both. Greater mass meant larger caliber weapons or heavier, denser material in a projectile of the same caliber. Thus, basic physics explains the general trend toward larger caliber weapons during World War II. This is true despite the fact that an increase in caliber alone would reduce the projectile's velocity unless the designer also took other steps. Velocity, in turn, would be increased through changes such as longer gun barrels, more effective propellants, and a better seal within the breech so that more of the propellant effect went to drive the projectile out of the gun tube.

In practical terms, World War II improvements in antitank guns had three consequences. First, the size and weight of those guns increased steadily as calibers increased, gun tubes lengthened, and stronger carriages were added to absorb the recoil of high-velocity weapons. Second, tanks needed increased armor protection, either by better materials, thicker armor plates, or different angles that caused the penetration to go through the armor diagonally. Third, the new antitank weapons were much more effective than those of the previous decade, but they were also more expensive and specialized. Such weapons formed the backbone of any antitank defense, yet no army could afford to have sufficient numbers of antitank weapons permanently assigned to every small unit that might need them. The kinetic energy antitank gun simply did not fulfill the battlefield requirement that every unit must have some protection in case it suddenly encountered enemy armor.

The alternative means of defeating armor was the chemical energy weapon. The detonation of an explosive charge usually had little effect against armor, because the blast effect would dissipate equally in all directions. Ordinary explosive artillery rounds had to be quite large before they could do more than damage the tracks and roadwheels on which a tank rode. Yet guns of sufficient size (usually 105mm or larger), like the antitank guns, were too large and specialized to be of general use by infantry, engineers, and other frontline troops. Moreover, using field artillery in an antitank role diverted it from its primary function of indirect fire.

The solution was to concentrate the effects of a relatively small amount of explosive on one particular point of the enemy's armor—the shaped-charge principle. Because the blast and not the momentum of the shell caused the destruction, a chemical energy weapon did not need the high velocity and elaborate gun carriage of a kinetic energy weapon. Here, at least potentially, was the means to give every soldier a cheap form of portable antitank defense.

By April 1942 the U.S. Army Ordnance Department had developed the 2.36-inch bazooka, which fired a shaped-charge warhead with a small rocket motor. Eventually, the United States concluded that this warhead was too small to penetrate thick German armor, and a larger, 3.5-inch super bazooka entered service. Meanwhile, later in 1942, the Germans had captured an American bazooka from the Soviets and from it developed the larger and more effective *Panzerschreck* antitank rocket launcher. The British PIAT (Projector, Infantry, Antitank) and the German *Panzerfaust* used the shaped charge propelled by a small conventional charge, similar to that of a grenade launcher. The same type of warhead enabled the Germans and Americans to develop experimental low-velocity recoilless rifles, light artillery pieces that eliminated the recoil by a controlled release of propellant blast behind the gun. Although recoilless rifles and rocket launchers lacked the long range and accuracy of conventional artillery, they gave the infantry, and indeed any unit, a much greater firepower and capability for short-range antitank defense.[9]

Tank Surrogates

By themselves, short-range antitank weapons were incapable of stopping a massed armor attack. Such weapons were most effective against the thinly armored flanks and rear of a tank that had already passed the defender. Indeed, a primary function of tank-supporting infantry, such as the German panzer grenadiers and the American armored infantry, was to protect the tanks from close-range ambush by enemy infantry. The towed antitank guns with which every army began the war presented a small target for the enemy to detect and engage; moreover, such guns could be maneuvered onto steep hills or river-crossing sites where a self-propelled gun could not go. The towed weapons, however, had very little armor; even if the enemy failed to score a direct hit on such a gun, a near-miss might cause casualties or at least disturb the gunner's aim. Many professional soldiers realized early in the war that the most effective antitank defense was a careful integration of obstacles, antitank mines, artillery, short-range antitank weapons, and some type of large caliber, longer range gun specially designed for antitank work. This requirement for mobile, large-caliber antitank guns in the defense matched the continued need for armor to support the infantry in the deliberate attack. Even if the nature of the enemy defenses did not always require tanks to overcome them, the presence of tanks or tanklike vehicles exerted a great psychological effect on both attacker and defender.

Armor experts in most armies, however, were determined to avoid being tied to the infantry, and in any event a tank was an extremely complicated, expensive, and therefore scarce weapon. The British persisted for much of the war on a dual track of development, retaining heavy tanks to support the infantry and lighter, more mobile tanks for independent armored formations.[10] The Soviets similarly ended the war with an entire series of heavy breakthrough tanks. Nevertheless, the widespread demand for tanks or tanklike vehicles outside the mechanized formations led to a number of tank surrogates, weapons designed to provide mobile antitank defense, close support of the infantry attack, or both. In the latter case, the surrogate needed considerable frontal armor and a dual purpose (antitank and antipersonnel) main gun.

The most original of these tank surrogates was the American tank destroyer. The German successes of 1940 evoked widespread American concern about antitank defenses, but none of the existing combat arms was interested in assuming the primary role in such a defense. As in so many other areas, Lesley McNair's solution to this problem was controversial.[11] McNair did not accept the extreme view, common in 1940 and 1941, that the armored division had rendered the infantry division almost obsolete. Instead, he agreed with the German concept that the best means to halt an armored attack was an antitank defense integrated with infantry units. McNair and Col. Andrew D. Bruce of the War Department staff sought highly mobile antitank guns that would end the psychological threat of the blitzkrieg by aggressive action against the attacking armored forces. In keeping with his concept of streamlining, however, McNair wanted most of these antitank forces concentrated in separate, nondivisional units. After a series of experiments in this regard during the U.S. Army's large-scale maneuvers of 1941, Bruce became head of a Tank Destroyer Center that developed its own doctrine for this new weapon.[12] Bruce took the concept even further than McNair, seeking a high-velocity gun mounted on a mobile platform that would sacrifice armor protection to achieve greater speed and gunpower. The tank destroyer, like the armored force, became a quasi-independent combat arm.

The 1942 tank destroyer battalions were combined arms forces in their own right, although they did not include a balance of all arms. Each tank destroyer platoon had four self-propelled guns, an armored car section for security, and an antiaircraft section; in addition to three companies of such guns, the tank destroyer battalion included a reconnaissance company of three reconnaissance platoons and an engineer platoon. Ideally, when an armored penetration occurred, the tank destroyer battalions would mass to ambush the enemy tanks in the depth of the American defenses. Within each tank destroyer battalion, the reconnaissance company selected likely anti-armor kill zones and emplaced minefields to impede the enemy advance through these areas. To reduce their vulnerability, the gun companies would move behind low hills so that only their turrets were exposed; from these protected or "hull down" positions they would engage the enemy armor.

When the U.S. Army first encountered the Germans in Tunisia during 1942 and 1943, the tank destroyers proved a dismal disappointment. Both tank destroyer doctrine and German armor design had outpaced the actual development of U.S. tank destroyers, which in 1942 were little more than improvised guns mounted on half-tracked, thinly armored vehicles. The early tank destroyers lacked mobility and effective penetration power, the very characteristics they were supposed to maximize. Moreover, most American units in North Africa were widely scattered with few roads to connect them, making it difficult to concentrate the tank destroyer forces according to doctrine. And much of the North African terrain was too open for tank destroyer vehicles to find effective hull-down positions. As a consequence, U.S. commanders in Africa tended to favor the British system of towed antitank weapons. They specifically asked that one-half of all tank destroyer battalions slated to participate in the 1944 invasion of France should use towed instead of self-propelled guns. Once in France, however, the Americans discovered that the towed antitank gun was almost useless in the more restricted terrain of Western Europe. Towed guns were not only slow to move but also too close to the ground to shoot over hedgerows and other obstacles. Furthermore, between the North African and Normandy campaigns the Tank Destroyer Center had procured much more effective, properly designed self-propelled guns. The M18 model with a 76mm gun and especially the M36 with a 90mm were excellent weapons, although even the 90mm had less armor penetration capability than the German 88mm. Beginning in July 1944, the U.S. Army therefore began to reconvert all tank destroyer battalions to self-propelled weapons. These newly converted battalions rarely followed Bruce's doctrine of concentrating at one point, however.

The original tank destroyer battalions had been created by removing antitank battalions from the standard infantry division. By 1944 improvements in German armor had rendered the standard 57mm antitank gun of the American infantry regiment largely ineffective. Tank destroyer units consequently became even more important for antitank defense. Prior to the Ardennes counteroffensive of December 1944, the German tank threat in the west was so decentralized that massed antitank defenses seemed unimportant. Instead, infantry commanders wanted a few effective antitank weapons distributed to every unit, where they could defeat the small German armored attacks that were common at the time. In most cases, therefore, corps and army commanders in the European theater habitually attached one tank destroyer battalion to each infantry division. In turn, division commanders parceled out the tank destroyer gun companies to individual infantry regiments. At that level, the self-propelled tank destroyer appeared to be a substitute tank and was used accordingly. The regiments used these weapons not only against enemy tanks but also as accompanying artillery and as substitutes for tanks to support their infantry attacks. Without adequate armored protection, such tank destroyers suffered significant casualties from German *Panzerfausts*. [13] At other times, tank destroyers acted as indirect-fire weapons to supplement the conventional field artillery. Thus, the American tank destroyer units became a classic case of an arm that rarely functioned according to its doctrine, because it was never articulated clearly to field commanders and because they had a pressing need for other types of weapon.

In keeping with their doctrine of maneuver, U.S. tank destroyers usually had their guns mounted in turrets and, in fact, resembled tanks so much that they were often mistaken for such. In European armies, however, relatively few tank surrogates had turrets, because a turretless vehicle was much simpler and cheaper to produce. The absence of a turret gave German and Soviet tank surrogates, both assault guns and antitank weapons, a lower profile that made them smaller targets on the flat, open battlefields of Eastern Europe. However, this apparent advantage was offset by the fact that the entire vehicle had to turn in order to traverse the gun more than a few degrees. Thus tank surrogates were at a disadvantage if they engaged tanks or infantry from anything other than an ambush position.

The Germans actually developed two series of tank surrogates—assault guns to support the infantry in situations where tanks were not available and tank hunters *(Panzerjaeger)* for the antitank role. Both were distinguished from self-propelled, indirect-fire artillery by considerably thicker armor protection and by a high-velocity, flat-trajectory gun intended for direct fire. Although armor purists criticized the expenditure of resources to produce these hybrids instead of true tanks, they performed a necessary role, particularly as the German towed antitank guns became progressively less effective against Soviet armor. The armored self-propelled tank hunter was much more survivable and mobile than its towed predecessor. The one drawback of such weapons was that, unlike the towed antitank guns, they had difficulty accompanying the infantry into inaccessible areas such as steep hills or bridgeheads across rivers.

The Soviet Union also produced outstanding, heavily armored assault guns during the second half of the war but tended to use them as one component of a three-way team in the deliberate attack. Medium tanks led the assault, using their mobility wherever possible to turn the flanks of German defensive positions. Heavy tanks, operating in pairs, advanced slightly behind the medium tanks, supporting the Soviet infantry and eliminating German strong points. Finally, the assault guns provided accompanying artillery support for both infantry and tanks. To accomplish this direct-fire role, the assault guns began the battle in camouflaged positions from which they could overwatch the advance of tanks and infantry. The assault guns engaged centers of resistance that had survived the massive Soviet artillery bombardment preceding the attack. This freed the assaulting forces to advance without halting to engage the enemy unless a counterattack appeared. At intervals, the assault guns bounded forward to new positions, trying to keep within 500 meters of the heavy tanks and infantry.[14] By staying behind in this manner, assault guns avoided meeting enemy armor in a maneuver battle at close range. In such a battle, tank turrets could traverse and fire much faster than the turretless assault guns could turn their entire vehicles to aim their guns. On many occasions, of course, the attacking Soviet unit did not have the three different types of armor, but the assault guns preferred to operate from an overwatch position where possible.

Tank Design and Production

These technological trends in antitank weapons and tank surrogates form a necessary background to the actual design and production of tanks during World War II. In general, both the armor and armament of tanks increased to keep pace with antitank technology, but different nations reached different design and production strategies. These factors exerted a considerable influence on the battlefield.

During the war, German tank design went through at least three generations, with constant minor variations.[15] The first generation included such unbattleworthy prewar vehicles as the Panzer I and II, which were similar to the Russian T-26 and BT series and the British cruiser tanks. The Germans converted their tank battalions to a majority of Panzer III and IV medium tanks after the 1940 French campaign, thereby stealing a march on the Soviets and the British, who still possessed obsolete equipment. The appearance of the T-34 and KV-1 tanks during the 1941 German invasion of the Soviet Union, however, compelled the Germans to begin a race for superior armor and gunpower. Simultaneously, their successes of 1939–1941 encouraged them to rely increasingly on armor rather than on infantry when conducting a rapid breakthrough attack.

The German solution was to design a third generation of tanks that combined greater armor protection with a shortened version of the 88mm antiaircraft gun that proved so successful in the antitank role. Design work began even before the invasion of the Soviet Union, but that experience increased the German need for these new vehicles. In August 1942 Adolf Hitler approved production of the Panzer VI, usually known as the Tiger I. By the time it entered production, however, its armor had been thickened to the point where the vehicle weighed almost sixty tons, with a top speed of less than twenty-four miles per hour. Moreover, the design was so complex that each Tiger cost more than 300,000 man-hours and 80,000 Reichsmarks to build. As a result, Germany produced only 1,350 Tigers in two years, and only 178 were available for the climactic battle at Kursk.[16]

Hitler's desire to mount the more lethal, long-barreled version of the 88mm gun in an armored vehicle also gave rise to the Elephant assault gun, sometimes called the Ferdinand after its designer, Ferdinand Porsche. This turretless assault gun was so massive that it weighed sixty-seven tons and could travel only eighteen miles per hour. Its innovative electrical transmission suffered significant problems when it was rushed into production. Moreover, the original Elephants lacked machine guns to defend themselves from enemy infantry armed with short-range antitank weapons. As a result, the seventy-six Elephants fielded for the battle of Kursk had to remain behind the front line, engaging Soviet tanks from overwatching positions.[17]

The third and most numerous of Germany's third-generation armored designs was the Panzer V or Panther tank. Designed as a direct response to the T-34, the Panther was equipped with a 75mm main gun. Again, however, its final weight of forty-one tons overburdened the available engine, limiting its speed. Moreover, in the spring of 1943 Germany rushed the Panzer V into production without adequate testing, which caused many mechanical problems both before and during the battle of Kursk.

Kursk was the objective of Operation Citadel, Germany's third great offensive in the Soviet Union.[18] Hitler had originally intended to pinch off the salient around Kursk by launching Citadel in May 1943, but various factors, including a desire to field the new armored vehicles, delayed the offensive until 5 July 1943. This delay allowed the German armored force, worn down by two years of struggle in the east, to rebuild and retrain. Even the older Panzer IVs received new side skirts that helped defeat shaped-charge weapons. Yet the same delay gave the Red Army time to construct massive, deep defenses at the expected points of German attack. Indeed, Citadel was the first instance in which the Soviets were able to predict and prepare for a major German offensive, a fact that goes far to explain the German failure.

At the time of the German attack, the Red Army had not yet fielded tanks equal to the Panzer V and VI, although it did produce its first effective self-propelled antitank gun. The SU-152 (*Samokhodnaya Ustanovka,* or mechanized mounting) was designed in only twenty-five days, marrying a 152mm field gun to an existing tank chassis. The few SU-152s that were available in time for Kursk were nicknamed "Zvierboy," or animal hunter, for the number of Tigers, Panthers, and Elephants they knocked out.[19]

Most of the Soviet success at Kursk, however, was due to careful preparation and courageous execution rather than to any technological solution. For days, the Germans struggled forward slowly through layer after layer of defenses, inflicting fearful casualties but failing to break through into the Soviet rear areas. Although the new German armor far outstripped its Soviet counterparts, the defenders slowly whittled down the panzer spearheads. Particularly on the southern side of the Kursk Bulge, Hermann Hoth's Fourth Panzer Army found its flanks constantly threatened by the determined, if sometimes ineffective, counterattacks of Soviet tank and mechanized corps (see Map 6, 2d Tank Corps and 2d Guards Tank Corps).

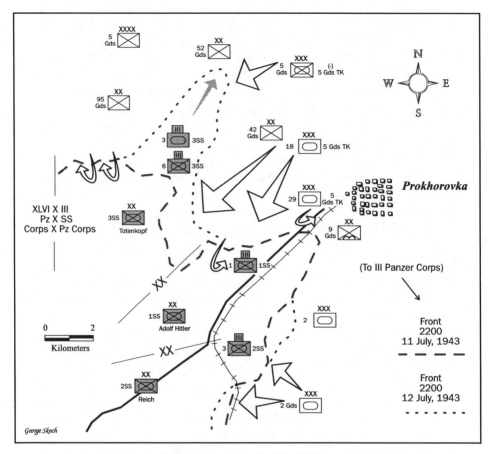

Map 6. Prokhorovka Action, Battle of Kursk, July 1943

The climax of this campaign came at the railroad junction of Prokhorovka, southeast of Kursk, on 11 and 12 July 1943. Prokhorovka has entered the realm of legend because of the supposedly huge number of tanks engaged at one spot. In fact, Hoth's spearheads, the XLVIII Panzer Corps and II SS Panzer Corps, were so hard-pressed on their flanks that they had only limited resources with which to continue the advance. Moreover, a railroad embankment and various shallow valleys effectively divided the battlefield into several different compartments. At Prokhorovka itself (Map 6), the 1st *(Liebstandarte Adolf Hitler)* and 2d *(Das Reich)* SS Panzergrenadier Divisions, supported by portions of the 3d *(Totenkopf)* SS, encountered a mixture of Soviet infantry units and three mobile corps—5th Guards Mechanized, 18th Tank, and 29th Tank—of the Fifth Guards Tank Army, the strongest of the new armored formations. Thus, perhaps 200 German tanks, including more Panzer IIIs and IVs than Tigers, fought 450 to 500 Soviet medium and light tanks at Prokhorovka. The commander of the Fifth Guards Tank Army, Pavel Rotmistrov, had concluded that the only way to overcome the technical advantage of the new German tanks was for the T-34s to rush forward to a range of 500 meters or less, where the Soviet 76.2mm guns could be effective against German armor.[20] Both sides attacked simultaneously, producing a confused melee that lasted for most of 12 July. Tanks, assault guns, and dive-bombers made a moonscape of the open, rolling terrain around Prokhorovka. Tactically, the Soviets got the worst of the exchange, and the *Totenkopf* Division was able to advance temporarily an additional two kilometers. Subsequently, *Das Reich* Division was able to advance farther, linking up with the German III Panzer Corps that had attacked from the southeast toward Prokhorovka. However, by this time the issue of Kursk had been settled. Coming on the heels of a week of desperate fighting, the struggle at Prokhorovka brought the Fourth Panzer Army to a halt. For the first time, a blitzkrieg offensive had failed to achieve a breakthrough, let alone an encirclement.

One reason for this and subsequent German defeats was the small number of German tanks produced. Hitler and his assistants were fascinated with technological improvements and frequently stopped production to apply the latest design changes to the existing tanks. Further, most German planners prized high quality and were suspicious of mass production techniques. Such problems, coupled with shortages of raw materials, meant that Germany could not compete with its foes in sheer numbers of tanks produced. In 1943, for example, Germany manufactured only 5,966 tanks of all types, as compared to 29,497 for the United States, 7,476 for Britain, and an estimated 20,000 for the Soviet Union.[21] A disparity in numbers of this magnitude would eventually overcome the highest quality of individual tank design, as Prokhorovka indicated. As early as 1941 production shortfalls forced the Germans to use captured French tanks and trucks to equip some units. Similarly, the presence of different versions of the same tank, often within the same company or battalion, sometimes made it difficult for the Germans to obtain spare parts and repair damaged equipment.[22]

The alternative to constant changes in tank design was to standardize on a few basic types and mass-produce them even though technology had bypassed them. The Soviet T-34, for example, was an excellent basic design. Although various minor changes were made, particularly to simplify mass production, the T-34 survived the war with only one significant improvement, the substitution (after Kursk) of an 85mm main gun for the 76.2mm. Even when the Soviets did introduce new designs, such as the heavier tanks and self-propelled guns of 1944, they did so without halting production of older types.

The United States had even more reason to standardize and mass-produce than did the Soviet Union. Unlike the other major combatants, the United States had to ship every pound of supplies and equipment over vast ocean distances in order to reach the war zones. By concentrating on mechanical reliability, America was able to produce vehicles that operated longer with fewer repair parts. This helped alleviate the chronic shortage of shipping space when the army moved to Europe and the Pacific. To further ease the shipping problems and to ensure that American tanks were compatible with existing military bridge equipment, the War Department restricted tank width to 103 inches and maximum weight to thirty tons. This restriction meant that navy tank transporters and portable bridges did not need to be redesigned in the midst of the war. The army relaxed these requirements only in late 1944.[23]

There was also a tactical purpose to these restrictions. General McNair wanted to ensure that American tanks were designed in accordance with the U.S. doctrine for employing armored divisions, which foresaw tank destroyers, not tanks, as the primary means of defeating enemy armor. Chance encounters between tanks would occur, but the principal role of the armored division was to exploit and pursue, not to fight enemy armor.

For all these reasons, the U.S. Army standardized on the M4 Sherman medium tank, an excellent compromise among reliability, mobility, armor protection, and gunpower. When the British first employed the Sherman in North Africa during late 1942, it proved to be at least equal, if not superior, to the German second-generation tanks, the Panzer III and IV. Once the Tiger tank appeared in Tunisia in early 1943, however, the Sherman tank and most of the U.S. antitank force seemed inadequate. It was this situation that gave rise to the common myth of the poorly designed, vulnerable Sherman.

The 103-inch width limitation further hampered the Sherman by forcing designers to give the tank narrow tracks. These meant much greater ground pressure per square inch, which in turn made the Sherman less mobile on muddy or swampy ground, at least by comparison to the wider tracks the Soviets and Germans used. The M4's only advantages over later German tanks were superior reliability and a power-driven turret. During unexpected meeting engagements at close ranges this latter feature allowed the Sherman's crew to traverse the gun and engage the enemy more rapidly than could German crews using hand-cranked turrets. Sherman tank crews often carried a white phosphorus round in the breech of their guns, ready to blind enemy tanks during such unexpected encounters.

Despite its drawbacks, the Sherman remained the main battle tank of the U.S. Army. In February 1945, apparently in response to the large-scale German armored attacks during the battle of the Bulge, the U.S. Army finally allowed a few M26 heavy tanks to be sent to Europe for combat testing. The army's Ordnance Department had developed a series of such tanks in 1943, but considerations of doctrine, shipping, and mass production had prevented its use in battle until the closing days of the war.[24]

Great Britain also used the Sherman during the latter half of World War II but was concerned by the limited penetrating power of the M4's 75mm, medium-velocity main gun. After considerable discussions with the Americans, the British finally modified some of the Shermans they received. The British version of the Sherman, called the Firefly, carried the third-generation British antitank gun, the seventeen pounder (77mm). Its long bore and higher velocity gave it much greater capability against German armor.[25]

Throughout World War II, the technological duel between offensive and defensive continued, often focusing on tank and antitank warfare. By 1943 infantry self-confidence, supplemented by technology, had placed severe limits on armored warfare, without completely eliminating the possibility of a successful blitzkrieg. At the same time, the most complex war in history brought with it many other developments that affected the battlefield.

CHAPTER 6

• • • • • • • • • • • • • • •

The Complexity of Modern Warfare, 1943–1945

To continue this survey of general trends in tactical practice during the second half of World War II, the more complex and specialized questions such as air–ground cooperation, airborne operations, amphibious landings, and special operations units bear closer examination.

Signals Intelligence and Communications

Signals Intelligence or SIGINT, the interception and analysis of radio signals, had flourished during the First World War, but it came of age during the Second. Closely related to SIGINT was Electronic Warfare, which included, among other techniques, radio jamming and deception by imitating enemy messages. Such tricks could disrupt an army's ability to communicate rapidly, rendering it almost impotent. SIGINT as a source of intelligence, and Electronic Warfare as a new means of disrupting enemy command and control, gave tactical commanders yet more considerations to coordinate in battle. Recent histories of World War II have overemphasized the strategic importance of high-level SIGINT concerning enemy intentions, however, while neglecting its role at the tactical, unit level.[1]

ULTRA, the British codeword for intelligence based on decrypting highly classified German radio teletype messages, gave the Western Allies only limited access to German military intentions and capabilities. Although the German Navy and Air Force sent most of their messages by radio, the German Army normally used landline (telephone cable) communications for higher headquarters messages, except when fluid operations forced them to make radio transmissions. Indeed, British intercept operators found that the most effective way to monitor Erwin Rommel's ground operations in North Africa was to monitor the SCORPION code messages sent by the Luftwaffe liaison officers attached to Rommel's panzer divisions.[2] Even then the Allies could not necessarily intercept, let alone decode in a timely manner, every German message. The Germans changed their codes every twenty-four hours and periodically made major shifts in codes or equipment. The Allies might go for days or even months without being able to decode transmissions on specific, highly lucrative radio networks. On 1 May 1940, for example, Germany changed virtually all its radio codes, blinding the Allied SIGINT effort until 22 May. By that time, the German offensive through the Ardennes had already succeeded.[3] Similar problems recurred during most of the war.

Nor were the deciphered messages of ULTRA always illuminating for the tactical and operational situation. The most senior German commanders communicated their specific plans only in unusual circumstances, such as when Hitler personally demanded such reports. Instead, Allied intelligence analysts had to piece together information by monitoring messages over long periods of time, or by inferring capabilities on the basis of requests for certain amounts of fuel and ammunition. Moreover, few Allied commanders even at field army level had access to ULTRA information, for fear that the Germans would learn of their vulnerability to decryption.

Perhaps the greatest drawback of ULTRA-level SIGINT was that it was usually so convincing that it discouraged the use of other sources of intelligence collection, sources that might confirm or deny ULTRA information. This blinded Allied commanders to threats that were not discussed in German radio traffic. In early 1943, for example, the Allied forces in Tunisia, North Africa, relied heavily on ULTRA; their other intelligence collection means were improvised and largely ineffective. The German offensive at Sidi Bou Zid and Kasserine Pass in February 1943 (Map 7) surprised the Allies because available ULTRA indicated that higher German headquarters had disapproved such an operation in favor of an attack elsewhere. Of course, SIGINT could not know that Rommel and other German commanders had met face-to-face on 9 February and had developed a plan that led to the attack on Sidi Bou Zid, which mauled a dispersed U.S. armored division before it was stopped by massed artillery fires at Kasserine.[4] Lack of SIGINT and misinterpretation of available intercepts also had a considerable effect

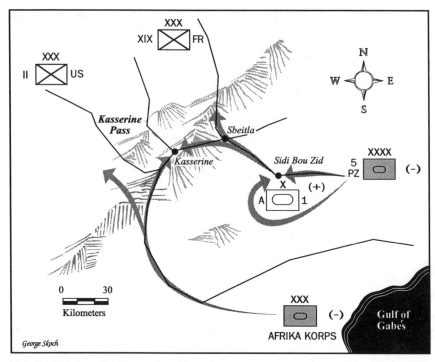

Map 7. Sidi Bou Zid–Kasserine Pass, February 1943

on Allied failure to predict the scale and intensity of the German counter-offensive in the Ardennes in December 1944.

Although the strategic-level SIGINT of ULTRA has been widely studied, historians have generally neglected the role of tactical-level SIGINT, derived from monitoring unit communications that used simple codes or no codes in the heat of battle. From 1940 to 1942, for example, a single German *Horch* (listening or intercept) company known as Unit 621 skillfully interpreted the unencrypted tactical communications of British units in North Africa. Erwin Rommel was always accompanied by a liaison team from this company, which gave him a complete picture of enemy dispositions and intentions during battle. When the British finally became aware of this unit's activities in July 1942, an Australian battalion raided and captured most of Unit 621, which was brazenly established on a hill less than one kilometer from the front lines. German replacements could not equal the priceless expertise of the analysts lost in this raid and thus had more difficulty detecting later British deception operations.[5]

Allied tactical SIGINT had a more difficult task, in part because of the nature of German communications. At division level and above, German commanders rarely used voice radio communications, preferring teletype or radio telegraph messages. These required less power output than did voice radios, making it more difficult for the Allies to intercept, let alone locate, the German transmissions. Moreover, because the message had to be written down for transmission by radio telegraph, it was simple and natural for the Germans to encode that message before sending it. In the heat of battle, the Germans discarded the Enigma codes that were the source of ULTRA intelligence, but even then they used simpler, three-letter codes.[6]

In the disastrous campaign of 1940, British tactical SIGINT units, known as Y units, had discovered that they needed more analysts to interpret the raw communications they were able to intercept and decode. The British therefore established a more centralized and effective Y service in the Middle East. They did, however, experiment by providing small intercept teams, mounted in armored cars, to support each armored division in the desert. Desert Y operators became so skilled that they could identify Rommel when he personally intervened in a tactical battle. Just before the 1942 battles at El Alamein, the British fielded three new direction-finding vans that could locate German transmitters to an accuracy of 100 yards. When Bernard Montgomery launched his offensive at the second battle of Alamein, airborne jammers disrupted German tactical radio communications for hours.[7]

The U.S. Army had to develop its own tactical SIGINT from a minuscule prewar basis and learned much from cooperating with the British. By an organizational quirk, however, the intercept units of what eventually became the Signal Security Agency belonged to the Signal Corps rather than to the intelligence staffs of the army. Intelligence officers found themselves competing for SIGINT and Electronic Warfare not only with the Signal Corps itself but also with the Army Air Forces, who wanted jammers for the bombing campaign against Europe. Despite such institutional wrangling, by 1945 the U.S. Army in Europe included a large and effective SIGINT organization, with Signal Service intercept companies supporting each corps.[8]

The U.S. Army Signal Corps was also at the forefront of another major change in World War II, the development of effective tactical radio communications. These were the basis for controlling fluid, mechanized operations as well as the raw material for tactical SIGINT. The military demand for such communications greatly accelerated research and development in this area. In particular, the U.S. Army pioneered the use of frequency modulation (FM) radios for short-range tactical communications and both very high frequency (VHF) and ultra high frequency (UHF) radios for longer ranges.[9] Unlike the European armies, the U.S. Army used FM extensively, because it provided static-free signals over a wide variety of channels without using a separate crystal for each frequency.

The combination of reliable radio communications and effective tactical signals intercept services also provided a new opportunity for senior commanders to follow the course of battle without delays in passing information up and down the chain of command. Both the British and American armies developed means for senior headquarters to receive battle reports by radio without waiting for the messages to be processed through the intermediate layers of command. That is, the senior headquarters could monitor tactical unit radio networks directly or else assign a radio-equipped liaison detachment to each forward unit to report the situation to the senior headquarters. The British General Headquarters Liaison (GHQ Phantom) units and the American Signal Information and Monitoring (SIAM) companies performed this service admirably during 1944 and 1945, and in the British case as early as 1942. The danger with such a monitoring system, as Gen. Dwight D. Eisenhower acknowledged after the war, was that the senior commander might be tempted to bypass the intermediate headquarters and interfere directly in the battle, using the system for command rather than as a source of timely operational and intelligence information.[10] In the latter role these monitoring services enabled much more effective coordination of the battle, allowing the commander to react through his subordinate commanders to situations as they developed.

Soviet Concepts and Practice, 1943–1945

Many of the foregoing technological considerations also manifested themselves in the German-Soviet conflict after the battle of Kursk. The Soviet Union held the initiative, although it was not always attacking the Germans and their Axis allies on all fronts. Before each new offensive, the Soviets used elaborate and usually successful deceptions, causing the Germans to concentrate their forces against imaginary attacks while leaving themselves exposed to the actual offensive.[11]

After such a deception, the Soviets exerted tremendous efforts to penetrate the German defenses. In the ensuing exploitation, logistical restrictions usually brought the Soviets to a halt even where there was little German resistance. In the course of the war, improvements in Soviet logistics led to steady increases in the depth of exploitation. Once the Germans gained a respite to reorganize their defenses, the cycle repeated itself. Accordingly, the Red Army developed a variety of techniques for both penetration and exploitation against the German defenders.

One significant development during 1944 was the change in Soviet reconnaissance techniques before a deliberate attack. Prior to that year, the Red Army had been very effective in conducting small, time-consuming long-range reconnaissance patrols. To shorten the time required to prepare for a new offensive, the Soviets in early 1944 sent out company- and battalion-sized units to engage the German outposts or to reconnoiter by fire. This process identified the main German defensive organization much more rapidly than conventional reconnaissance and attacks. In the process, the Red Army received an unexpected bonus. Soviet reconnaissance units were often able to seize control of outposts that the Germans were defending only lightly, as part of the long-standing German doctrine of flexible defense-in-depth. By late 1944 the Soviets had transformed their reconnaissance units into the first wave of the deliberate attack. Company and larger units on reconnaissance missions attacked up to twenty-four hours before the main offensive, seizing the German outposts and thereby unmasking or even disorganizing their main defenses. Then the main attack could focus on the remaining defenses.[12]

Although Soviet commanders massed their forces on relatively narrow breakthrough frontages, their successes were due to more than sheer numerical superiority. Whether in the reconnaissance echelon or during the main attack, the Red Army used a variety of procedures to overcome German defenses. First, artillery units fired under centralized control and according to elaborate plans. The Soviets also used a variety of deception measures, such as sending the assault infantry forward during a lull in the firing in order to lure the Germans from their bunkers so that renewed Soviet artillery fire could destroy them. Heavy tanks to support the infantry and eliminate strong points, medium tanks to penetrate rapidly and suppress enemy infantry fires, and assault guns for direct-fire support against antitank guns and strong points cooperated. Combat engineers or specially trained infantrymen frequently rode on each tank. Their mission was to eliminate obstacles and to provide protection for the tank against German short-range antitank weapons.[13] The tank might temporarily take up a hull-down position behind a low hill, in order to provide covering fire while engineers cleared minefields and infantry eliminated enemy short-range antitank weapons.

Soviet commanders reluctantly accepted the high casualties produced by this technique in an effort to accelerate their rate of penetration. Given the meticulous German defensive preparations and the lack of Soviet armored personnel carriers, the Red Army had to combine engineers, infantry, and tanks in this manner, regardless of losses. By 1944 casualties were a subject of great concern for the Soviet generals. The best means to reduce casualties were deception, concentration, speed of penetration, and careful task organization of the attacking forces. Instead of advancing on-line in masses of infantry, the Soviet attackers operated in tailored assault groups of platoon to battalion size (Fig. 13). When time allowed, Soviet air and ground reconnaissance collected complete information about specific German strong

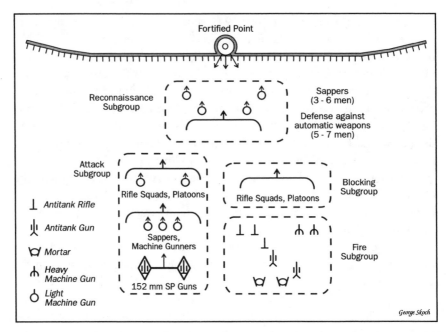

Figure 13. Soviet Assault Group Formation, 1944–1945

points to be reduced. Then each Soviet assault group would be trained to eliminate one of these, dislocating the German defensive organization by removing a key position. Assault groups normally included four subgroups: a reconnaissance subgroup to clean an approach route to the objective; a blocking subgroup to engage and pin down the defenders; a fire subgroup whose guns would isolate the strong point from reinforcements; and an attack subgroup, including engineers and heavy tanks or assault guns, to eliminate the objective by attacking it from the flanks or rear.[14]

Once the Soviets completed their penetration, their commanders sought to sustain the momentum, moving rapidly from encirclement to renewed exploitation and pursuit so that the defenders had no opportunity to reorganize a coherent defense. The German exploitations of 1940–1942 had normally been centrally controlled to ensure that all elements moved in the same general direction and were available to support each other in the event of counterattack. By contrast, Soviet exploitation tended to be more decentralized and diffused, particularly after the initial encirclement was completed. Notoriously poor Soviet radio communications may have been partially responsible for this decentralization; but more to the point, the Soviets retained their belief in the interwar theory that rapidly moving forces could fan out and confuse as well as disorganize the defender. Contrary to the postwar stereotype of rigid, unimaginative Soviet operations, relatively junior Red commanders were encouraged to use their initiative. Such decentralization allowed leading Soviet units to seize targets of opportunity, such as bridges and river crossings, that were not immediately obvious to the senior planners. These targets were especially important, because achieving bridgeheads across a river denied the Germans the opportunity to use it as a defensive obstacle.

The same decentralization made the Soviets more vulnerable to defeat in detail by German counterattacks. However, beginning in 1943, a combination of factors, including declining German combat effectiveness, growing Soviet tactical experience, and better close air support of the exploitation forces allowed the Soviets to defeat most German counterattacks and to continue their mission.

The most common formation for Soviet exploitation was the forward detachment, a combined arms organization of great mobility and firepower that was sent ahead of the main units to seize key objectives and to disrupt enemy efforts to reorganize an effective defense.[15] During the course of the war, both the size of the typical forward detachment and the distance it operated ahead of the main body increased steadily. In the last two years of the war, a forward detachment typically was a tank brigade reinforced by batteries or battalions of field and antiaircraft artillery, heavy tanks, assault guns, and engineers. For advances that were of great importance to senior commanders, an air controller accompanied the detachment to direct close air support, and air units were dedicated to support specific detachments. This reinforced brigade of perhaps 2,500 men operated as much as ninety kilometers (fifty-six miles) ahead of the rest of its parent tank corps, which in turn might be acting as a forward detachment for a tank army. A forward detachment did not necessarily follow the same routes as the main body of troops and was not responsible for providing security for it. Frequently, an efficient forward detachment commander could brush through hasty German defenses along the way, allowing the following troops to continue their exploitation and pursuit without deploying to attack the scattered Germans. Thus the main body could remain in a column on the roads, saving considerable time and increasing the overall tempo of pursuit. When the forward detachment ran low on fuel, ammunition, and troops, its commander attempted to seize a bridgehead over the next river obstacle as a starting point for a renewed offensive at a later date.

Of course, the forward detachments sometimes paid for their audacity, being halted and even wiped out by a large German armored force. Nevertheless, the forward detachments led the mobile groups envisaged by prewar

Soviet doctrine and thereby accelerated the speed and distance of exploitation and pursuit.

The German Decline, 1943–1945

While the Red Army grew in both equipment and tactical proficiency, its German opponent declined not only in numbers but also in overall training and combat ability. When the Soviets achieved an overwhelming superiority by massing their forces on a narrow breakthrough frontage, German defenders naturally ascribed their successes to their numerical advantage. In reality, the quality of the German armed forces declined as a result of their declining quantity. As early as the summer of 1942, those German divisions that were not involved in the second great offensive in the east were deliberately filled to only 55 percent of their authorized personnel. Even spearhead units received only 85 percent of authorized equipment. By 1943 a typical German division could accomplish its supply and construction tasks only by using a thousand noncombat "helpers," former prisoners of war who acted as laborers in return for food and better treatment.[16]

In order to maintain their armies in the field, the German leaders progressively reduced the amount of training given to replacements. Frequently, the field training units that belonged to combat divisions were pressed into service to stop a Soviet breakthrough. This became a vicious cycle, in which poorly trained German soldiers survived for only short periods of time at the front and had to be replaced even more rapidly than before.[17] This decline in the quality of their infantry prompted German commanders to seek ever-increasing amounts of firepower in the form of assault guns, antitank rockets, automatic weapons, and artillery.

Given the shortages of personnel, many German infantry divisions operated with only six of the nine authorized infantry battalions from 1942 onward; the other three battalions existed only on paper. In 1944 the German General Staff formally changed the division structure to reflect this reality. According to the 1944 reorganization, an infantry division consisted of three infantry regiments of two battalions each. This configuration allowed each battalion to have a larger share of the weakened regimental artillery and antitank companies than had been the case with a three-battalion regiment. On the other hand, such a structure retained the large overhead of three regimental staffs and support elements yet denied the regimental commander the luxury of a local reserve force. With two battalions to cover a wide defensive frontage, the regimental reserve was often only one company or less. In practice, some infantry divisions organized themselves into two regiments of three battalions each. In either case, the 1944 German infantry division retained the four artillery battalions of the previous structure, so that, at least on paper, the declining ability of the infantry was offset by a larger proportion of fire support. The habit of manning only three guns in each battery meant that many German infantry divisions had only twenty-seven tubes in their division artillery, less than one-third of their British and American counterparts. Recognizing enemy air superiority, the 1944 divisional organization also included a battery of self-propelled antiaircraft guns.[18]

After 1943 the German defenders found themselves increasingly hard-pressed to contain, let alone halt, Soviet offensives. The basis for the German doctrine of defense-in-depth was to absorb enemy attacks and separate attacking armor from its supporting infantry, in order to defeat each element independently. By 1944, however, improved Soviet cooperation among the combat arms nullified German efforts to isolate those fighting components from one another. Many German commanders experimented with the idea of a preemptive withdrawal, pulling back their troops just before a Soviet deliberate attack in order to save lives and to force the Soviets to reorganize for another attack at a point a few kilometers to the west. Yet such a withdrawal under pressure required not only excellent intelligence but also high morale and well-trained troops, the very commodities that were declining most rapidly in the German Army.[19]

While the infantry divisions gradually wore down, the Germans made a belated effort to rebuild their panzer forces. Heinz Guderian dedicated himself to this task as Inspector General of Panzer Troops (1943–1944) and then as Chief of the General Staff (1944–1945). However, his continued insistence on the panzer arm as a force separate from the rest of the German Army was no longer appropriate. It was true that panzer divisions were the principal German instrument for counterattacking enemy penetrations and encirclements. Yet these divisions were so few in number compared to the great distances on the Russian front that they were often counterattacking singly or in pairs, wearing themselves down as fast as Guderian could rebuild them. By removing armor training and doctrine from the appropriate branches of the General Staff, Guderian only increased the estrangement between the panzer and infantry forces and made training between the arms more difficult.[20]

Despite these problems, the balanced panzer division remained an extremely effective force at the tactical level. Only minor changes in organization and tactics occurred after 1941. Production requirements for tanks, assault guns, and other tracked vehicles meant that the panzer grenadiers (accompanying infantry) remained largely truck-mounted rather than being mounted in vehicles that could protect them from small-arms fire and artillery fragments. As a result, the panzer grenadiers were much more vulnerable than were the tanks they supported. Even at its peak in the fall of 1943, the German Army panzer force had only 26 of 226 panzer-grenadier battalions, or 11 percent, mounted in armored half-tracks.[21] Thus, except in certain elite units such as the SS, no more than one of the four to five infantry battalions in a panzer division was actually mechanized.

Generally speaking, one or two companies of half-track mounted panzer grenadiers accompanied each panzer battalion in the advance, with the motorized infantry following later to consolidate and defend the area seized by the first attacks. Artillery forward observers in tanks or half-tracks were also in the first wave. Where only motorized infantry was available, these troops went into battle dismounted, following in the lee of the tanks until they were needed to clear obstacles or defend against enemy infantry. To avoid being tied to this dismounted infantry when the attackers met with effective antitank fire, the German tanks sometimes bounded forward, assumed hull-down positions that reduced the target they presented to the enemy, and provided suppressive fires to cover the infantrymen hurrying to rejoin the tanks. To protect the attacking panzer force from the enemy armored counterattack, antitank guns leapfrogged into a series of overwatching positions on the flanks of the advance. Assault guns remained with the motorized infantry reserves to consolidate gains or to engage an enemy counterattack force that got past the panzer spearheads. Because of Allied air superiority on all fronts, German armored forces needed much greater air defense protection in 1944 and 1945 than in 1940. Truck-mounted panzer-grenadier battalions therefore included the 20mm antiaircraft guns that had proven so effective earlier in the war, and battalions of tanks and half-track mounted infantry received self-propelled antiaircraft guns. In some cases such guns were attached to each company.[22] Such, at least, was the theory of panzer organization and tactics; in practice, of course, the declining strength of such units produced a variety of improvised battle groups.

American Concepts and Practice, 1943–1945

The initial contact of American forces with Axis troops did not fulfill the promise of previous U.S. developments in doctrine and organization. During the 1942–1943 invasion of North Africa a variety of factors, including inexperience, led American commanders to scatter their forces in regimental or smaller units, thereby depriving them of the advantages of the U.S. centralized fire-control system. The U.S. armored divisions had stressed decentralized, mobile combat by direct tank fire so often in training that their self-propelled artillery battalions had neglected the study of indirect-fire techniques. Inadequate logistics forced the Americans to leave their corps artillery far behind the Tunisian front lines, further reducing available fire support when the Germans counterattacked in February 1943. In the crisis of Kasserine Pass, however, the artillery of the 1st and 9th Infantry Divisions was finally able to operate on an organized basis, with devastating effect on the Germans (see Map 7).[23]

Similar problems arose in the Southwest Pacific, where in 1942 Gen. Douglas MacArthur committed the 32d Infantry Division to battle in Papua with no artillery and only a few mortars. Despite the protests of the 32d Division's commander, MacArthur's staff mistakenly thought that artillery would be ineffective in the jungles. Moreover, the local air commander, Gen. George C. Kenney, assured the division that "the artillery in this theater flies" but then failed to provide effective air support throughout a long and bitter campaign.[24] Weather and terrain prevented such air support on many occasions, and there was so little communication between air and ground that Kenney's pilots attacked Americans by mistake on a weekly basis. Based on the bitter experience of assaulting Japanese bunker complexes without appropriate fire support, the 32d Division learned at great cost the need to coordinate artillery and air support with the infantry.

Two years later, the 32d Division had the opportunity to apply these lessons in the battle of Ormoc Valley (Map 8).[25] In October 1944, the U.S. Sixth Army invaded the island of Leyte as the first step in the reconquest of the Philippines. In response, the Japanese reinforced their Leyte garrison with a number of units, including the well-equipped, full-strength 1st Infantry Division. As part of a general counteroffensive, the Japanese division attempted to advance northward and eastward out of the Ormoc Valley. It ran directly into the U.S. 24th Infantry Division, soon replaced by the 32d, which was attempting to secure the island by a sweep in the opposite direction. The result was a classic confrontation, on nearly equal terms, between American and Japanese divisions. As such, Ormoc Valley was a fair test of American doctrine and practice in World War II.

Despite its experience, the 32d Division had a number of disadvantages in this struggle. The jungles of Leyte often restricted movement to 100 yards per hour and made resupply difficult. Because of the monsoons of November 1944, much of the battle was fought without air support on either side. Moreover, like other U.S. divisions in combat, the 32d lost at least one-third of its troops before and during the battle, due to illness and casualties. One advantage that the Americans did possess, however, was the support of Filipino resistance forces, which provided scouts, laborers, and supply porters throughout the battle.

For six weeks, from mid-November through late December 1944, the opposing armies struggled with one another, the rains, and the jungle for control of the northern end of the Ormoc Valley. At the very beginning of the operation, two battalions from the 24th Infantry Division's 19th and 34th

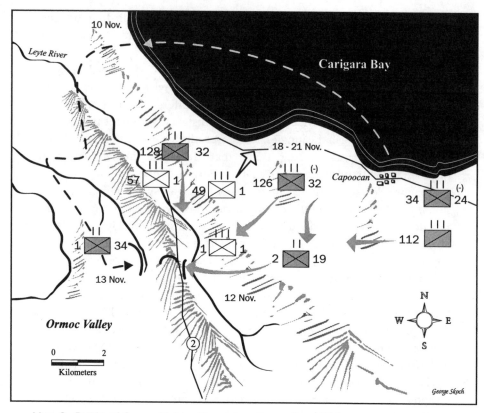

Map 8. Battle of Ormoc Valley, November–December 1944

Infantry Regiments hacked their way across country and took up positions overlooking but not blocking Route 2 in the Ormoc Valley. These two battalions were able to direct artillery fire and otherwise hamper Japanese use of the key supply route throughout the battle. The 32d Division assumed control of these two battalions but still had to slug its way southward through the persistent defenses and ill-timed counterattacks of the Japanese 1st Division's 57th, 49th, and 1st Regiments. The initial American attacks were launched hastily, without adequate support, but eventually settled down to the unavoidable but costly effort to push the Japanese southward.

Throughout these struggles, poor maps and observation continued to limit artillery support on either side, unless the infantry pulled back and allowed a deliberate artillery preparation. Instead, the U.S. troops had a few tanks accompany the lead elements at all times to provide direct-fire support. However, tank–infantry cooperation was often lacking. Because of enemy sniper fire, the tanks began by operating with their hatches closed. These "buttoned up" tankers could not see the infantry and suspected that they had been abandoned, and the infantry could not communicate with the tankers. Finally, on 21 December, three successive tank platoon leaders rode on the back decks of their tanks, fully exposed to enemy fire so that they could talk both to their tank crews and to the accompanying infantry. Two lieutenants were wounded in this manner, but the American infantry finally learned to trust and cooperate with tanks. This effort, in cooperation with other U.S. attacks from the east (112th Cavalry Regiment, fighting as infantry) and south (77th Infantry Division) finally broke the back of the Japanese defenses.

To some extent, the U.S. troops who invaded Normandy in 1944 had to learn the same lessons about cooperation. Many of the U.S. infantry divisions used in that invasion had not been in combat before and had not had the opportunity for extensive tank–infantry training with the separate tank battalions that supported them. Furthermore, the radios issued to infantry, tank, and fighter aircraft units had different frequency spectra, making communications among the arms impossible. Even when the infantry commander was riding on the outside of a tank or standing next to it, the noise of the tank engine made it difficult for him and the tank commander to communicate face-to-face.[26]

The U.S. Army gradually corrected these problems and developed more effective combined arms teams during the breakout from Normandy. The need for close tank–infantry cooperation reinforced the habitual association of the same tank battalion and infantry division. Communications personnel installed improvised external telephones on the rear of tanks so that the accompanying infantry could communicate with the tank crews. In July 1944 the commander of IX Tactical Air Command, Gen. Elwood A. Quesada, provided VHF aircraft radios for installation in the leading tanks of each armored battalion task force. When the United States broke out of the Normandy beachhead, these tanks could communicate with fighter-bombers overhead. For a few high-priority advances, the IX Tactical Air Command flew "armored column cover," providing on-call fighter-bombers for close air support. It is true that this tactic was wasteful of air resources, but aircraft could keep pace with armored advances even when the field artillery fell behind. The resulting high tempo of exploitation saved lives and justified the expenditures involved.[27]

Advancing on parallel routes also facilitated American exploitation and pursuit across France. Where the road network allowed, U.S. armored divisions and combat commands advanced with two or more task forces moving along parallel routes. Frequently, a German strong point would halt one column, only to find itself outflanked by another American column a few kilometers away. The Allied forces usually found their progress hindered as much by logistical factors as by enemy defenses. Strategically, logistics hampered the Allies throughout 1944 and 1945. Tactically, some armored units found it more secure to travel with their supply and maintenance units in the midst of the column, instead of following behind where they might encounter bypassed enemy resistance. Of course, such a tactic was appropriate only when exploiting against limited enemy defenses. When logistics elements moved on their own, they often required small antiaircraft, tank destroyer, and infantry escorts for local security.[28]

This dispersion of antiaircraft units in small detachments exemplified the fate of specialized American forces when their particular function was not in demand. U.S. antiaircraft units conducted a number of air defense operations, most notably the protection of the captured bridge at Remagan during the conquest of Germany. In general, however, overwhelming Allied air superiority made an integrated air defense system increasingly unimportant during 1944–1945. Instead, senior commanders used antiaircraft weapons to engage targets on the ground and inactivated some antiaircraft units to provide much-needed infantry replacements during the fall of 1944. Similarly, chemical smoke generator companies, intended to provide smoke screens to conceal installations and activities from the enemy, were pressed into service to repair asphalt roads when line units did not need smoke. This misuse developed a set of false attitudes and priorities among combat commanders, but the shortage of manpower was so severe that no unit could stand idle. During the German counteroffensive at the battle of the Bulge, many specialized units found themselves fighting as infantry. Their excellent performance in this role justified the American policy that support troops should be trained and equipped to defend themselves and to fight when necessary. Even if, for example, the engineers had been employed to construct barriers in front of the German advance, there were no other infantry forces available to provide the firepower needed to defend these obstacles. At that point, the situation was so desperate that local commanders were fully justified in using all available soldiers as riflemen.

Air–Ground (Non)cooperation

Although Allied commanders steadily evolved techniques for combining the ground combat arms, cooperation between air and ground was a much more difficult problem.[29] The apparent intimacy of German air and ground forces in the 1940 campaigns jolted the British and American forces into addressing this issue, but the results were painfully slow.

The 1940 British performance in this area was abysmal; the army and Royal Air Force headquarters had been located at separate positions with uncertain communications between the two. In late May, after all Royal Air

Force units had been withdrawn from France, the remaining British Army officers in battle resorted to telephoning the War Office in London to arrange for air support.[30] While the battle of Britain raged in August, the British took the first steps toward developing an effective air–ground system. RAF Group Captain A. H. Wann and Army Colonel J. D. Woodall conducted a series of tests in Northern Ireland. The resulting Wann-Woodall Report advocated the creation of a separate tactical air force and a number of mobile communications teams, known colloquially as tentacles, to accompany ground maneuver divisions and brigades. On the basis of this report, in December 1940 the RAF created an Army Co-Operation Command and began developing tentacle equipment and procedures that would put an Air Liaison Officer (ALO) at each brigade headquarters. Despite this promising start, however, the RAF was understandably concerned about maintaining control over its aircraft, as many British ground commanders made little effort to understand the complexities of air operations.

Meanwhile, British forces in North Africa had to improvise their own solutions to the air–ground issue. In 1941 two senior RAF commanders—Sir Arthur Tedder and Arthur Coningham—struggled to overcome the problems by collocating various RAF headquarters with the equivalent ground headquarters. In October 1941 they created joint RAF–Army Air Support Control staffs at each corps and armored division headquarters; at brigade level, an RAF Forward Air Support Link could forward requests for air support. In 1942 the first trained tentacle team arrived in North Africa from the United Kingdom and demonstrated procedures that cut response time on air requests to thirty minutes. Indeed, by the end of 1942 the United Kingdom and desert methodologies had been fused into a single procedure. However, the resulting system was still unwieldy.[31]

American forces suffered from many of the same problems and learned much from their British counterparts. Throughout the war, the U.S. Army Air Forces (AAF) operated almost independently from the other elements of the army. Soon after Pearl Harbor, Pres. Franklin D. Roosevelt gave the AAF the enormous mission of precision strategic bombing of Germany and eventually of Japan. This unprecedented task not only strained AAF resources to the limit but also fit in perfectly with its doctrine. This assignment greatly encouraged the tendency of AAF leaders to distance themselves from the ground arms. The result was near disaster on the battlefield, retrieved only by the common sense of the tactical commanders on the spot.

Army Air Force doctrine defined three priorities for tactical aviation: first, air superiority; second, isolation of the battlefield, which in effect meant air interdiction of enemy supply lines; and third, attack on ground targets in the zone of contact between opposing armies.[32] Throughout the war, the

AAF phrase for close air support was "third phase" or "priority three" missions, reflecting a belief that such targets were an uneconomical, inefficient, and unimportant use for airpower and rightfully belonged to the field artillery. Centrally directed interdiction of the enemy by tactical air assets, the AAF argued, was the most efficient use of this weapon. In practice, most AAF interdiction campaigns fell well short of their goal of isolating the battlefield.[33] Moreover, ground commanders valued the psychological effects of close air support on both friend and foe, but the unseen interdiction attacks had no such effects. The more air leaders opposed the decentralized use of their aircraft for close air support, the more ground commanders felt the need to control at least some air assets to ensure their availability when needed.

After the fall of France, the AAF commander, Lt. Gen. Henry (Hap) Arnold, responded to the apparent success of Stuka dive-bombers. He turned to the U.S. Navy, which had pioneered dive-bombing, and by 1941 had established two groups of A-24 dive-bombers, the army version of the Navy's SPD-2. The AAF later acquired the A-36, a modified P-51 fighter with dive brakes attached, which did considerable damage in Italy and Burma before being phased out.[34]

As commander of the Army Ground Forces, Lesley McNair led a vain effort to change Army Air Force priorities. Hard-pressed to meet its many obligations, the AAF was unwilling to provide aircraft even for major ground maneuvers, let alone for small-unit training. Six months before the Normandy invasion, for example, thirty-three U.S. divisions in England had experienced no joint air–ground training, and twenty-one divisions had not even seen displays of friendly aircraft for purposes of recognition in battle. In 1943 the AAF changed the radios in fighter-bombers to a frequency that was incompatible with ground radios. In short, air and ground units went overseas with little understanding of the tactics and capabilities of their counterparts.[35]

The results were predictably poor. During the 1942–1943 North African invasion, ground forces received little air support, and ground commanders with no experience in the employment of tactical air support misused the little that was available. The AAF dedicated most aerial photography missions to its own intelligence needs, compounding the intelligence failures of ground units. U.S. troops saw so few friendly aircraft that they fired on anything that flew. One American observation squadron lost ten aircraft in North Africa—two to enemy air attack, three to enemy ground fire, and five to American ground fire. Gradually, both sides learned to recognize and cooperate with each other, but the process was painful.[36]

When the British Commonwealth and American forces worked together in the long Italian campaign (1943–1945), their proximity led to cross-fertilized ideas about close air support. The AAF's XII Air Support Command collocated its headquarters with the Fifth U.S. Army, meeting every evening to plan strikes for the next day and improvising a common network of liaison officers and radios. Within the air resources allocated by higher headquarters, the ground operations officer established priorities that the air operations officer rejected only when the proposed use was a technical impossibility. As a concession to speed up response time, the British established a few Forward Control Posts (FCPs), each with a senior RAF officer and several army staff officers. These FCPs, colloquially called Rover, were able to direct close air strikes, although the British did not mount air controllers in tanks until the final months of the war.[37]

Thus, by early 1944 the British and Americans had improvised a close air support system. However, as one historian remarked, "It was manned by an air force whose high command was fundamentally opposed to the role and an army unused to utilizing air support, and whose General Staff was, for the most part, too unfamiliar with air power to fully appreciate its potential, and its drawbacks."[38]

The Normandy invasion of 1944 brought much larger Allied armies and tactical air forces into the field. The British ground commander for the invasion, Montgomery, brought his experiences in North Africa to the air support question in the British Army, but most innovations occurred in the U.S. Army. There the Ninth Air Force provided tactical support to the Twelfth Army Group, with the headquarters of a tactical air command collocated with each of the three U.S. field army headquarters. As commander of IX Tactical Air Command, supporting the First U.S. Army, Gen. Elwood Quesada worked to provide the greatest possibility in close air support and air interdiction. In addition to developing the armored-convoy cover technique, Quesada used a modified antiaircraft radar to track fighter-bombers in flight so that they could be directed to drop bombs on targets they could not see or redirected for immediate air support missions as necessary.[39] The AAF also experimented with placing tactical air control parties with armored combat commands and assigning fighter pilots to short tours as ground-based air controllers, in order to increase their understanding of the problems of close air support.

In practice, however, even Quesada emphasized close air support only when his ground counterparts were launching a major offensive. On a day-to-day basis, the British and American tactical air forces continued to devote more of their missions to interdiction, including armed reconnaissance to engage targets of opportunity. These interdiction operations remained a priority even though subsequent analysis showed that they were often twice as dangerous as close air support missions.[40]

Moreover, the problem of mistakenly attacking friendly troops persisted throughout the war. In Italy, for example, A-36s continued to bomb U.S. tanks because the pilots did not know the recognition signals, until finally, in self-defense, the ground forces shot down one of their own aircraft. In a three-day period in August 1944, the First Canadian Army suffered fifty-two separate attacks by RAF and AAF aircraft, resulting in 72 killed and 191 wounded.[41] The ground forces were equally likely to fire on their own aircraft, a problem that prompted the broad black and white stripes painted on all Allied aircraft prior to the Normandy invasion.

With the partial exception of the A-36, no British or American aircraft was specifically designed for the close support role. In a forerunner of the debates during the Cold War, air commanders preferred to have fighters or fighter-bombers that could, in a pinch, provide close support. Some of these aircraft, notably the P-47 Thunderbolt, performed superbly in ground attack. However, most multirole aircraft had considerable limitations. The drag of carrying external bombs and other ordnance burned up fuel at a ferocious rate; furthermore, many fighter engines were optimized for high-altitude operations, facts that severely restricted the operating range of these aircraft when flying ground support. Thus, when such aircraft arrived over the battlefield, they had to expend their ordnance immediately and return to base instead of loitering while specific targets were identified. Moreover, in air-to-air combat, machine guns and cannon on fighters had to be sighted so that their bullets converged a few hundred meters in front of the aircraft. This was inappropriate for the ground-attack role, where the bullets should converge at a greater distance and at a point somewhat below the line of flight, in order to strafe effectively. Quesada and others also had to develop ordnance that was appropriate for the ground-attack role instead of dropping smaller versions of the weapons carried by strategic bombers. One of the most significant innovations of 1944–1945 was the introduction, pioneered by the RAF, of unguided rockets that could hit a target with slightly greater accuracy than "iron bombs." Even then, however, ground-attack aircraft were still more effective against large numbers of thin-skinned trucks than against entrenched troops and dispersed armored vehicles.

While the Allies struggled to create an effective air–ground system, even the Germans experienced interservice misunderstanding and rivalry. As late as November 1941, for example, the Luftwaffe refused Erwin Rommel's request for even one air liaison officer to arrange on-call aircraft for the Afrika Korps, because such an arrangement "would be against the best use of the air force as a whole." Given these attitudes, it is not surprising that German Stukas dive-bombed their own armored divisions on at least one occasion.[42]

On the Eastern Front, of course, German air–ground cooperation reached its peak during the period 1941–1943. Thereafter, the number and quality of German tactical aircraft declined steadily. This was caused not only by the growing strength of the Red Air Force but also by the redeployment of Luftwaffe assets to defend Germany against American and British strategic bombardment. Moreover, from 1942 onward the improved quality of Soviet tanks caused the Luftwaffe to experiment with better air–ground antitank weapons, including 30mm automatic cannon and shaped-charge armor-piercing bombs.[43] However, the Germans still found it difficult to deliver such ordnance against individual tanks. Thus, although the Luftwaffe developed adequate procedures for air–ground cooperation in most respects, the lack of sufficient aircraft and fuel to conduct such support and the Luftwaffe's technological decline in comparison to its opponents made that support rare after 1943.

Air Transportation and Air-Landing Forces

One often-overlooked aspect of air–ground operations during World War II was the use of air transportation to move supplies and even nonparachute troops within a theater of operations. Just as railroads and trucks had changed the logistical and operational mobility of earlier armies, so air transportation promised to eliminate the historical vulnerability of all ground forces— their land-based lines of communication. Leaving aside for the moment the true airborne forces, the techniques of air transportation and resupply bear closer examination.

Their most significant use was in Asia, where vast distances, poor road networks, and scarce railroads made aerial supply almost a necessity. In order to understand the British use of air transport in Burma, however, we must digress briefly to consider the tactics of Britain's opponent, Japan.

Japanese industry could not hope to compete with the industrial production of weapons by its enemies. Much as the Japanese Army would have liked to have possessed such weapons, it often had to rely on unorthodox tactics to make up for lack of equipment and firepower. In particular, surprise attacks by night or from unexpected directions seemed to allow the Japanese to move rapidly to close contact, "hugging" their enemy. Paradoxically, the closer the Japanese troops were to their enemies, the safer they were, because the British and Americans then could not use artillery, air strikes, and other technological advantages for fear of hitting their own troops. In hand-to-hand fighting, Japanese leaders believed that their superior morale and training would compensate for shortages of equipment and manpower.[44]

During the conquest of Malaya and Burma in 1942, the Japanese tactics made a virtue of their lack of heavy weapons. Generally speaking, British and Commonwealth defenders were tied to the few available roads for supply purposes and considered the surrounding hills and jungles almost impassable. Upon contacting the enemy, the Japanese therefore used a small demonstration attack on the road to fix the attention of their opponents and then sent a lightly armed infantry force on a long flank march through difficult terrain into the enemy rear. Once in position, this flanking force would attack British supply installations and set up roadblocks behind the bypassed defenders. The British response was predictable—they turned their combat forces around to fight through the roadblocks behind them and rejoin their supply trains, allowing the Japanese to defeat both in detail. As the war continued and their supplies became thinner, many Japanese commanders acquired a habit of planning to live off captured enemy supplies. Having achieved their objectives, the Japanese would then establish elaborate bunker defenses that were difficult to locate, let alone destroy, when the British counterattacked.

Some of the British responses to these tactics were simple and effective. Divisions reduced their establishment of wheeled vehicles and trained to secure their flanks and move through "impassable" terrain, just as the Americans did on Leyte. To destroy Japanese bunkers, the British Fourteenth Army developed two tactics, which incidentally represented partial solutions to the continuing problem of how to keep the defender pinned down by fire while the attacker covered the final few meters in the assault. First, British tanks accompanying the attack fired a careful sequence of ammunition at the bunkers—simple explosives to clear the jungle, then high explosives with delayed-action fuses to break into the bunkers, and finally solid armor-piercing shot as the infantry made the final assault. So long as the infantrymen stayed out of the tank's direct line of fire, they could safely close with the Japanese because this solid shot had no explosive effect. Later in the war, the extremely high degree of cooperation and mutual confidence between air and ground elements in Burma allowed the British close air support to fly a final, "dummy" bombing pass against the enemy, causing the Japanese to stay under cover until the Allied infantry and tanks were on top of them.[45]

The key to defeating Japanese infiltration tactics was air transportation, which would allow bypassed British installations to survive almost indefinitely. In March 1944 Gen. William Slim, the Fourteenth Army commander, correctly predicted a major Japanese offensive against the logistical base area around the town of Imphal (see Map 9). Using large numbers of RAF and U.S. transportation aircraft, Slim was able to parachute or air-land supplies to all his bypassed elements, thus allowing them to fight without being tied to their threatened lines of communications. Furthermore, Slim air-landed most of the 5th Indian Division on the airfields around Imphal, and these fresh troops went straight into battle without a debilitating overland march.

By 1945 the victorious advance of the Fourteenth Army in the more open country of central Burma was made possible only by a combination of air and surface transportation. Two of Slim's divisions reorganized in an unusual manner for this advance. Two of the three infantry brigades in each division reequipped with a full complement of trucks so that they could accompany attacked army tank brigades in a mechanized advance down

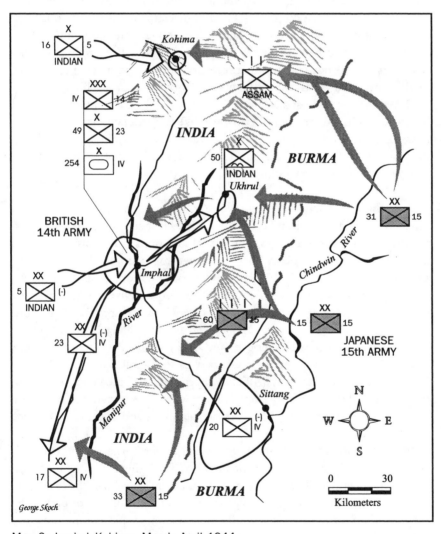

Map 9. Imphal–Kohima, March–April 1944

major roads. As each objective fell, one of these two brigades paused long enough to construct an air strip for resupply. The third brigade in each division was specially equipped with small, light trucks and narrow artillery gun carriages that would fit onto the transport aircraft of the day. Thus the entire brigade could be air-landed onto airstrips or captured airfields to reinforce the ground elements when they encountered significant resistance. Until that time, however, the brigade was in essence a divisional reserve that did not burden the logistical system in the combat zone. This combination of armor, wheeled infantry, and air-landed infantry established a tempo of advance that the poorly equipped and foot-mobile Japanese could not hope to match. In this situation, British Commonwealth engineers often suffered the heaviest casualties, removing land mines to facilitate the motorized advance. The only other drawback to this form of aerial resupply and redeployment was the need for air superiority or at least air parity with the enemy, to allow hundreds of transport flights into forward areas each day.[46]

Other nations also used air transport for resupply and limited movement of troops. For the Germans, air transport, like close air support, was a promising concept that the Luftwaffe was too weak to sustain in many cases. Thus the surrounded German forces in encirclements like that of Stalingrad rarely received adequate air resupply.

Airborne Operations

All the considerations and difficulties of close air support and of air transportation loomed even larger when ground troops used parachutes and gliders to land behind enemy lines. Consider, for example, the Allied invasion of Sicily in July 1943. Without proper coordination, the U.S. Army Air Force transports carrying part of the 82d Airborne Division flew over the invasion fleet. Having just been attacked by German aircraft, the nervous gunners of this fleet opened fire on their own aircraft, killing 229 paratroopers and aircrew.[47] In fact, the Americans and British finally decided that the only solution to such coordination problems was to establish a joint and combined organization—the First Allied Airborne Army, which controlled both the troops and the transport aircraft. Even with close integration of air and ground assets, the potential for error in planning and executing airborne operations was great.

In theory, airborne operations appeared as an answer to the difficulties of penetrating prepared defenses—the attacker simply flew over them and assaulted the enemy rear areas. Sudden assault from above had the same psychological effects as early armored penetrations, confusing and disorganizing the defending army. In practice, of course, planning and communications between the air and ground elements of such an operation were complicated in the extreme. The effects of the defender's antiaircraft fire, the inaccuracies of air navigation, and the difficulty of controlling early parachutes and gliders during landings meant that most airborne drops were widely scattered. Para-troopers had to land prepared to fight as individuals or in ad hoc small groups, without the advantages of organization that made any military unit so much more effective than the sum of its individual members.

In the early hours of such a confusing battle, the paratroopers could have profited greatly from close air support. However, the scattered and confused nature of airborne drops multiplied a pilot's difficulties in distinguishing between enemy and friendly forces. German airborne forces used signal lights to communicate with aircraft, but the Allies relied on the fragile radios of the day. Under such circumstances, the loss or damage of a single special-ized radio could make air support of the paratroops impossible. Moreover, at least in the U.S. Army, some developments in air–ground cooperation were not passed on to airborne troops because they were not part of the conventional, front-line command structure.[48]

In a few operations, such as the German capture of the island of Crete in 1941, airborne troops took and held an objective almost unsupported, but only at great cost in men and equipment. Generally, airborne operations were best conducted in conjunction with a conventional ground offensive, so that the paratroopers could link up with the attacking ground forces within a few hours or days of the initial airdrop. Finding such an ideal situation was dif-ficult. Commanders had to abort many planned airborne operations because, by the time the decision was made and planning completed, the advancing ground troops had overrun the proposed drop zones.

Because of the difficulties of transporting heavy weapons and vehicles even in gliders, airborne units could not be equipped like conventional infantry forces. Furthermore, the parachuting personnel often found themselves separated from the gliders and cargo parachutes carrying their heavy weapons. Thus, an airborne unit lacked much of the firepower, protection, and ground mobility of ordinary infantry divisions. In the air, airborne forces had great operational mobility; once on the ground, an airborne division could move only at a walk and was extremely vulnerable to enemy mechanized attacks. Airborne troops needed to seize their objectives as soon as possible, before the defender could react, and to defend those objectives until help arrived overland. Gen. James Gavin and other U.S. airborne leaders concluded that it was better to accept heavy casualties and parachute injuries by landing on or close to the objective than to descend on a safer drop zone that was sev-eral miles from the objective.[49]

The poor firepower and mobility of an airborne division was especially significant for the British and Americans. The shortage of combat troops of all kinds in 1944 and 1945 meant that airborne divisions frequently remained committed to ground combat alongside conventional divisions even after the two had linked up, because there were no other divisions to relieve the paratroopers. Ultimately, U.S. airborne commanders urged that their divisions be organized and equipped like conventional infantry divisions, with the heavy weapons and vehicles rejoining the airborne division overland after the drop zone had been secured.[50]

Many of the same problems plagued the Soviet efforts in airborne warfare. Despite an initial lead in airborne concepts and training during the 1930s, by 1941 the Red Army's higher ranking paratroop commanders suffered from the same problems as their ground-bound peers—poor leadership and staffwork, inadequate intelligence, and lack of key equipment, including sufficient transport aircraft. Of the two division-sized Soviet airborne operations of World War II, the Vyazma landing in early 1942 was at best a partial success, because attacking ground elements never established firm contact between the airborne pockets and the main Soviet lines. The Dnepr River landing of September 1943, on the other hand, was a disaster because the troops landed on an unsuspected concentration of German troops, just as the British airborne suffered at Arnhem in 1944. Smaller Soviet airborne drops, often launched to divert German attention from ground attacks, were more successful. Overall, however, the mixed airborne experiences of the Red Army prompted Soviet dictator Joseph Stalin to virtually ignore airborne tactics and development in the years immediately after the war.[51]

Amphibious Operations

Airborne operations required meticulous cooperation and coordination between air and ground services, but amphibious operations were even more complex. The opposed amphibious landings—that is, landings in the teeth of prepared enemy defenses—of World War II were among the most elaborate operations ever undertaken. These landings foreshadowed the nature of future wars, when sea, air, and land forces would have to be integrated and coordinated with one another and often with the forces of other nations.

Many of the largest such landings, including those in Normandy, Leyte, and Luzon, were conducted by the U.S. Army instead of by the Marine Corps. With only a rudimentary doctrine for such joint operations, the army and navy had to develop effective planning and command relationships. When airborne insertions were added to the equation, the situation became even more difficult to coordinate. Similar problems arose even when the landing was conducted by the Marine Corps. Rear Adm. R. K. Turner, who commanded many of the amphibious operations in the central Pacific, was instrumental in developing the procedures by which the amphibious force commander remained in control of all forces, regardless of service, until the senior army or marine commander moved his headquarters ashore and assumed responsibility.[52]

The U.S. Marine Corps (USMC) had developed the tactical procedures for amphibious landing during the interwar period, at a time when most professional soldiers considered such operations impossible. When World War II broke out, the marines were still struggling to resolve the problems of air support. An amphibious assault against prepared enemy defenses had all the problems of a deliberate attack, plus the inability of the attacker to bring his own artillery onto the beach immediately. The attackers also faced the difficulties of wind, tide, and underwater obstacles. The solution to these problems, besides careful organization and command and control, was fire support from naval and air units. Yet as late as 1940, the USMC's own aviators followed the familiar argument that air strikes should be used only when conventional artillery was unavailable. Even during the invasion of Saipan in June 1944, there was only one radio frequency available for forty-one air liaison teams to control marine close air support, causing considerable delays in air strikes. Still, by the end of the war the USMC had extremely effective and responsive air support, and even naval gunfire became so refined that it could provide a rolling barrage in front of the attackers on the beach. Only the flat trajectory of naval guns limited their ability to provide fire support inland. To control both air and naval gunfire support, the services eventually developed the Joint Assault Signal Company (JASCO), a combination of up to 600 army, army air force, and navy communicators.[53]

In addition to coordinating the elements of fire support, there was the question of moving the assault infantry and support forces across the beaches and through enemy shoreline defenses. In many respects, the invading forces faced the same disadvantages of exposure and slow movement that World War I attackers had faced in trench warfare. The amphibious tractor, followed later by armored amphibious vehicles, gave the attackers the means to overcome such problems even when the water was too shallow for ordinary landing craft. Although these developments were most prominent in the U.S. landings in the Pacific, similar developments also occurred in Europe. The British Army developed an entire armored division, the 79th, which was equipped with specialized weapons such as amphibious Sherman tanks or mine-roller or flail tanks to clear obstacles. This equipment proved invaluable, not only during the invasion of Normandy in June 1944 but also in the assault river crossing of the Rhine in 1945. Both of these operations, with the combination of ground, air, amphibious, and parachute troops of several nations, were models of the steps required to combine many different weapons and units into an effective whole.

Special Operations

World War II also brought new refinements to special operations. Generically, such operations can be divided into two groups, depending on the type of organization involved. Unconventional, ranger, or commando operations involve a specially trained force that is self-contained—it penetrates enemy rear areas to gather intelligence or to conduct sabotage, ambushes, and other combat operations. By contrast, guerrilla or partisan warfare depends on the indigenous population of an enemy-controlled area, although special teams of soldier-instructors may be sent to organize, train, and lead the local population. Inevitably, such guerrilla operations require much longer time periods to come to fruition and are heavily dependent on the support of the local population, both armed and unarmed.

German and Japanese forces faced widespread guerrilla resistance. Dozens of German divisions were involved in rear-area security against partisan forces in the Soviet Union and the Balkans. In France and again in the American reconquest of the Philippines, these guerrilla forces were much more than an additional irritant to the occupying army. On a number of occasions, U.S. and British forces used the guerrillas as an economy-of-force tool, bypassing enemy positions and leaving the guerrillas to protect friendly flanks and rear. This approach, along with the great intelligence and sabotage potential of guerrillas, made them a significant weapon.

The principal drawback to the Allied use of guerrillas was largely one of perception. Because most military planners regarded the guerrillas as an auxiliary force, dependent on the conventional armies for weapons and training, they tended to underestimate the capability of guerrillas for independent action of the type that dominated the 1950s and 1960s. Few, if any, Western soldiers studied the effectiveness of Chinese Communist guerrillas against the Japanese or against the Chinese nationalist government.

Paradoxically, though professional soldiers underestimated the role of guerrillas, nonmilitary intelligence agencies overestimated it. The British Special Operations Executive (SOE) and the U.S. Office of Strategic Services (OSS) controlled thousands of highly trained soldiers and civilians who secretly entered occupied territories and organized resistance to the Axis foe. Based on this experience, the OSS and its descendent, the Central Intelligence Agency (CIA), acquired an exaggerated idea of what a few well-trained operations could do to overthrow a hostile power. Such successes, however, depended on widespread popular opposition to the regime in power, a situation that did not always recur in the postwar world.

Meanwhile, ranger and commando units flourished. Small, elite forces, primarily light infantry, could reconnoiter and in some instances assault enemy headquarters and other installations. Guerrillas were traditionally local inhabitants, perhaps trained by a cadre of professional soldiers, but these special-purpose troops were formal units of the organized armed forces. Such units avoided conventional combat on the front lines, seeking to enter the rear areas by covert parachute drops, amphibious landings, or ground infiltration. Like guerrillas, special-purpose forces were essentially weak, relying on the enemy's surprise and poor organization for rear-area defense. If surprise were lost, such units were vulnerable to counterattack by larger, more heavily armed conventional forces. Under such circumstances, only prompt air or naval gunfire support would allow the commandos to extricate themselves.

These tactics had been evident at least as early as the German infiltration units of 1918. During World War II, virtually all the combatants had special-purpose units. The Red Army and Navy repeatedly sent such forces to seize key positions in the German rear.[54] The U.S. Army formed a multitude of such specialized units, of which the most famous were the Ranger battalions. Given careful training and reasonable missions, the Rangers could and did seize key positions, rescue prisoners of war, and otherwise disrupt Axis rear areas. On other occasions, however, they were misused as shock troops, suffering heavy casualties while assaulting strongly fortified positions.[55] Under the eccentric genius Orde Wingate, the British Commonwealth created numerous light infantry brigades of Chindits, which were only partially successful, losing many troops while attempting to disrupt and distract the Japanese by rear-area operations in Burma.[56]

Such a mixed record, coupled with the perception that specialized units diverted the best soldiers from more conventional assignments, made many senior officers as skeptical of special-purpose units as they were of guerrillas. Across the board, the major powers in World War II dedicated some of their most highly trained combat leaders to these two forms of special operations but achieved mediocre results at best.

Urban Combat

Guerrilla warfare, like mechanized combat, traditionally evokes images of operations in rural areas. In fact, many of the most desperate struggles of World War II occurred in urban areas, where buildings made it easy for defenders to ambush attackers at close range and where air and artillery strikes might create an impassable pile of rubble. Tanks had to be carefully escorted and used in a stand-off role to avoid exposing them to short-range antitank weapons.

The most famous city battle was, of course, Stalingrad, where the Germans attacked the Soviets in the fall of 1942 and then had to defend the ruins after a Soviet counteroffensive. In such a battle, the Red Army in particular adopted the same type of hugging tactics that the Japanese had used to neutralize enemy fire support. Often only a single building wall separated the opponents, who fought in small groups of infantry, engineers, antitank troops, and, on occasion, tanks.[57]

As the Allies converged on Germany, they found themselves increasingly involved in city fighting in the heavily urbanized terrain of Central and Western Europe. Perhaps the most serious such conflict in the west was the American recapture of Metz, France, in November 1944. Here, the usual difficulties of urban warfare were complicated by a river crossing, extensive flooding, and a number of obsolete French fortresses manned by the Germans. Fortunately for the Americans, the German defense proved disjointed, and both sides avoided the horrors of house-to-house combat.[58] Meanwhile, the Red Army used its experience at Stalingrad to conduct much more extensive attacks on the defended German cities, especially Berlin itself.

PART II CONCLUSION

To some extent, the experience of the German Army reflects the evolution of all armies in World War II. Initially, Germany had advantages in training and experience, advantages that allowed its soldiers to integrate the different weapons on the battlefield and to move so rapidly that their opponents became disoriented and incapable of rapid response. As the war lengthened, the Germans tended to rely increasingly on their air support and on high-quality armored formations to perform missions that were beyond the capacity of such forces, such as penetration of a prepared defense or halting a major enemy offensive. Heavy tanks took precedence in production plans over half-tracks for the accompanying infantry, and thus German production was never able to support a fully mechanized force. Simultaneously, Germany's opponents were learning how better to integrate their forces at a tactical level and how to organize an effective defense-in-depth against armor, as the Soviets demonstrated at Kursk. From 1943, moreover, improvements in both the quantity and quality of Allied air and ground forces dissipated the early German advantages of training and weaponry. The twin issues of quality and quantity became even more acute for the Japanese, who were never able to compete in manpower and production with their enemies, especially because hundreds of thousands of Japanese troops were tied down in China.

Sheer mass was not sufficient to defeat the Axis forces on the battlefield, however. The Soviet, British, and American armed forces also gained greater skill in combined arms and adjusted their organizations to improve this combination. By 1945 these armies had developed true combat effectiveness at the small-unit level, even though their effectiveness was sometimes the product of field improvisation rather than of careful institutional development. At that point, the problem of combined arms integration shifted, at least temporarily, to a higher level of organization. The lingering problems of combining the arms in 1945 were not so much at battalion and division levels as they were between the armies and the other services. Air support in particular was a critical link in the success of most offensives in World War II, yet the U.S. Army achieved only a temporary truce with the Army Air Forces on this issue. Once the war was over, the practical lessons of small-unit integration and interservice cooperation were frequently forgotten.

• • • • • • • • • • • • • • • •

HOT WARS AND COLD, 1945–1999

Task Force Smith, 1950

At 0400 on Sunday, 25 June 1950, the North Korean People's Army (NKPA) attacked South Korea without warning. The Soviet Union had trained and equipped the NKPA as an imitation Soviet field army, with 135,000 men, 150 T-34 tanks, and 110 World War II–era, propeller-driven combat aircraft. Thirty-eight thousand of these troops had fought in the Chinese Civil War, and a number of their leaders, such as the dictator Kim Il-Sung, had served in the Red Army during World War II.

Faced with this sudden attack, South Korea's army did not run away; but it lacked the weapons to halt tanks, and it was quickly outmaneuvered and forced to retreat. In the context of the global Cold War, U.S. President Harry Truman felt compelled to commit American troops to slow down and eventually halt the invasion.

The initial task of delaying the North Koreans fell not on the U.S. Army as a whole, but on one group of 440 men, flown from Japan to Korea on 1 July 1950. This group, built around the 1st Battalion, 21st Infantry Regiment, was known as Task Force Smith, named after its commander, Lt. Col. Brad Smith.* Smith was a

*The story of Task Force Smith has been told many times, most notably by Roy K. Flint, "Task Force Smith and the 24th Division," in Heller and Stofft, *America's First Battles,* 266–

veteran of the Pearl Harbor attack; in 1950 he literally led the U.S. Army into combat for the second time in nine years. In those hectic early days of July 1950, Colonel
Smith must have felt that he had traveled backward in time by at least ten years,
to the days before the U.S. Army's combined arms experience of World War II. He
had no tanks, no air support, and almost no effective antitank weapons. Task Force
Smith consisted of only one-half an infantry battalion—two companies, a small
headquarters, and a handful of mortars and other heavy weapons—plus one battery of six 105mm howitzers. Smith had borrowed people and equipment from many
parts of the 24th Infantry Division to bring this unit up to strength, but this meant
that the leaders did not know many of their men. The entire division had only thirteen high-explosive antitank (HEAT) artillery rounds capable of penetrating the North
Korean tanks; these thirteen went with Task Force Smith.

When he arrived in Korea, Brad Smith went to see the local American commanders for instructions. He was told to head northwest along the main road and to try to
slow down the enemy advance toward the port of Pusan. He drove forward with a
few assistants to find a suitable place to do this and finally picked a low ridge that
straddled the road just south of the town of Suwon. Smith went back to pick up his
troops, who moved northward through the night, having to dodge refugees and fleeing South Korean soldiers and to persuade jittery Koreans not to blow up the bridges
over which they traveled. Just after midnight on 5 July 1950, Task Force Smith reached
its assigned position south of Suwon. Smith did what he could to organize the position, setting up his infantry along the ridge with their short-range antitank weapons
forward and the artillery behind. One of the howitzers was given all the HEAT ammunition and positioned to shoot at any tanks coming down the road. In a cold, misty
rain the task force began to dig foxholes and prepare their defenses. At dawn, the
troops were only half dug in, but they test-fired their weapons.

Many of the troops, and even their officers, thought that the North Koreans would
back off when they realized they were fighting Americans, but that did not happen.
At 0700 on 5 July, Colonel Smith spotted eight North Korean tanks moving down
the road toward him—the leading edge of a rudimentary Soviet-style forward detachment. When the tanks were well within range, the American infantry fired their ba-
zookas and recoilless rifles, but the shells bounced off the tanks. The one American
howitzer knocked out two of these tanks but then in turn was hit, and the remaining
special ammunition was destroyed. The other tanks kept coming, firing their guns
and passing through the Americans to continue their advance south.

All was quiet for an hour, while the Americans dug their holes deeper. Then at
1000 the main body of the enemy appeared. Three tanks led a long line of trucks

99; T. R. Fehrenbach, *This Kind of War: A Study in Unpreparedness* (New York, 1963), 97–
108; and Clay Blair, *The Forgotten War: America in Korea, 1950–1953* (New York, 1987),
94–103.

and dismounted infantry. When the enemy came within 1,000 yards, Smith opened fire with all his weapons. The enemy infantry took a terrible beating but deployed and sent its troops around the two open flanks of the American force. Smith had lost communications with his supporting artillery so that he could not tell them to fire at this new threat. By 1430 he was almost surrounded, low on ammunition, and without any communications to higher headquarters. He decided that he had done what he could and began to pull back. But withdrawing under fire, in daylight, is extremely difficult in the best of circumstances. The enemy was so close that many Americans simply fled, leaving their equipment and weapons behind. The American wounded and dead also fell into North Korean hands. The trucks to tow the artillery howitzers had already been destroyed, so the gunners could only remove vital parts and start walking to the rear.

At a cost of 153 dead and captured, Task Force Smith had slowed the enemy advance down by perhaps five hours. It took a further two months of desperate fighting for the U.S. Army to rebuild the artillery–infantry–airpower team of 1945 and to halt the invaders at the Pusan Perimeter in southeastern Korea.

CHAPTER 7

• • • • • • • • • • • • • • • •

"Limited" Warfare, 1945–1973

By 1945 the victorious armies of the United Nations had developed a highly sophisticated, equipment-intensive form of combined arms mechanized warfare. Even in the Pacific, the Americans and British used generous amounts of airpower, specialized landing craft, and armored vehicles to support their infantry operations. Yet during the immediate postwar years, the armies of the major powers faced two trends that argued against the mechanized, armored solution to the problems of combined arms combat. First, the destructive power of the atomic bomb convinced many strategists that traditional land warfare was obsolete and caused others to expect radical modifications to any future land combat. The atomic bomb made dense concentrations of ground forces on narrow frontages extremely dangerous and caused the airpower advocates of the world to regard air–ground cooperation as even less important than they had previously viewed it, because the superweapon seemingly made close air support unnecessary. Especially during the late 1940s, when the United States had a nuclear monopoly, the future role of armies appeared to be to secure the bases for strategic bombers before a war and to mop up and occupy shattered enemy territory after a nuclear bombing. Until the early 1950s, technological limitations restricted the design and production of truly small-yield, tactical nuclear weapons. Thus by definition nuclear warfare meant using large-scale, strategic nuclear weapons; ground combat fell into neglect.

Later, when it became evident that land armies still had a role to play, the nuclear option continued to militate against large conventional armed forces. Particularly to the United States and its European allies, nuclear weapons appeared to be a cheap alternative to the type of armed forces needed to meet potential adversaries on even terms. Moreover, nuclear deterrence made it seem suicidal for two nuclear powers to fight directly; instead, the United States and the Soviet Union fought a series of proxy conflicts, in which one or both sides provided aid to lesser allies while maintaining the fiction that the two superpowers were at peace with each other.

Such proxy wars were also largely responsible for the second challenge to the mechanized armies of 1945, the so-called wars of national liberation that employed guerrilla tactics and emphasized political rather than military objectives. During the later 1940s, insurgencies in China, Indochina, Greece, and Malaya made conventional armies appear too expensive and too musclebound to compete efficiently against the politicized peasant outfitted with a rifle and a bag of rice. In retrospect, most of these insurgents were more nationalistic than Marxist, but the Communist powers did provide aid to the rebels in order to discomfit and destabilize the West. It is worth noting, however, that these guerrilla or people's wars worked only where the population was well-disposed to the guerrilla; when the Chinese, Vietnamese, and other insurgents tried to fight outside their borders, they quickly found that they needed many of the logistical and transportation structures of Western armies.

To meet the challenge of the wars of national liberation, Western armies had two choices—they could attempt to adapt their conventional forces to a style of war for which they were not intended, or they could neglect the development of new generations of armored weapons in favor of a renewed interest in light infantry forces. The French in Indochina and Algeria and the British in Malaya, Kenya, and Aden were clearly distracted from the mechanized trends of 1945. Of course, some developments in the counterinsurgency wars may have had application in a more intense, mechanized environment, but these developments were exceptions to the general trend. In the 1960s the Europeans were again willing and able to focus on defense of their homelands in an intensive mechanized war. Almost simultaneously, however, the United States became involved in Vietnam, which distracted American attention and resources from the grave threat of the Warsaw Pact. Not until the mid-1970s were all the members of the North Atlantic Treaty Organization (NATO) actively studying and developing doctrine for their own defense in Europe. In the interim the Soviet Union, which did not fight a significant war between 1945 and 1979, had gone far to make up for its previous technical disadvantages in conventional combat.

Most major armies, including that of the Soviet Union, had to adjust to the challenges of nuclear warfare, guerrilla insurgency, or both. The only major exception was Israel, and even there persistent terrorism at home and guerrilla warfare in Lebanon posed difficult problems for the mechanized Israeli forces. Thus, major themes in combined arms since World War II are difficult to identify. Different armies faced the same problems, but rarely at the same time. I will begin with the period from the end of World War II in 1945 to the end of the second Vietnamese conflict and the eve of the fourth Arab-Israeli War in the early 1970s. The focus of this analysis is on three different developments: the evolution of organization and doctrine in the Soviet Army, the confusing experience of the United States and its European allies, and finally the rapid development of the Israeli Defense Forces from guerrillas to armor-heavy conventional soldiers.

The Soviet Army, 1945–1966: The Decline of Conventional Forces

The Soviet Army, as it was renamed after World War II, experienced four distinct periods of doctrine and organization after 1945. First, from the end of the war to the death of Stalin in 1953, the Soviets demobilized a portion of their forces but continued with the same tactical and operational doctrines and organizations developed during the war. Second, from 1953 to approximately 1967, the ground forces took a back seat to the nuclear-equipped arms of the Soviet state. During this period, the Soviet Army shrank in size and neglected its historical experience in combined arms in favor of an armor-heavy force designed to survive and exploit nuclear strikes. Third, from the late 1960s to the mid-1980s, the Soviet Union reversed this decline of land forces, restudied the experience of the Great Patriotic War, and prepared for the possibility of an extensive, combined arms mechanized conflict with or without the use of nuclear weapons.[1] Finally, the experience of Afghanistan and the development of various new weapons influenced the structure of the Soviet Army in the later 1980s (see chapter 8).

Immediately after World War II, the Soviet Union had no nuclear weapons and therefore sought to refine its mechanized conventional forces for any European eventuality. At the time, this was the only possible counterweight to the U.S. nuclear monopoly. Stalin demobilized his forces from a total of over 500 division-sized units to approximately 175 divisions by 1948. Within the peacetime Soviet Army, however, the number of armored and mechanized units of division size actually increased during this period from thirty-nine to sixty-five. In the process, the wartime tank corps became tank divisions, and the mechanized corps became mechanized divisions (see Fig. 14).[2] Each of these new division structures reflected the experience of World War II, including the integration of tanks, self-propelled guns, infantry,

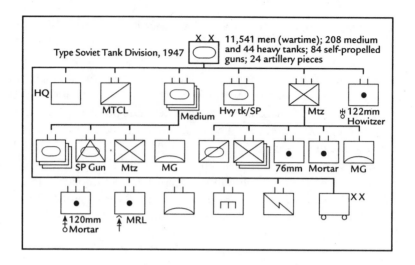

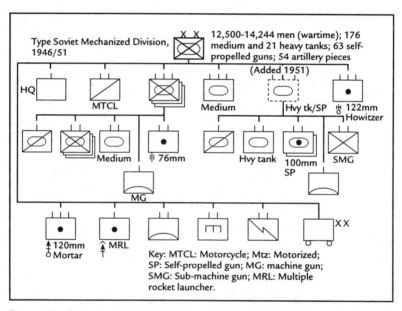

Figure 14. Type Soviet Tank Division, 1947, and Mechanized Division, 1946–1951

artillery, and air defense at regimental level. Indeed, the addition of a combined heavy tank/self-propelled gun regiment to the mechanized division in 1951 made the latter almost too unwieldy for a small Soviet staff to control.

Simultaneously, the Soviets motorized their remaining rifle divisions. The demobilization of 1945–1949 allowed them to equip all remaining units completely with motor transportation, as evidenced by a three-fold increase between 1944 and 1946 in the authorized number of trucks for a rifle division. The first primitive Soviet armored personnel carriers, the BTR-152 series (*Brontetransportr*, armored transport), came into production in late 1945, but even the motorized rifle regiment of a tank division was still truck-mounted until well into the 1950s. Eventually, however, reviving industrial production permitted the Soviets to mount virtually all their infantry in armored carriers that provided some protection from shell fragments and small-arms fire. The BTR-50, the first fully enclosed, tracked armored personnel carrier, came into production in the mid-1950s to equip the riflemen of mechanized and tank units. Apparently the lower priority rifle units then inherited the BTR-152. Despite the advent of the tracked BTR-50, the Soviet Army also produced large numbers of the wheeled BTR-60 series. Not only were tracked vehicles more expensive to build and operate, but they were also slower than the wheeled versions. As a result, the Soviets decided to maintain a mixture of tracked and wheeled personnel carriers. By 1963 all rifle divisions had been mounted in some form of armored vehicle. In their haste, however, the Soviets had standardized on very large vehicles. With each BTR carrying one-half a platoon, any maintenance failure or combat loss would seriously reduce the effectiveness of the platoon as a whole.[3]

This mechanization occurred against the background of significant changes in Soviet doctrine. The Soviet development of nuclear weapons made conventional ground forces seem less vital to national strategy, a trend that climaxed in 1959 with the creation of the Strategic Rocket Forces as a separate branch of the armed services. Beginning with the death of Stalin in 1953, Moscow began to recognize what came to be called a "Revolution in Military Affairs," involving not only nuclear firepower but also improvements in electronics and communications. Two figures—Marshals Georgi Zhukov and Vasily Sokolovsky—were central to this change in conceptions. As Chief of the General Staff from 1952 to 1960, Sokolovsky provided the intellectual framework for efforts to develop new concepts.[4] Marshal Zhukov, whose political popularity had led to his disgrace after 1945, was able to return to power after Stalin's death. By 1955 Zhukov had won government approval for a major reorganization. His primary goal was to adjust the ground forces to the realities of nuclear warfare. All units had to become smaller, for better command and control, and better armored, for protection against the effects of nuclear weapons. The conventional field-gun artillery preparations of the Great Patriotic War declined in significance, giving way to a doctrine that viewed mechanized, armor-heavy forces as the exploitation element after nuclear strikes had shattered the enemy defenses.

In the realm of organization, Zhukov abolished the rifle corps headquarters, the unwieldy mechanized division, the foot-mobile rifle division, and the remaining horse-cavalry units. The motorized rifle division replaced both the mechanized and the rifle division. By 1958 only three types of divisions remained within the Soviet force structure: tank, motorized rifle, and airborne rifle. In contrast to the large, multicorps field armies of 1943–1945, a combined arms army now consisted of two or three motorized rifle divisions and one tank division; the tank army included three or four tank divisions. Missile-equipped artillery and air defense units replaced much of the field-gun, conventional artillery of the Soviet Army.[5]

At the same time, the influx of new equipment and the reduction in the overall size of the army meant that all units, with the exception of airborne divisions, were at least motorized and in many instances mechanized. The term "mobile group," which for three decades had designated mechanized and cavalry forces that were more mobile than conventional infantry, lost its meaning and fell out of use. The function of exploiting penetrations remained, however, becoming a role for the tank and motorized rifle divisions.

During the early 1960s, this continuing need for rapid exploitation and Deep Battle prompted Soviet commanders to investigate the use of airmobility on a limited scale. As early as 1951 Stalin had ordered the design of a helicopter capable of transporting up to twenty-four men. When these design efforts matured, helicopter lift offered the Soviets another option to increase the tempo of exploitation by seizing river crossings and other key terrain. The helicopter promised to eliminate the problem, so prevalent in parachute operations, of units becoming scattered and disorganized during insertion. Initially, however, the Soviet Army did not dedicate ground forces to this type of operation. Instead, helicopter lift became an additional mission for conventional motorized rifle battalions. Leaving its armored vehicles behind, such a battalion with attachments of antitank, air defense, artillery, and engineer troops would be inserted into the enemy rear area. Eventually, it became standard practice for the helicopters to remain with this battalion task force at its landing zone, providing additional tactical mobility as necessary. Yet armed helicopters were confined to a narrow role of escorting the transport helicopters.[6]

With this exception, however, the entire concept of combined arms seemed less important once the Soviet Army had decided that any future war would be a nuclear war, with ground forces generally mopping up after nuclear strikes. In particular, infantry as well as conventional artillery shrank within existing organizations. In 1947, for example, a typical mechanized army consisted of two tank and two mechanized divisions. Because the maneuver regiments in these divisions had integrated infantry units, there was a total of thirty-four motorized or mechanized infantry battalions in the mechanized army. By contrast, the 1958 tank army consisted of only four tank divisions, and they had lost the motorized rifle battalions previously included in their tank regiments. Consequently, the tank army had only twelve infantry battalions, all mounted in armored personnel carriers that would provide some protection from the blast and radiation of nuclear weapons.[7]

Nikita Khrushchev's emphasis on Strategic Rocket Forces meant that the conventional units of the Soviet Army were relegated to an unusually low profile. Individual organization size, as well as the total strength of that army, declined to a postwar low of 140 small divisions. Moscow appeared totally committed to the concept of the single option, the expectation that any major war must be a nuclear war.

Rebirth of Soviet Combined Arms After 1967

Following Khrushchev's ouster in late 1964, a debate began within the Soviet military about the general direction of military affairs. Its exact cause remains unclear, although to some extent it may have been a response to the American doctrine of flexible response. This doctrine called for military forces that would be capable of fighting along the entire spectrum of possible conflicts, from terrorism and guerrilla warfare to full conventional and even nuclear war. Khrushchev had become associated with the emphasis on nuclear warfare, and his departure freed the Soviet military to consider the same options as their American counterparts.[8] Regardless of the causes of the Soviet reappraisal, by 1966–1967 the Kremlin had determined that the single option was too simplistic. In January 1968, for example, an article by the leading theorist Maj. Gen. S. Shtrik announced that "a situation may arise in which combat operations begin and are carried out for some time (most probably for a relatively short duration) without the use of nuclear weapons, and only subsequently will a shift to operations with these weapons take place."[9]

To meet this possibility, Soviet officers renewed their study of conventional combined arms warfare. The government permitted many senior commanders of World War II to publish their memoirs, openly identifying the operational and tactical errors that they had made while fighting the Germans. More important, these memoirs focused on the continuing relevance of certain techniques of the Great Patriotic War. In particular, Soviet commanders paid attention to the concepts of the mobile group and the forward detachment, both of which were key to Soviet methods of mechanized exploitation and pursuit. Although the term mobile group itself no longer applied to a fully mechanized Soviet Army, the functions involved remained relevant to conventional Soviet tactics.[10]

Soviet organization reflected these doctrinal and historical concerns. During the 1970s Soviet tank regiments gradually regained the mechanized infantry and conventional artillery battalions that they had lost under Zhukov's regime. Perhaps most important, some Soviet divisions received a "new" formation, the separate tank battalion. Western observers regarded this as a pure tank battalion that might serve as an additional reserve for the division commander. Within the context of renewed Soviet interest in the Great Patriotic War, however, the separate tank battalion might well be the nucleus of a task-organized forward detachment in any future exploitation and pursuit.

Thus, by the mid-1970s the Soviet Union had come full circle in the doctrine and organization of combined arms combat. While the United States lost a decade of mechanized development because of its involvement in Vietnam, the Soviet Union had developed new generations of armored fighting vehicles and even helicopters to implement fully its long-standing doctrine of Deep Battle and mechanized combined arms.

The U.S. Army: Demobilization to Korea

In contrast to Soviet leaders in 1945, American field commanders were only partially satisfied with their organization and equipment. In 1945 and 1946 the General Board of the U.S. European Theater of Operations conducted an exhaustive review of past and future organization and tactics. It recognized the actual practices of the army in 1944–1945, thereby departing from Leslie McNair's concepts to a considerable extent.

In reviewing the performance of the triangular infantry division, for example, both the General Board and subsequent War Department studies concluded that armor should be organic to that division in order to provide support for infantry attacks and to act as the primary antitank weapon of the army. The infantry's 57mm antitank gun seemed ineffective, and the entire tank destroyer branch was too specialized to survive in the shrunken peacetime force structure. In a reversal of previous doctrine, the U.S. Army concluded that "the medium tank is the best antitank weapon."[11] Although such a statement may have been true, it ignored the difficulties of designing a tank that could outshoot and defeat all conceivable enemy tanks. Moreover, even if the tank were the best antitank weapon, using it in that role might not be the best employment of available tanks, which found themselves tied to their own infantry instead of attacking and exploiting enemy vulnerabilities. In any event, each infantry regiment in the postwar U.S. Army received authorization for an organic tank company, with the division as a whole acquiring an additional tank battalion.

By the time the War Department finally approved a new infantry division structure in November 1946, a variety of changes had occurred based on wartime experience (Fig. 15, top). The self-propelled antiaircraft machine guns and 4.2-inch mortars that had frequently provided fire support to the World War II division became organic to that division. Regimental cannon companies and antitank companies disappeared, but each infantry battalion received recoilless rifles, the new low-velocity antitank guns, to perform these missions. Even the infantry squad and platoon changed. After a conference at the Infantry School in Ft. Benning, Georgia, in 1946, the army reduced the rifle squad from twelve to nine men. This change not only facilitated the squad leader's control of his squad but also released personnel for a separate weapons squad in each reorganized platoon. This squad included a light machine gun and an antitank rocket launcher. The resulting platoons had a greater capacity for independent fire and maneuver than their wartime predecessors. On the other hand, the nine-man squad had little staying power once it suffered casualties or other absences.[12]

In the armored division, similar modifications occurred (Fig. 15, bottom). The limiting factor in most armored operations during 1944–1945 had been the shortage of armored infantry, even in the smaller 1943 divisions. At the end of the war, Gen. George S. Patton estimated that the armored infantry had suffered 65 percent of all casualties in these divisions while inflicting only 29 percent of the German casualties.[13] Conventional infantry and armored engineers had found themselves pressed into service to perform the infantry's close security and urban combat functions for armored task forces. In 1946 the War Department therefore increased the armored infantry in each armored division from three battalions of three companies each to four battalions of four companies each.

Just as in the infantry division, the postwar armored division acquired a number of units that had previously been habitually attached to it. A "heavy" tank battalion, actually equipped with M26 medium tanks because of their 90mm high-velocity guns, replaced the departed tank destroyers as the antitank element of the armored division. Battalions of self-propelled 155mm artillery and self-propelled antiaircraft machine guns also became organic. The three armored engineer companies of the World War II division had proved inadequate for mobility missions, let alone for doubling as armored infantry. As a result, the postwar armored engineer battalion received a fourth line company and a bridge company. The two truck companies normally

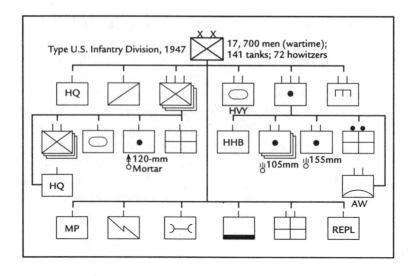

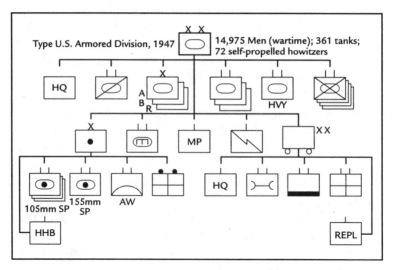

Figure 15. Type U.S. Infantry and Armored Divisions, 1947

attached to any armored division were not added as separate units, but the division's available wheeled transportation certainly grew during the post-war reorganization. For example, the number of 2.5-ton cargo trucks in an armored division increased from 422 in 1943 to 804 in 1947.[14]

Most of these notable improvements in the combination of arms were stillborn because of postwar demobilization. The U.S. Army shrank to a garrison force occupying Germany and Japan, with only a few skeleton units at home. Given America's nuclear monopoly, few people outside the army saw any need for combat-ready forces. Except for one division in Germany, the U.S. Army had no formations that even approached the 1946–1947 tables of manning and equipment. Each of the four divisions occupying Japan in 1950 had only two-thirds of its wartime authorization in men and weapons. Each of these divisions had only one tank company and one antiaircraft battery and was missing one out of every three infantry battalions and artillery batteries. Moreover, budget cuts, personnel turnover, and shortages of training areas severely hampered unit training.[15]

The Korean Conflict

When the Soviet-equipped North Korean People's Army invaded South Korea in June 1950, the understrength American divisions occupying Japan entered combat in a matter of days. The budgetary shortages of previous years were immediately evident. World War II–era ammunition failed to detonate, and units lacked even boots for their soldiers and batteries for their radio transmitters. As each successive division left for the war zone, it was brought up to strength at the expense of the units left behind in Japan, causing serious damage to unit cohesion.[16]

This sudden commitment to combat revealed more than a simple lack of training and combat power, however. It also demonstrated that the U.S. Army had a force structure that did not fit its doctrine. Regimental commanders were deprived of their preferred antitank weapon, the medium tank, and had only the obsolete 2.36-inch rocket launcher for short-range antitank defense. Such weapons were largely ineffective against the Soviet-supplied T-34 medium tanks used by the North Koreans. Moreover, with only two infantry battalions instead of three, a U.S. regiment had no reserve if it tried to defend on a normal frontage of two battalions. The shortage of manpower and the hilly terrain of the Korean peninsula increased the dispersion and isolation of defending units. Such dispersion allowed the North Koreans to practice tactics that were a combination of Japanese offensive operations in 1942 and the Soviet forward detachment. A small unit of T-34s led each column as the North Koreans moved south along the roads. If this tank force encountered a strong point that it could not overrun, light infantry forces bypassed it through the surrounding hills, cut the defender's line of communications behind him, and forced him to withdraw or be cut off. American battalion and regimental commanders, most of whom were veterans of the road-oriented battles of Europe in 1944–1945, found it difficult to adjust to these tactics.[17]

Later in the war, the Americans, like the British a decade before them, learned to accept being cut off and under attack from the flank and rear.

Throughout the conflict, the most common American defensive position was a company entrenched for all-round defense of a ridge or hilltop, separated by hundreds or even thousands of meters from the companies on its flanks. This type of dispersed, strongpoint defense has become increasingly common in most armies since 1945, but it requires excellent fire support and, if possible, active patrolling to provide an effective defense. In the case of Korea, U.S. infantry frequently had to forego patrols and outposts, relying on superior firepower to defeat sudden enemy attacks delivered at close range. When such attacks occurred, a combination of artillery, heavy weapons, and the organic weapons of the infantry proved effective in halting them.[18]

The initial contacts with the Chinese Communist Force (CCF) in October and November 1950 were not deliberate attacks or small-unit defense but a series of chance encounters, meeting engagements in which both sides were trying to use the same roads and streambeds as avenues of movement. By late 1950 the U.S. divisions had built up to their full tables of organization and were oriented on the few roads in an effort to occupy and pacify North Korea rapidly. Although much more lightly equipped, the CCF was also in the habit of using the low ground, moving southward in solid columns with security screens out, and then hiding in villages or woods when aerial reconnaissance searched the area. Once the initial surprise encounter was over, the CCF, many of whom were veterans of the guerrilla wars in China during the 1940s, shifted their attention to the high ground, moving around the U.S. and allied forces tied to the roads. American firepower soon made any daytime movement dangerous for the Communists, and the establishment of company and battalion perimeter defenses on high ground further hampered the CCF movements. Thus, during the later years of the Korean conflict, the preferred CCF maneuver once again became the advance along the low ground at night, seeking to bypass enemy strong points and attack from unexpected directions.[19]

Infiltrators often attacked the artillery batteries that supported the forward infantry. If, despite these efforts, the Chinese still found it necessary to attack a defended hilltop, they rarely resorted to the costly human wave attacks so often described at the time. Instead, the attackers used their machine guns and mortars to pin down the UN defenders while assault groups crept forward, looking for weak points. Persistent attacks would then gradually deprive the defenders of ammunition and manpower. Like other outgunned armies in the past, the Chinese in Korea attempted to "hug" the UN forces, remaining so close to their opponents that the defenders could not call in fire support without hitting themselves.[20]

When the front began to stabilize in 1951, the Korean War became a war of attrition, each side launching limited attacks to destroy enemy personnel. The United States used its World War II doctrine for combining the different arms in such attacks, modifying it slightly to maximize the available firepower and to minimize friendly casualties. One example of this operational technique was the second phase of Operation Punch, a multibattalion limited attack conducted by the 25th U.S. Infantry Division during early 1951. Two large task forces advanced along parallel roads to reduce CCF resistance, withdrew at night to avoid infiltrations, and then returned to inflict additional casualties after the enemy had reoccupied his defenses. One of the two U.S. elements was Task Force Dolvin, which consisted of a battalion headquarters with two companies of medium tanks, a battalion of infantry, a 4.2-inch mortar platoon from a regimental mortar company, a self-propelled antiaircraft machine-gun platoon, a combat engineer platoon, and elements for communications, medical aid, and tactical air control. Because the purpose was to clear bunkers in the area of Hill 300, the infantry commander controlled the entire force. Communication between tank crews and the infantry riding on the tanks was difficult, however, because the newer M46 tanks, like the early model M4 tanks of 1944, had no external telephones mounted on them.

On 5 February 1951 the entire task force moved up the highway and deployed around the base of Hill 300. The self-propelled antiaircraft guns, with the enormous firepower of multiple heavy machine guns, deployed behind the tanks, the two lines of vehicles staggered so that all could aim at the hill to engage the enemy defenses. For thirty minutes, the 4.2-inch and 81mm mortars, the infantry recoilless rifles, the antiaircraft machine guns, and the tank weapons methodically blasted Hill 300, trying to suppress and if possible destroy enemy resistance. Then the infantry, which was sheltered behind the tanks during this preparatory fire, advanced cautiously up the hill. One man in each platoon deliberately exposed himself by wrapping a colored panel, originally intended for signaling aircraft, around his body. Whenever these leading men took cover because of enemy fire, all supporting weapons knew exactly where the friendly troops were, together with the approximate area of enemy resistance. The result was considerable damage to the CCF without major American casualties.[21]

In November 1951 the United Nations and its Communist opponents tentatively agreed to a demarcation line for the armistice they were negotiating. Thereafter, the United States and its UN allies had little reason for maneuver attacks even as small as that of Operation Punch, because there seemed no purpose in losing soldiers to capture ground that would be surrendered at the armistice. Except for patrols, raids, and counterattacks in response to Communist advances, the war became largely a matter of holding defensive positions.[22] Many observers compared this phase of the Korean

War to the artillery and trench struggles of World War I, but in fact there were notable differences. Instead of a defense-in-depth along relatively narrow unit frontages, UN units in Korea formed a thin crust of discontinuous strong points on high ground. Centralized fire control and artillery fuses that detonated in proximity to the ground, rather than on impact, gave the UN defenders unprecedented firepower in the defense. By contrast, the attacking Communists often had only limited artillery or mortar support. In 1951 the U.S. Army further improved its fire-direction capability by introducing rotating plotting boards, allowing a fire-direction center to adjust fire on a target without knowing the observer's precise location. Upon report of a Communist attack, a preplanned, horseshoe-shaped concentration of artillery and mortar fire, or flash fire, would descend around a UN outpost. This firepower isolated the area from further enemy reinforcement for hours while providing illumination to assist the defenders. Within the horseshoe of exploding shells, the defending infantry had to deal with attackers who had come within close range of the strong point. A defending infantry company often had up to a dozen machine guns above its normal authorization and, in some cases, could call on self-propelled antiaircraft machine guns for ground fire support. On occasion, the artillery of an entire corps would fire in support of one such outpost under attack. During a twenty-four-hour period in April 1953, nine artillery battalions fired a total of 39,694 rounds to protect one infantry company.[23] Although justified by the limited objectives of the time, this avalanche of shellfire tended to make the infantry too dependent on artillery support.

Airpower in Korea

Artillery fire, even on such a lavish scale, could stop a determined enemy only while the shells were actually falling. By contrast, air support had a tremendous psychological impact on both sides in a ground action. Recognizing this, the U.S. Marine Corps in the Korean War maintained the tradition of intimate air–ground cooperation. This was especially important for the marines, who had less nondivisional artillery and fire support than the army.

Unfortunately, the other armed services could not match the marine record. During the budget cuts of the later 1940s, the newly independent U.S. Air Force had focused on expensive bombers and jet interceptors, virtually abandoning the ground support role that remained the last priority in its doctrine. In the crisis of 1950 the limited range of available air force fighters and the initial necessity of operating from Japanese bases meant that the USAF was at a severe disadvantage in fighting for air superiority, let alone ground support, in Korea. Repeated army demands led the air force to divert B-29 heavy bombers to attack enemy supply lines immediately behind the front line, a mission that one USAF officer rightly compared to "trying to dam a stream at the bottom of the waterfall."[24]

Eventually, the air force concentrated sufficient pilots, aircraft, and forward air controllers to provide battlefield support to the army, but misunderstandings and mistrust continued. The doctrine of centralized control of air assets meant that ground commanders had to deal with new and unfamiliar pilots on a daily basis, with long waits while the aircraft arrived and found their targets. As one regimental commander (and future Chief of Staff of the Army) observed concerning close air support in Korea, "If you want it, you can't get it. If you can get it, it can't find you. If it can find you, it can't identify the target. If it can identify the target, it can't hit it. But if it does hit the target, it doesn't do a great deal of damage anyway."[25]

Such ill-feeling was part of the reason that, when Douglas MacArthur launched the combined marine-army invasion of Inchon (September 1950), USAF aircraft were excluded from the airspace over the invasion area; and the invasion force relied on Marine Air Group 33 to support all air operations.[26] Later in the war, a number of senior army officers advocated giving operational control of all ground support aircraft to the ground commander. In December 1951 the commander of the Eighth U.S. Army, Lt. Gen. James Van Fleet, formally asked the overall UN commander, Gen. Mark Clark, permanently to attach an air force fighter-bomber squadron to each of the four army corps in Korea. This move would ensure that the pilots were familiar with the terrain and units in a specific area and would respond rapidly when needed. General Clark reluctantly rejected the proposal because it would divert scarce aircraft from other missions such as the often ineffectual interdiction of supply lines. He did, however, get both the navy and the air force to provide a much larger proportion of available aircraft to close air support, culminating in 4,500 sorties in October 1952.[27]

To its credit, the USAF made major efforts to improve the responsiveness and effectiveness of ground support aircraft during the war. Eventually, pilots had to serve eighty days as ground forward air controllers (FACs), a requirement that greatly improved understanding and coordination when they returned to the air. The air force also used a variety of aircraft to provide airborne FACs at critical locations. Gradually, the air and ground leaders became more familiar with one another's operations and capabilities. For example, the army learned that firing high explosive rounds with proximity fuses just before an air strike would help protect the aircraft by suppressing enemy antiaircraft fire in the target area.[28]

One new area of air operations in Korea was the use of helicopters. At the end of World War II, both the U.S. Marine Corps and the U.S. Army had purchased a few primitive helicopters and studied their employment. The marines organized an experimental helicopter squadron in 1947 and used it in small assault landings during amphibious exercises. Interservice agreements meant that the USAF controlled the design and procurement of helicopters for the army, significantly impeding development of this capability. Moreover, the U.S. Army's senior officers stressed parachute and glider mobility at the expense of newer concepts. Still, by 1953 both the army and the marines had used helicopters not only for medical evacuation and liaison but also for limited movement of troops and supplies.[29]

Special Operations Forces

If helicopters made considerable progress in Korea, special operations units had at best a mixed record.[30] At the time of the Inchon invasion in 1950, an enterprising lieutenant in the U.S. Navy organized a mixed group of U.S. Army and South Korean soldiers, local volunteers, and Royal Marines into an ad hoc force that gathered extensive information and seized the lightly held islands near Inchon. During this same operation, however, the U.S. Army's hastily organized 1st Special Operations Company failed to seize the nearby Kimpo Airfield when the river current proved too swift for the rafts being hand-paddled by the invaders. Still, a few days later this company did conduct an effective diversionary attack.

With U.S. leadership, a number of effective guerrilla groups operated from the islands off the coast of North Korea. Efforts at airborne insertion into the depths of North Korea were much less successful. Some of these "donkey teams" became scattered by the parachute drop. Others succeeded in gathering intelligence and organizing local resistance to the Communist regime but made the mistake of using standard infantry techniques when they became stronger, allowing their opponents to track and destroy them. Overall, these operations suffered from a confused chain of command and uncertain methods of insertion and resupply.

The Korean experience prompted the U.S. Army to reverse its previous decision against the value of specialized forces to train and support guerrillas. A few veterans of the World War II operations in the OSS had worked tirelessly to create such units in the army. In March 1952 Chief of Staff J. Lawton Collins achieved the creation of the first Special Forces group at Ft. Bragg, North Carolina, although these troops were not employed in Korea until early 1953 and then were sometimes misused as shock troops. Indeed, this was almost unavoidable, considering that the personnel authorization for the Special Forces had been acquired by eliminating the Ranger companies that had previously conducted special operations at a tactical level.[31]

The Asian Communist Powers

The United States was not the only participant who tested special operations forces in Korea. Because Kim Il-Sung and other Korean Communist leaders were veterans of anti-Japanese partisan campaigns, it was inevitable that the North Korean Army would lay great stress on such operations. During the later 1940s, northern-trained agents had attempted to provoke insurrection in South Korea. These operations generally failed, and so the North Koreans neglected traditional guerrilla concepts in favor of self-contained commando or sniper units that would attack targets in rear areas. For more than forty years after the 1953 armistice, these North Korean special forces worked strenuously to undermine the Republic of Korea. In particular, between 1966 and 1968 the North Koreans launched a series of often unsuccessful special operations, including an attempt on the presidential palace in Seoul. Although thwarted in these attacks, the North Koreans continued to expand their special operations forces, approaching 100,000 men by the 1980s.[32]

By contrast, the rest of the North Korean Army bore a close resemblance to the Soviet structure and doctrine of the 1940s. Infantry divisions followed the standard Soviet pattern, although the mountainous terrain of Korea dictated that these units have more high-angle weapons, such as mortars, than did their Soviet counterparts. During the 1970s, the North Koreans even replicated the Soviet structure of mobile exploitation forces from World War II, building a tank corps and several mechanized corps that were obviously intended to exploit an infantry breakthrough in the event of renewed hostilities with the UN and South Koreans.

The Chinese People's Liberation Army (PLA) also learned hard lessons in Korea.[33] The conflict showed the PLA that it could not fight modern war without modern equipment and training. After their initial successes, the Chinese discovered that they were heavily outgunned by their opponents and that projecting power beyond their own borders called for much greater supply efforts than had the people's wars of the previous two decades. The enormous effect of U.S. airpower prompted a rapid expansion in the fighter arm of the People's Air Force. More generally, beginning in 1951 the PLA eagerly embraced Soviet equipment, tactics, and to some extent doctrine. By 1957 the Soviet Union had provided at least 3,000 advisers and $2 billion worth of military aid to China. These advisers, together with selected Chinese officers who attended Soviet schools, accelerated the conversion of the PLA from an extremely light, guerrilla-oriented force to an underequipped version of the Soviet Army. Field and antiaircraft guns, as well as other heavy weapons, became standard in at least the highest priority PLA units, and tank and truck regiments were organized to give the

Chinese more offensive power and logistical capacity. Like their North Korean allies, the Chinese did preserve some aspects that were unique to the mountainous, light infantry combat of Korea, such as a large number of recoilless rifles. By 1952 the PLA's infantry divisions were no longer light and resembled those of many Western powers. The infantry still used infiltration tactics, but it did so with much greater logistical and fire support.

In the process of converting, the People's Liberation Army became Soviet in discipline as well as in appearance. The officer corps became more professional and acquired distinctive uniforms and rank insignia. This professionalism inevitably clashed with the ideological, politically oriented wing of the Chinese Revolution. Moreover, the Chinese leadership eventually concluded that Soviet military doctrine, like Soviet domination of the Communist world, was incompatible with the revolutionary traditions of China. During the Cultural Revolution of the later 1960s, Mao Tse-tung abolished officer ranks and punished some of his most effective commanders. This upheaval and the difficulties of obtaining and maintaining complex weapons for the huge Chinese Army inevitably delayed its development. Still, the typical Chinese infantry division included a small armored regiment and significant numbers of field guns and supporting weapons (Fig. 16).[34]

In Search of a Mission: U.S. Army Organization from Triangle to ROAD

The genuine success of the U.S. Army in the Korean War caused a temporary increase in its size and budget. Armored forces especially profited from the example of North Korean tanks in 1950, and the army increased its armored strength from one combat command to four armored divisions by 1956.[35]

At the same time, the Eisenhower administration chose to base its national strategy on massive retaliation with nuclear weapons. In order to justify its existence and mission, the U.S. Army had to develop a doctrine and structure that would allow ground forces to function effectively on a nuclear battlefield. Concentrated, fixed defenses of the type used in both world wars appeared to be vulnerable targets for nuclear attack, and so the army had to find a means of greater dispersion and flexibility that would still preserve efficient command and control. The Soviet Army had to fight only in the terrain of Europe and northern Asia—terrain that was contiguous to the Soviet homeland and favorable to mechanized operations. By contrast, the U.S. Army had to remain relatively lightly equipped so that it could deploy rapidly to any trouble spot in the world.

These strategic considerations greatly influenced the tactical structure and concepts of the army. Units had to be sufficiently small and dispersed over

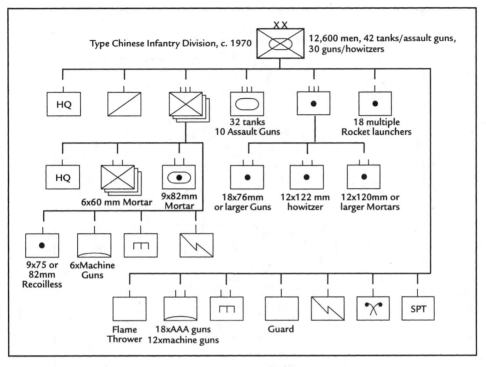

Figure 16. Type Chinese Infantry Division, c. 1970

such large areas that they would not present a lucrative nuclear target, sufficiently balanced among the combat arms so that they could defend themselves when isolated, and sufficiently self-supporting so that they could fight without vulnerable logistical "tails." Army commanders also wanted to streamline the command structure in order to speed the passage of information and decisions. The need for dispersion and for fewer command echelons prompted some theorists to consider increasing the span of control at each level from three subordinate units to five. Five units, spread over a greater area, would report to one higher headquarters, thereby reducing their number at any level.

The result of these concerns was the Pentomic Division, a public relations term designed to combine the concept of five subordinate units with the idea of a division that could function on both an atomic and a conventional battlefield. Five battle groups replaced the three regiments at the core of the pentomic infantry division (Fig. 17). Each battle group was an infantry formation that was smaller than a regiment but larger than the battalion in the older triangular organization. The authors of this design believed that

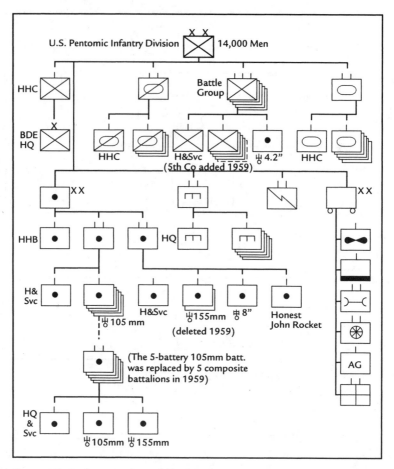

Figure 17. Pentomic Infantry Division

they were eliminating the battalion level of the chain of command while retaining the reconnaissance, heavy weapons, and command and control elements of the triangular infantry regiment. In retrospect, however, a battle group appeared to be an oversized battalion, consisting of a headquarters and service company, four infantry companies of four rifle platoons and a heavy weapons platoon each, as well as a 4.2-inch mortar battery. The battle group's headquarters was much closer in size and complexity to that of a regiment than to the small tactical command post of the old battalion. Within the headquarters and service company, a variety of specialized arms was available. The reconnaissance platoon, for example, integrated light tanks, an 81mm self-propelled mortar, and an armored infantry squad. The assault-gun platoon, equipped with the unarmored, self-propelled M56 gun, pro-vided both antitank and limited offensive gun support for the infantry. The infantry companies, which included the 81mm mortars and 106mm recoil-less rifles previously located at battalion level, proved to be too large for effective control. In 1959 each battle group therefore acquired a fifth rifle company headquarters, but each company was reduced to three rifle pla-toons and one weapons platoon. Even the squad changed, increasing from nine to eleven men and officially acquiring a second automatic rifle. As a result, the pentomic infantry squad was formally divided into two fire teams. These could alternate, one providing covering fire while the other maneu-vered, just as all marine corps and some army squads had done during and after World War II.[36]

The pentomic division structure allowed the division commander to at-tach to each battle group, if necessary, one tank company, one engineer company, and one 105mm howitzer battery. This fire support proved inad-equate, and in 1959 the division's five direct-support batteries gave way to five composite direct-support battalions, each consisting of a 105mm bat-tery and a 155mm battery. Such a composite battalion posed notable prob-lems in training, ammunition supply, maintenance, and fire control of two dissimilar weapons. Even the differing ranges of the two weapons argued against collocating the two batteries. Because mortars had again proved unsuitable as an artillery weapon, the 1959 modifications also reduced the number of 4.2-inch mortars in a battle group and returned their responsi-bility from artillery to the infantry.

Fire support was not the only difficulty with this organization. The divi-sion commander had only one brigade headquarters, commanded by an assistant division commander, to help control the five battle groups, the tank battalion, and the armored cavalry (reconnaissance) squadron. Even with a new division trains headquarters to control logistical support, the division commander and headquarters risked being overwhelmed by the number of subordinate units reporting directly to them. Similar problems existed at the battle-group level, where a colonel and his staff had to control four or five rifle companies, a mortar battery, reconnaissance and assault-gun platoons, and attached tanks, engineers, and artillery. By eliminating one level of head-quarters, the pentomic structure left all other headquarters with an exces-sive span of control. The loss of any one of the remaining headquarters could be disastrous in battle.

Battlefield mobility was another concern. The pentomic structure in-cluded both a helicopter company and, for the first time, a large number of armored personnel carriers (APCs) in an infantry formation. Early APCs gave only limited protection from direct hits. However, unlike the half-track carriers of World War II, most postwar APCs had armored roofs, protecting the troops from shrapnel if artillery shells burst overhead. These carriers, grouped together in a transportation battalion, were sufficient to move one battle group at a time. Because the carrier drivers belonged to one unit and the infantry to another, close cooperation between the two was difficult. Any battle group without these armored carriers had only limited protection and mobility. Moreover, many senior commanders anticipated that their divisions would be deployed for nonnuclear struggles in various parts of the world. Such a deployment could well mean leaving the tank battalion, armored personnel carriers, and other heavy equipment behind for lack of strategic transportation.

In fact, this separation of personnel carriers from their infantry represented a significant shift in infantry doctrine. In an effort to standardize organization, tactics, and training, the Infantry School sought to eliminate the specialized armored infantry of World War II. In its place would be regular (foot-mobile) and mechanized infantry. The tactics of these two groups were supposed to be identical; only their means of mobility between battles would differ. The APC was simply a battle taxi from which the troops still dismounted to fight. This doctrine tended to retard the later development of infantry fighting vehicles in the U.S. Army (see chapter 8).[37]

The effects of the Pentomic concept on the rest of the U.S. Army were much less drastic than on the infantry. The armored division retained its three combat command headquarters, four tank battalions, and four mechanized infantry battalions. It acquired an aviation company to centralize existing aircraft and received the same general support artillery battalion, a combination of 155mm howitzers, eight-inch guns, and Honest John nuclear rockets, as did the infantry division. This new battalion replaced the previous 155mm battalion. As in the infantry division, the armored signal company grew to a complete battalion, reflecting the ongoing growth in communications and staff coordination.

The pentomic changes also brought the nondivisional armored cavalry regiment, the descendant of the World War II cavalry reconnaissance group, to the structure that it retained into the 1990s. Each of three reconnaissance squadrons in this regiment received sufficient logistical support elements to enable it to operate semi-independently, reflecting the large distances over which the cavalry frequently operated. Such a squadron consisted of a headquarters and headquarters troop (equivalent to a company), three armored reconnaissance troops, a tank company, and a self-propelled howitzer battery. A reconnaissance troop represented an ideal of combined arms organization at a low level, because each of its three platoons integrated tanks, infantry, scouts, and a mortar. Thus armored cavalry lieutenants had to deal with the type of combined arms questions that concerned much more experienced officers in other organizations.[38]

This organization of cavalry reconnaissance units served two purposes. First, the variety of combat vehicles in such units made it difficult for an opponent to distinguish between U.S. cavalry forces and other combined arms units. The attacker might be unable to determine whether the U.S. force in question was simply a strong screen of cavalry or an entire U.S. division. This uncertainty would supposedly force the attacker to slow his advance, deploying his forces for a formal attack against the cavalry covering force. A thinly spread armored cavalry could thus provide great security for the division or corps behind it. Second, this combination of weapons and vehicles allowed U.S. reconnaissance forces to fight, if necessary, to develop intelligence about the enemy. As the Soviets had discovered in 1944, a reconnaissance force that is able to fight will be much more effective even in its primary role of intelligence collection and protective screening.

By 1959 the U.S. Army had a radically new structure and operational concept to meet the changing demands of nuclear warfare. This structure and concept differed markedly from the armor-heavy solution of the post–Stalin Soviet Army, but the American commanders were no happier with the results than were their Soviet counterparts.

During the same period, the possibility of nonnuclear conflict increased. The Kennedy administration came into office in 1961 committed to the concept of "flexible response," fielding forces that could fight a broad spectrum of wars from terrorism to full-scale conventional or nuclear war. Despite the army's original intent, the pentomic division was heavily oriented for nuclear warfare and therefore did not appear to be in line with the needs of flexible response. The new administration quickly approved ongoing army studies for a different divisional organization, known as the Reorganization Objectives Army Division, or ROAD (Fig. 18). The different types of ROAD divisions shared a common base, including a cavalry reconnaissance squadron of some type, three brigade headquarters, an artillery headquarters, a support command, battalions of engineers and signal troops, and eventually an air defense battalion. The brigade headquarters, like the combat commands of the World War II armored division, could control a varying number of combat and combat-support elements, depending on their specific missions. The battalion—whether infantry, mechanized infantry, airborne infantry, or armor—replaced the battle group as the largest maneuver organization with a fixed structure. However, this battalion retained many of the battle group's elements, including reconnaissance, mortar, and service-support units. Thus the ROAD structure completed the process of transfer-

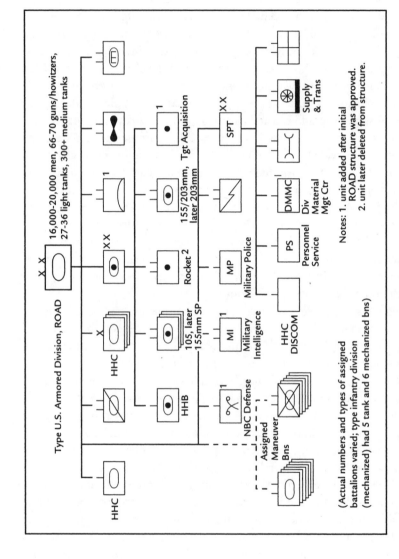

Figure 18. Type U.S. Armored Division, ROAD, 1965–1983

ring many staff, combat support, and logistical functions from the regimental/brigade level to the battalion level.

The unique aspect of the ROAD division was its ability to "task organize," that is, to adjust or tailor structures at any level, based on the anticipated mission. Strategically, the army could choose to form and deploy armored, mechanized, conventional infantry, airborne, and later airmobile divisions, depending on the expected threat. Although there were recommended configurations of each division type, in practice planners could further tailor these different types by assigning various numbers and mixes of armored, mechanized infantry, infantry, airborne infantry, and airmobile infantry battalions, for a total of anywhere from seven to fifteen maneuver battalions per division. The division commander and his staff had considerable flexibility in attaching these battalions to the three brigade headquarters. A brigade making a main attack might have as many as five battalions under its control, but in a subsequent operation the same brigade might have only two battalions to perform a less important mission. Finally, within the brigades and battalions, commanders could form tailored combined arms forces for a specific operation by temporarily cross-attaching infantry, mechanized, and armored companies and platoons as well as attaching small elements of engineers, air defense, and other arms. Thus a battalion task force or company team might receive a variety of subordinate units of different types, allowing integration of the arms as the mission required. In practice, of course, such tailoring and task organizing was prey to the same problems that the World War II system of pooling and attachment had suffered. Constantly shifting units resulted in some inefficiency and poor coordination between subordinate elements that were unfamiliar with one another. A captain or first lieutenant commanding a company team had to be capable of optimizing employment of a variety of different weapons under his control, including tank, mechanized infantry, and later antitank platoons as well as engineer and air defense teams. As a result, battalion and brigade commanders tried to keep the same elements "habitually attached" to one another unless a radical change of mission or terrain occurred. Nevertheless, the ROAD structure gave the U.S. Army the span of control and flexibility of organization it had lacked under the pentomic model.[39] It remained to be seen, however, whether the ROAD or any structure designed for conventional, mid- to high-intensity conflict could adapt successfully to the amorphous requirements of counterinsurgency.

Air Assault

The Kennedy administration's dedication to flexible response also resolved the long-standing question of helicopter mobility. The result was a noteworthy new capability in air–ground interaction and in tactical operations in general.

During the 1950s the British (in Malaya and Suez) and especially the French (in Algeria) had begun to use helicopters extensively to move troops, supplies, and casualties.[40] In the United States, the Marine Corps continued to lead the other services in the application of helicopters for battalion and larger unit assaults. While the army struggled with the pentomic structure, the marines reconfigured their divisions and regiments to eliminate much heavy equipment. They relied on mortars, naval gunfire, and aircraft rather than on howitzers for direct-support artillery. The assault elements of a marine division became completely air transportable as a result.[41]

The more limited innovations by the U.S. Army focused on helicopters in a cavalry role, with small aviation units for screening, raids, and reconnaissance. Brig. Gen. Carl I. Hutton, commandant of the U.S. Army Aviation School from 1954 to 1957, conducted extensive experiments to improvise gun and rocket armament for helicopters and then to use armed helicopters tactically. The army's Infantry School made similar efforts, and the director of Army Aviation, Maj. Gen. Hamilton H. Howze, attempted to popularize the concept of completely heliborne units. The U.S. Air Force adamantly opposed any expanded role for army aviation, especially any form of armed helicopters, because it would impinge on air force missions. Thus only limited progress was possible during the 1950s.[42]

Then in 1962, following the suggestions of several army aviation advocates, Secretary of Defense Robert S. McNamara asked the U.S. Army to study the bold use of aviation to improve tactical mobility for ground forces. The result was the so-called Howze Board. For much of 1962, General Howze and his staff tested everything from dispersed fuel stockpiles for helicopters to close air support bombing by fixed-wing army airplanes. Howze recommended the formation of a number of air assault divisions depending almost entirely on army aircraft, as well as separate air cavalry brigades for screening and delay roles and air transport brigades to improve the mobility of conventional divisions. He noted that an air assault division could maneuver freely to attack a conventional foe from multiple directions and could use both artificial and natural obstacles to delay or immobilize an enemy while itself remaining free to fly over them.[43]

After a considerable internal struggle, the Defense Department authorized the creation of a division for further testing. From 1963 to 1965, the 11th Air Assault Division (Test) at Ft. Benning acted as the vehicle for tactical training and experimentation. The 11th itself was not a full-strength division and often had to borrow elements of another division to conduct exercises. When the air assault division first formed, army regulations still forbad aircraft to fly in formation, and thus many techniques had to be developed with little or no background experience. To make the division's supply system as mobile as its maneuver elements, the division commander, Maj. Gen. Harry Kinnard, developed refueling and rearming points that could be camouflaged and dispersed near the battle area. Artillery, aviation, and infantry had to cooperate closely to suppress enemy resistance during an assault landing. Artillery and any available air force aircraft fired on the proposed landing zone (LZ) until one or two minutes prior to landing, when the assault helicopters began their final approach. The last artillery shells were smoke, to signal armed helicopter gunships that it was safe for them to enter the area and begin firing machine guns and rockets to suppress any remaining defenders. Meanwhile, the troop helicopters landed and discharged their infantrymen. Early helicopter weapons were rather inaccurate, but this fire was certainly impressive to friend and foe alike. Airmobile artillery and infantry changed locations frequently by helicopter and often conducted false, temporary landings in multiple locations so as to confuse the enemy as to their actual dispositions and intentions.

The division's air cavalry squadron combined elements for aerial observation, insertion and recovery of ground reconnaissance teams, and helicopter gunships within each company-sized air cavalry troop. The air cavalry conducted the traditional cavalry missions of reconnaissance, screening, and raids almost entirely from the air.

After a number of tests, the air assault division had clearly demonstrated its potential. Its two most obvious vulnerabilities were the loss of mobility and resupply capability in darkness or extremely poor weather and the much-debated effectiveness of enemy air defense on helicopters.[44]

During the same period of the early 1960s, U.S. Army helicopter units, both armed and unarmed, supported the Army of the Republic of Vietnam (ARVN). This provided a combat test of the concepts developed by Howze, Kinnard, and others, and personnel and ideas passed frequently between Vietnam and the 11th Air Assault Division at Ft. Benning. Initially, American helicopters in Vietnam did little more than transport troops from one place to another. By 1964, however, American helicopter gunships and transports formed small air assault units with Vietnamese infantry on a semipermanent basis.[45]

Inevitably, the U.S. Air Force protested the U.S. Army's use of armed helicopters and even fixed-wing aircraft in a close air support role in Vietnam. The government of South Vietnam was so concerned about possible disloyalty in its own forces that it further complicated the already cumbersome procedures for requesting air support from Vietnamese air force elements. Despite USAF protests, therefore, American and Vietnamese ground commanders felt compelled to use any aircraft that were available, including army aviation when air force channels proved unresponsive.

By 1967 the massive U.S. involvement had reversed this situation, providing large amounts of air force close support for ground troops. Because there was no enemy air threat over South Vietnam, the USAF supported the ground forces to such an extent that Congress held hearings about the neglect of its air superiority mission. This artificially high level of air–ground cooperation temporarily buried much of the rivalry between the two armed services. Yet even then, critics estimated that of all USAF sorties over South Vietnam, only 10 percent were actually close air strikes in support of ground forces; the rest were interdiction attacks, with uneven success.[46] In any event, no air force would have been able to provide such sustained tactical ground attacks if it had been required to struggle simultaneously for air superiority against a comparably equipped enemy air force.

In the interim, the U.S. Army fully integrated the helicopter and its tactics. In the summer of 1965, the 11th Air Assault Division became the 1st Cavalry Division (Airmobile) and deployed to Vietnam (see Fig. 19). General Howze's plan to use fixed-wing army aircraft in a ground attack role had failed, but many of his other recommendations were reflected in the new airmobile division. Because its operational area exceeded the range of conventional artillery, an aerial artillery battalion armed with rocket-firing helicopters replaced the general support artillery battalion found in other ROAD division structures. A division aviation group, including two light and three medium helicopter battalions and a general support aviation company for command and control flights, could carry several infantry battalions simultaneously, moving them dozens of miles without regard to obstacles on the ground.

Entering combat in the fall of 1965, the 1st Cavalry found itself fighting North Vietnamese conventional light infantry regiments much more often than it encountered small guerrilla bands. On 14 November 1965, for example, a U.S. battalion landed by helicopter in what proved to be the base camp of the North Vietnamese 66th Regiment, forcing the enemy to turn and fight in his own rear area. Superior mobility and firepower of this type temporarily halted a North Vietnamese invasion of the south, although the air cavalry suffered heavily in the process.[47]

One key to the airmobile or air assault concept was the close integration, within the same unit, of helicopter and ground forces. By contrast, using helicopter gunships and transports from one major unit to airlift infantry or artillery elements of another unit was much less efficient, requiring more time and effort to ensure coordination and mutual understanding between the parties involved. In practice, the U.S. Army lacked sufficient helicopter

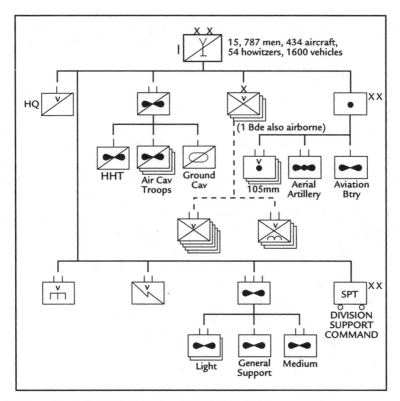

Figure 19. 1st Cavalry Division (Airmobile), 1965

assets to make all the American, Korean, and Vietnamese units fully airmobile with their own organic aviation. Instead, the 1st Aviation Brigade controlled up to 100 company-sized aviation units of various types. Battalions were habitually associated with different divisions. Even the two airmobile divisions, the 1st Cavalry and 101st Airborne, frequently had to lend their assets to support neighboring units.[48]

Airmobility did more than put the enemy off balance and neutralize conventional obstacles. It also forced the U.S. Army to change many procedures to accommodate operations over a large territory without a defined front line. In conventional operations, for example, field artillery and communications units ordinarily oriented their support in the direction of a particular front line or axis of advance. By contrast, in Vietnam these branches had to operate on an area concept, providing fires and communications in any direction from a network of mutually supporting hilltop bases. Even this system did not always give sufficient artillery support for a large-scale operation, and thus the 1st Cavalry Division controlled a nondivisional 155mm artillery battalion that could be lifted by heavy transport helicopters.[49]

As time passed, some of the well-trained American units that arrived in Vietnam experienced so much personnel turnover that they had difficulty maintaining tactical proficiency. More important, the very success of the American machine-intensive form of warfare often became an obstacle in itself. Airmobility allowed U.S. troops to cover large areas and to concentrate their forces rapidly. As Andrew Krepinevich has argued, however, this air movement meant that friendly troops were not on the ground, asserting government control by their constant, visible presence.[50]

Moreover, the almost unlimited availability of artillery and air force firepower led to two understandable but regrettable practices. First, at least some American units made it standard practice to call in artillery, helicopters, and close support aircraft in response to any enemy contact, however small. The purpose was to maximize enemy casualties while minimizing friendly losses, an important consideration for any commander even without the American political reluctance to suffer additional casualties. In fact, until the M16 automatic rifle and M60 light machine gun became plentiful in U.S. units, those troops were outgunned by an enemy armed with AK-47 automatic assault rifles, a situation that naturally prompted the U.S. commander to seek additional fire support. Yet this habit of depending on firepower meant that American troops lost their tactical momentum, pausing to assemble fire support while the enemy maneuvered or fled. The company commander on the ground could not direct the battle when his attention was consumed with adjusting fire support and responding to the inquiries of senior commanders orbiting overhead in command helicopters.[51] Thus American and ARVN troops tended to neglect the fundamentals of fire and maneuver, relying on the artillery and air force. This, in turn, bred unrealistic expectations about fire support in future conflicts.

The abundance of firepower had another effect. Up to 90 percent of the air force sorties over South Vietnam and as much as 60 percent of artillery fire were expended not against specific targets to provide support for troops but in a vain effort to interdict and harass the unseen guerrilla enemy. Harassment and interdiction (H and I) fires were a nightly practice in many areas, the troops firing blindly at suspected areas of enemy activity. On occasion, unattended sensors or small reconnaissance teams made such attacks effective. In most instances, however, this lavish expenditure of firepower served only to convince the populace and the enemy that U.S. and Saigon troops were frightened of their opponents and unconcerned about any civilian casualties.[52]

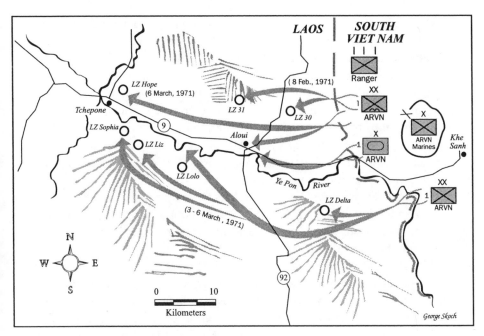

Map 10. Operation Lam Son 719, February–March 1971

Lam Son 719

When the 1st Cavalry Division deployed to Vietnam in 1965, it used the tactic of terrain flying—hugging the ground with helicopters—to present a fleeting target for ground-based air defense. This procedure worked well over jungle and other rough terrain, but in more open areas the enemy on the ground had more time to react and to fire on helicopters. Because the principal air-defense threat in South Vietnam was rifle and automatic weapons fire at low altitudes, many aviation units began to fly above the effective range of such weapons. Many observers argued that such high altitude, level flight would be suicidal against an enemy with larger and more sophisticated air defenses. One battle in 1971, known as Lam Son 719, became the center of the debate on the vulnerability of helicopters in combat.[53]

The purpose of Lam Son 719 (Map 10) was to destroy the North Vietnamese base areas in Laos, specifically the large logistical installations around the town of Tchepone. This move would forestall a major North Vietnamese offensive to take control of the northern provinces of the Republic of Vietnam. The First ARVN Corps planned to make the main effort with the 1st ARVN Airborne Division conducting airmobile operations north of the Ye Pon River, while the 1st Armored Brigade, which was attached to the airborne division, advanced westward along Route 9 into Laos. The 1st

ARVN Infantry Division would conduct a secondary attack south of the Ye Pon River, providing fire support and flank protection for the main attack. Finally, a three-battalion force of Vietnamese Rangers was responsible for the northern (right) flank of the 1st Airborne Division.

The plan had problems even before the offensive began. First, the U.S. government would not permit U.S. forces to operate on the ground inside Laos, and thus the ARVN units had to fight for the first time without their American advisers. Although many ARVN units were capable of such operations, the absence of advisers made coordination of air support and airmobile transport much more difficult. The ARVN units depended on American helicopters and fighter-bombers for their mobility and firepower. U.S. Army aviation and ARVN ground unit commanders had to plan each operation as equals, which inevitably slowed down the planning process even though both sides tried to cooperate.

Terrain was another major handicap. The Ye Pon River, including Route 9 that paralleled the river, was the natural avenue of approach between Vietnam and Tchepone. This valley was so narrow that the 1st ARVN Armored Brigade lacked maneuver space for its three armored cavalry squadrons. The valley was also a natural air corridor, especially when clouds reduced visibility over the high ground on either side. The Ye Pon River was the most prominent terrain feature for helicopter navigation. As a result, much air traffic was channeled down the valley; once the ARVN forces began their attack, their future axis of advance would be immediately obvious to the defending North Vietnamese. Huge ARVN convoys near the border gave the enemy ample warning of the projected attack.

For several years prior to 1971, the Communists had established an integrated air defense system oriented on the valley and on the few natural helicopter pads in the surrounding hills. Nineteen anti-aircraft artillery battalions were in the area, including 23mm, 37mm, 57mm, and 100mm antiaircraft guns as well as 12.7mm heavy machine guns. The antiaircraft coverage was thickest around the Tchepone supply dumps. In addition, the defenders had preplanned artillery fires to strike all likely landing zones. The North Vietnamese reinforced their defenses during the battle, reaching a total of twelve infantry regiments, two tank battalions, and considerable artillery support. By the end of the operation, the ARVN attackers were in many instances outnumbered by the defenders.[54]

The result was a "midintensity war" rather than a counterinsurgency campaign. The ARVN began its attack on 8 February 1971 but had to delay operations the next day because of poor weather. Throughout the offensive, U.S. and Vietnamese air support was often unavailable because of low cloud cover. Even single helicopters on medical evacuation and supply flights needed armed helicopter escorts to suppress enemy air defense. This in turn strained the available resources of AH-1 attack helicopters and forced the U.S. Army to use the slower, more vulnerable, and generally obsolete UH-1C gunships.

Cobra attack helicopters fired on North Vietnamese light tanks, destroying six and immobilizing eight. Still, a score of fourteen out of sixty-six tanks suggested a significant weakness in attack helicopter armament. T-34 medium tanks overran the ARVN firebase at LZ 31 after repeated attacks. Because the U.S. and ARVN forces had rarely needed large-caliber anti-tank weapons before this battle, they had few effective defenses. The U.S. Army aviation commander for Lam Son 719 urged the army to renew its study of specialized helicopters for the antitank role.[55]

After several weeks of limited success and occasional defeat, the ARVN commander abandoned plans for a ground advance west of Aloui, twenty kilometers inside Laos. Instead, during the first week of March 1971, the 1st ARVN Infantry Division established a series of temporary firebases on the escarpment overlooking the southern side of the river. On 6 March, two battalions of the 1st ARVN Airborne Division air-assaulted into LZ Hope, northeast of the ultimate objective at Tchepone. This LZ was in the center of the enemy air defense umbrella, but the two battalions lost only one helicopter of 120 involved in the attack. These air assaults succeeded because they were carefully planned and supported. Strategic and tactical bombers suppressed local enemy defenses and often created clearings to be used as new, unexpected LZs. Gunships and air-delivered smoke screens protected the infantry during their landings.

The ARVN accomplished its mission, destroying the support facilities around Tchepone, before withdrawing with considerable losses. This operation delayed a major North Vietnamese offensive for a year, but the cost seemed excessive. In addition to several infantry battalions that were virtually destroyed, the U.S.–ARVN attackers lost a total of 107 helicopters shot down in six weeks. Many observers cited Lam Son 719 as proof that airmobile operations were too vulnerable to enemy air defense and could not be conducted in complex, machine-intensive wars.

Yet these helicopter losses must be evaluated carefully. One hundred seven helicopters represented perhaps 10 percent of the U.S. aircraft involved in the operation but only a small loss in an offensive during which the U.S. Army flew more than 100,000 sorties, although many of these were only short "hops." The terrain neutralized many of the advantages of an air assault force, allowing the defender to focus his attention on a few critical areas through which the advance and withdrawal had to pass. This concentration of antiaircraft fires, in combination with poor weather, forced the helicopters to avoid terrain flying by increasing their altitude to about 4,000 feet above ground level. Finally, after 1971, helicopters acquired improved navigational equipment, night flying devices, and more survivable mechanical designs. Similar circumstances of weather and terrain might still hamper air assault operations, but Lam Son 719 by itself did not definitely prove such operations to be impossible.[56] Certainly the other NATO powers and the Soviet Union used the airmobile experience of Vietnam to help in the development of their own army aviation doctrines.

The NATO Powers

For fifteen years after 1945, the military policies and posture of Western European states resembled those during the same period after 1918. The war had exhausted the Europeans, who were reluctant to finance major new weapons systems for their armed forces. The British, French, and Americans allowed West Germany to rearm only after a decade of occupation, and even then only because of the conflict between East and West. The new *Bundeswehr* (Federal Armed Forces) could not afford to mechanize all its formations in accordance with the experience of World War II, and so the first-line units had different equipment and tactics from other German ground units. France and Britain had even greater difficulties, developing three elements within their armies: a fully mechanized force committed to the defense of Central Europe, a less well-equipped conscript and reserve force at home, and a lightly equipped but well-trained and strategically mobile element for conflicts outside Europe. Such conflicts and the demands of strategic mobility encouraged British and French interest in light tanks and armored cars that might be used both at home and abroad. This interest echoed the interwar British desire to maintain lightly armored forces for imperial police work.

In the 1960s the end of conscription in Britain and the gradual termination of counterinsurgency wars abroad caused both the British and French armies to reorient their attention on defense in Europe. Even then, democracies remained naturally suspicious of "offensive" weapons such as tanks, preferring to develop "defensive" weapons such as the antitank guided missile (ATGM). The French SS-11 was the first effective ATGM in NATO, and many nations, including the United States, adopted it during the early 1960s.

Britain, France, and West Germany accepted the concept of combined arms or "all-arms cooperation" as a principle of tactics. This similarity of concept was reflected by some similarity in large-unit organization. The three armies converged on fixed, combined arms structures that in U.S. terms were of brigade rather than divisional size. By contrast, within the U.S. ROAD division, brigades remained flexible, changing their configuration to adjust to different situations and missions. The evolution of the fixed European brigade may have been a result of orientation on the single mission of mechanized operations in Europe. In any event, this evolution deserves a brief review.

At the end of World War II, the British Army retained its two-brigade armored division and three-brigade infantry division with only minor changes. The mixture of three tank and one motorized battalion in an armored brigade, and the reciprocal one tank and three infantry battalions in an infantry brigade, allowed for cross-attachment of tanks and infantry at battalion and company level. The resulting combinations would be in the proportion of three companies or platoons of one arm with one of another. During the 1950s the British Army of the Rhine (BAOR) developed a square brigade structure that was more balanced, more suitable for a variety of tactical situations. Each brigade consisted of two tank and two mechanized infantry battalions. These brigades came to have a fixed organization of other arms and services, generally including a 105mm artillery battalion, two engineer companies, and more generous logistical support than any other NATO brigade. Some of these supporting arms nominally belonged to the division as a whole, but they were habitually assigned to specific brigades. Outside the BAOR, the brigade level of command was even more important. Although division structures existed on paper in the United Kingdom, the basic unit for deployment outside Europe was usually the infantry brigade, consisting of three to five infantry battalions and other arms.[57]

As late as 1954 the French Army, whose Free French divisions had been equipped by the United States during World War II, retained the equipment and organization of the 1943 U.S. armored division. After the Algerian War ended in 1961, the French Army renewed its study of mechanized operations and organizations. This review culminated in the 1967 or Type-67 mechanized division, consisting of three mechanized brigades. Each of these, like their German and British counterparts, had a permanent structure. The brigade included one main battle tank battalion, two mixed mechanized battalions, a self-propelled artillery battalion, and an engineer company. As in the case of Britain, this French structure for operations in Europe was so fixed that the brigade and division levels of command were somewhat redundant. As a result, in the mid-1970s the French Army began to convert its units to a new organization, labeled a division, that was in fact an oversized brigade. The armored division, for example, consisted of only 8,200 men, organized into two tank, two mechanized, one artillery, one engineer, and one headquarters and service battalion. The French infantry division became even smaller, totaling 6,500 men in three motorized infantry and one armored car battalions, and other arms as in the armored division. The French hoped that this smaller division structure would be more responsive and fast-moving on the nuclear battlefield. For the French Army, the function of armored divisions in such a battle was to cause the enemy forces to mass, thereby presenting a vulnerable target for French tactical nuclear weapons.[58]

One of the unique aspects of French Army structure during the 1960s and 1970s was the organic combination of different arms within one battalion. The French began experiments with combined arms battalions in the early 1960s, culminating in the mixed or "tank–infantry" battalion of 1967. Within this battalion, each of the two light tank companies contained four tank platoons and an antitank guided missile platoon; two mechanized infantry companies had three platoons each. The two types of companies cross-attached platoons for tactical operations. The battalion headquarters of such a unit included other arms as well, such as communications, reconnaissance, and mortar platoons. Use of the same basic vehicle chassis reduced the maintenance problems of the battalion and ensured that all elements had uniform cross-country mobility. First the AMX-13 and later the AMX-10 family of armored vehicles consisted of compatible vehicles for light armor, ATGM launchers, and infantry. The French had to extend greatly the amount of training given to junior leaders to enable them to train and control three types of platoons. This problem helped force the French Army to reduce the size of both tank and mechanized infantry platoons to three vehicles each, a size that was easier to supervise and control. Finally, because these tank–infantry battalions could no longer provide infantry support for the pure tank units, the medium or main battle tank battalion in each mechanized brigade acquired an organic mechanized infantry company. In practice, this tank battalion often had to support the tank–infantry battalions because of their limited armor protection against a massed enemy attack.[59]

While France led the Western powers in the integration of different arms within a battalion, West Germany pioneered the development of mounted infantry integrated with armor. In this, West German commanders were following the World War II tradition of panzer grenadiers. While the U.S. continued to build armored carriers to transport infantry that would dismount to fight, the Germans wanted effective infantry fighting vehicles (IFVs) in which the infantry could fight as well as ride. The first generation of such a vehicle was the SPz 12-3 (*Schuetzenpanzer*, armored infantry vehicle), fielded in the early 1960s. Intended as a true fighting platform, the SPz was armed with a 20mm automatic cannon and carried four more tons of armor than the contemporary American APC, the M113. To achieve this load, the SPz had to give up any amphibious capability, but the Germans reasoned that it was pointless if the SPz were to cooperate intimately with nonamphibious tanks. By 1963 the first grenadier brigades were formed, consisting of one tank battalion, three infantry battalions, and a self-propelled artillery battalion. Only two of the three infantry battalions were equipped with SPz, however; the third battalion (*Manschaftstransportwagen*, MTW) was at first truck-mounted and later equipped with the U.S. M113 APC. The Germans made a virtue of this variation, however, because the MTW battalion provided additional dismounted troops to fight in urban areas, conditions of reduced visibility, and other circumstances where true mounted infantry was not essential.[60]

The second-generation German IFV was even more impressive. The *Marder* (marten) was the first true IFV in NATO. It had a turret-mounted automatic cannon and gun ports that allowed the infantry to shoot from inside the vehicle. Unlike the SPz, the *Marder* had a protective air circulation system to keep nuclear radiation, biological, and chemical (NBC) weapons out of the crew compartment. German commanders intended the mechanized infantry to fight from their IFVs, to dismount only when necessary for special operations such as patrols or urban combat. The trade-off to this mounted capability, however, was that the German panzer grenadiers had the smallest dismounted squad size—seven men—of any Western army. The *Marder* itself became the base of fire around which the dismounted squad maneuvered as an assault team, much as light machine guns had provided the base in earlier generations.

The German concept and design for an IFV drew considerable attention and imitation both in the other NATO nations and in the Soviet Union. Yet if tanks and mounted infantry operated as a team under all circumstances, the IFV required the same mobility and protection as a tank, becoming in essence another less heavily gunned tank itself. The British Army had recognized this at the end of World War II, when it had used a limited number of Sherman tank chassis without turrets as Kangaroo heavy personnel carriers. The *Marder* itself went a long way in the same direction, but its weight at 27.5 tons made crossing obstacles difficult, and its production cost prevented the *Bundeswehr* from equipping all German infantry with it.[61]

The Germans were also the only power to field new armored tank destroyers during the 1960s, although a decade later the *Bundeswehr* replaced them with tanks. In part, the tank destroyer filled a perceived gap when the United States and its allies abandoned the creation of true heavy tanks, choosing to standardize on medium-weight main battle tanks. The *Jagdpanzer* (tank hunter) was organic to German brigades and sometimes carried

ATGMs as well as a 90mm high-velocity gun. A gun-equipped antitank vehicle of this type seemed too specialized to maintain in peacetime, especially when ATGMs were so much more effective and flexible. In the 1970s, however, new forms of ceramic and other specialized armor protection greatly reduced the effectiveness of the shaped-charge chemical energy warheads used on most ATGMs and low-velocity, infantry-portable guns such as the recoilless rifle (see chapter 8). The shaped-charge round was not totally useless, because no nation could afford to use ceramic armor on all its combat vehicles, or even on all surfaces of main battle tanks. Still, by the late 1970s the higher velocity gun in a tank or tank surrogate was again the most effective weapon against enemy tanks. As a result, infantry units were potentially more vulnerable to armored attack than at any time since 1943, a situation that had a significant effect on the development of new antitank weapons during the 1980s.

From Home Defense to Blitzkrieg: The Israeli Army to 1967

In five wars and numerous undeclared conflicts since 1948, Israel has become famous as an expert practitioner of highly mechanized combined arms warfare. Israeli experience had a profound impact on other armies, especially after 1973 (chapter 8). Yet to understand the strengths and weaknesses of the Israeli Defense Forces we must remember their origin.

In 1948 the Israeli portions of Palestine declared independence from Great Britain while under attack from their Arab neighbors. At the time, the Israeli armed forces were a loose confederation of self-defense militia, anti-British terrorists, and recent immigrants. A number of Israelis had had training as small-unit leaders, both in the local defense forces and in the British Army of World War II. What Israel lacked were commanders and staff officers with experience or formal training in battalion or larger unit operations and logistics. Even after independence, Great Britain would allow only a few Israelis to attend British military schools. Moreover, until the 1960s Israel could find neither the funds nor the foreign suppliers to purchase large quantities of modern weapons.

As a result, the Israeli Army of 1948–1956 was an amateur force, poorly trained and equipped except as light infantry. The honored elite of the Israeli Army were the paratroopers of 202d Brigade, who conducted reprisal raids into Arab territory. Indeed, throughout its history Israel has always assigned the cream of its army recruits to the airborne brigades. In general, the Israeli Army relied on its strengths in small-unit leadership and individual initiative, strengths that were sufficient for self-defense until 1955. In that year, the Soviet Union changed the Middle Eastern equation by promising to supply Egypt with large quantities of more modern heavy weapons.

The colorful Moshe Dayan became chief of staff of this unusual army in 1953. In 1939 Dayan had been one of a number of Jewish self-defense leaders who received unauthorized small-unit training from Capt. Orde Wingate, the erratic British genius who later founded long-range British Chindit attacks in the jungles of Burma. During the 1948 War of Independence, Dayan commanded the 89th Mechanized Commando Battalion, a ragged collection of half-tracks and light vehicles that conducted daring raids into Arab rear areas. Subsequently, while visiting the United States, Dayan by chance met Abraham Baum, a U.S. tank company commander who, during World War II, had led a small raiding party behind German lines to release American prisoners of war at Hammelburg, Germany. Baum's account of American armored tactics in 1945 reinforced Dayan in his own belief in speed, mobility, and commanders going forward to make decisions on the spot. Thus, Dayan discovered that his own ideas were in part a reinvention of the principles used by both Americans and Germans in World War II.[62]

By contrast, the Egyptian Army of the 1950s was almost a caricature of the British Army that had trained it. Egyptian tactics relied on the centralized, methodical staff planning that had made British units slow to react against the German attacks in France in 1918 and in North Africa in 1941 and 1942. Dayan's success in the 1956 war lay in his recognition of this vulnerability:

> The Egyptians are what I would call schematic in their operations, and their headquarters are in the rear, far from the front. Any change in the disposition of their units, such as forming a new defense line, switching targets of attack, moving forces not in accordance with the original plan, takes them time—time to think, time to receive reports through all the channels of command, time to secure a decision after due consideration from supreme headquarters, time for the orders then to filter down from the rear to the fighting fronts.
>
> We [the Israelis] on the other hand are used to acting with greater flexibility and less military routine.[63]

The Egyptian defenders of the Sinai Desert in 1956 occupied a string of positions at key terrain points, positions that lacked both depth and flank security. These defenses were vulnerable to outflanking Israeli movements and lacked a large counterattack force to support them. Despite the promise of newer Soviet equipment, the forward Egyptian units were armed principally with World War II–vintage equipment. Dayan planned to disorganize and ultimately collapse this defense by rapid thrusts at Egyptian lines of communication.

Still, the instrument that Dayan planned to use for the 1956 campaign was not a mechanized force. On the contrary, he depended on the Israeli strengths in small-unit leadership and light infantry operations. An airborne drop to seize the critical Mitla Pass would assist the ground infantry columns, which moved across the desert in commandeered commercial trucks, accompanied by a few light tanks and artillery pieces. Initially, Israel's only full-strength armored brigade, the 7th, remained in reserve, with no mission except to use its tank guns as additional indirect-fire weapons.

The 7th was a fairly typical armored brigade of the immediate post–World War II period.[64] It consisted of a battalion of Sherman medium tanks, a battalion of AMX-13 light tanks, a battalion of half-track mounted infantry, an additional motorized infantry battalion, a reconnaissance company mounted in jeeps, and two battalions of artillery. Each of the tank battalions included an organic company of mechanized infantry, but parts of the brigade were attached to other units, leaving the brigade itself with three understrength battalion-sized task forces for much of the campaign.

The 7th Brigade commander, usually identified as Col. Uri Ben-Ari, was dissatisfied with his allotted passive role in the campaign. Early on, however, his reconnaissance company penetrated the poorly guarded Dyka Pass on the southern flank of the key Egyptian position of Abu Agheila–Um Katef (see Map 11). Although this reconnaissance indicated that the pass would support only a few vehicles, Ben-Ari committed his three cross-attached task forces on three different axes inside the Egyptian defenses. Task Force A attacked in vain against the southern side of the Um Katef defenses, where two other Israeli brigades made half-hearted frontal assaults that suffered significant losses. Task Force C moved to the southwest, seeking to intercept relief forces coming from the Suez Canal area. Ben-Ari sent Task Force B, consisting of one company of Sherman tanks and one company of mechanized infantry, through the Dyka Pass into the middle of the Egyptian position. This task force's commander, Lt. Col. Avraham Adan, held his position against Egyptian attacks from two directions and strafing by Israeli aircraft. In fact, when the main Israeli force finally penetrated through Um Katef, Adan mistook the arriving Israeli tanks for Egyptians and destroyed eight of them.

This was the last of a series of errors on both sides, errors that muddied the waters of the apparent Israeli victory. Indeed, Egyptian sources later insisted that their troops at Abu Agheila–Um Katef had held out for four days against superior Israeli forces and withdrew only when Pres. Gamal Abdel Nasser decided to regroup to defend the Suez Canal against a British-French invasion.[65] Had the Egyptian command been more responsive to the threat, the 7th Armored Brigade might have suffered a severe defeat. Still, for whatever reason, Dayan's "collapse theory" proved to be correct; the Egyptians were unable to respond effectively to an opportunistic, improvised Israeli attack.

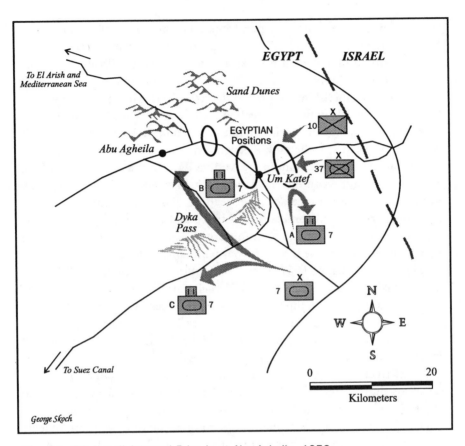

Map 11. 7th Israeli Armored Brigade at Abu Agheila, 1956

The 7th Armored Brigade did not win the 1956 war by itself, but its apparent success prompted a serious reevaluation within the Israeli armed forces. Dayan and other leaders became convinced that armored forces were a superior instrument for the type of maneuver warfare they had always preferred. During the decade after 1956, the Israeli Defense Forces gave the armored corps almost as high a priority for men and materiel as the air force and paratroops received. As deputy commander of the armored corps from 1956 to 1961, and commander after 1964, Gen. Israel Tal shaped Israeli armor into an effective force with superior tank gunnery. Tal soon discovered, however, that complicated armored tactics and equipment required the same discipline and methodical maintenance that had long been common in Western armies but that were rare in Israeli forces.

The principal problem was that Israel lacked the resources to maintain a superior air force and an elite paratroop element while still developing a balanced mechanized army. Tal persuaded the government to purchase modern American and British tanks and to add larger caliber guns to the older Shermans, but the rest of the armored force suffered. In 1967 most of the Israeli mechanized infantry still rode in the 1941–vintage M3 American half-track, a vehicle with no protection from overhead artillery bursts, limited side armor, and increasing maintenance and mobility problems as it aged. Tal insisted that the tank-mechanized infantry team was a European tactic that was less important in the Middle East. In the open spaces of the Sinai peninsula, he contended, Israeli tanks needed less infantry security against enemy short-range antitank weapons. To Tal, infantry was useful for reducing bypassed centers of resistance and for mopping up after the battle. Otherwise, he agreed with the British tankers in North Africa, who had regarded infantry as more of a burden than a partner.[66]

The Six Day War of 1967 seemed to confirm these arguments. The set-piece attacks conducted by teams of Israeli paratroops, infantry, artillery, and tanks to break open the Egyptian border defenses were forgotten in the euphoria of another armored dash to the Suez Canal. Once again, Egyptian command and staff procedures, now heavily influenced by their Soviet advisers, proved too slow and rigid to counter the Israeli thrusts. 1941–vintage half-tracks could not keep pace with 1961–era tanks, either under fire or across difficult terrain. The close and constant assistance of the Israeli Air Force made army air defense and field artillery seem unimportant, especially in fluid operations when the Air Force could arrive more quickly than the artillery could deploy. At the very start of this war, Israeli fighter-bombers had destroyed the Egyptian Air Force on the ground, achieving an air superiority that Israeli ground commanders took for granted. Consciously or otherwise, Israel came to rely largely on the tank–fighter-bomber team for its victories. In doing so it was simply echoing the German successes of 1941 and 1942 and inviting the same disillusionment that the Germans had suffered at Kursk.

5th Regimental Combat Team, 1st U.S. Cavalry Division, Kumchon, Korea, September 1950. (U.S. Army Military History Institute)

UH-1 helicopters with door-mounted machine guns. (U.S. Army Military History Institute)

M2A2 Bradley Fighting Vehicle and dismount team. (Steve Zaloga)

Russian Federation BMP-2 fighting vehicle, with T-90 tank. (Steve Zaloga)

CHAPTER 8

• • • • • • • • • • • • • • • •

Combined Arms, 1973–1990

Israel: The Failure of Combined Arms, 1967 to 1973

Many of the trends evident in the 1967 Israeli campaign continued after that victory. Higher budgetary priorities, combined with the wealth of Arab armor captured in 1967, allowed the Israeli armored force to grow from nine armored and two mechanized brigades in that year to an estimated sixteen armored and at least four mechanized brigades by 1973. The rest of the army remained relatively stable in size. Because Israeli doctrine regarded the tank as the best means of defeating other tanks, the Israeli Defense Forces refused an American offer to supply the newly developed Tactical, Optically Guided Weapon (TOW) ATGMs.[1]

Armor became the main avenue for promotion in the Israeli Army. Aside from the small number of paratroop units, no mechanized infantry officer could expect to command above company level without first qualifying as an armor officer. Israel distinguished between paratroop, conventional, and mechanized infantry, the latter being part of the armor branch but having the lowest priority for quality recruits. Most conventional and mechanized infantry units were in the reserve, where they received less training and priority than tanks. For example, the tanks and crews of the three armored brigades located in the Sinai Desert when the 1973 war began were at a high level of availability and training, but their mechanized infantry components were still in the unmobilized reserve. These brigades went into battle as almost pure tank forces.[2]

As commander of the armor corps from 1969 to 1973, Maj. Gen. Avraham Adan, a task force commander at Abu Agheila in 1956, had tried to reverse these developments. He assigned higher quality recruits to the mechanized infantry units, only to have them seek reassignment away from such an unprestigious branch. Adan also tried to obtain large numbers of M113 armored personnel carriers to replace the dilapidated M3 half-tracks. On becoming chief of staff in 1972, Gen. Israel Tal opposed this purchase. Tal argued that the true role of mechanized infantry, if it had a role, was to fight mounted, as in the West German doctrine. Although the M113 was a considerable improvement over the M3, neither vehicle had enough armor protection and firepower to act as the IFV that Tal sought. The chief of staff therefore opposed spending scarce funds on a good but not perfect vehicle.[3] Israel continued to emphasize the tank and the fighter-bomber to the neglect of other arms. This neglect was also apparent in Israeli unit structures. Despite the great increase in the size of the Israeli Army, all echelons above brigade remained ad hoc task forces *(Ugdah)*, rather than deliberate designs to integrate an appropriate balance of arms and services.

By contrast, the Egyptian Army carefully analyzed its weaknesses and strengths between 1967 and 1973. Indeed, one reason for its initial success in the 1973 conflict was that, for the first time, the Arabs initiated a war with Israel according to a detailed plan instead of Israel's launching a preemptive attack. Moreover, Pres. Anwar Sadat recognized that a holy war to destroy Israel completely was impossible in the foreseeable future. In 1972 he overruled his commanders, appointing new senior leaders and staff to plan a rational, limited war. Sadat's goal was a limited success, a creditable attack that would shake Israeli confidence and permit diplomatic negotiation on a more equal basis. To improve the chances of success, Syria agreed to attack at the same time.[4]

Sadat's staff recognized the same problems that Dayan had exploited since 1948. Egyptian leadership and control procedures could not react quickly to sudden changes in mission. In this respect, Egypt preserved the British method of warfare long after the British Army had resolved its problems, and the growing Soviet influence reinforced this pattern. Having a low opinion of their Arab students, Soviet advisers encouraged a lockstep, phased advance based on fixed timetables. Moreover, the Egyptian troops became demoralized rapidly in a maneuver battle where Israeli troops could bypass them and attack from unexpected directions.

The classic World War II solution would be to prepare the troops psychologically to continue fighting when cut off and surrounded and then to develop a defense-in-depth to absorb Israeli armored attacks before they could penetrate. Yet the Egyptian leaders recognized the lack of cohesion and mutual trust in their units. Nor could the Egyptian Air Force hope to achieve even air parity in battle with the Israelis, a necessary precondition for an effective defense-in-depth.

Instead, the Egyptians sought a different solution to their problems. They planned to force the Israelis to attack Egyptian positions at a time and place of their own choosing. By penetrating no more than fifteen kilometers to the east of the Suez Canal, a limited Egyptian offensive would remain under the protection of a Soviet-built integrated air defense umbrella, an elaborate series of overlapping range fans for fixed SA-2 and SA-3 surface-to-air missiles (SAMs), supplemented by SA-6 missiles and ZSU-23-4 self-propelled guns that would accompany the forward troops. This plan would allow the Egyptian soldier to fight at his best, stubbornly defending his own position against frontal attack without worrying about his flanks or his air cover.

The Egyptian plan therefore called for a surprise attack across the Suez Canal, the line of contact between Egypt and Israel since the 1967 war (Map 12). This attack would isolate the small Israeli outposts known as the Bar-Lev Line, located along the eastern bank of the canal. Egyptian units that were not involved in the initial attack relinquished their Sagger antitank guided missiles and air defense weapons to the assault echelons, who therefore had three times their normal complement.[5]

Egypt succeeded in achieving strategic surprise—Israeli military intelligence did not inform its government until less than ten hours before the attack, and mobilization had barely begun when the two Arab states attacked.[6] The first waves of well-equipped Egyptian troops rushed about four kilometers to the east of the canal and then set up defensive positions. When the local Israeli armored reserves (first a battalion and later an armored brigade) counterattacked to relieve the Bar-Lev outposts, the missile-armed Egyptian infantry faced perfect targets of pure tank units with little infantry or artillery support. Israeli aircraft suffered heavily when they tried to support their armor inside the umbrella of the Egyptian surface-to-air missiles.

The Egyptians also profited from the famous Israeli method of command, which depended on leaders operating well forward and communicating with one another on the radio in a mixture of slang and code words. Using Soviet-supplied equipment, the Egyptian Army jammed many of the Israeli command nets and captured codebooks that enabled them to interpret messages they could not jam. Moreover, Israeli commanders committed the classic mistake of Rommel, becoming personally involved in local clashes instead of directing their troops. On the night of 8 October 1973, the third day of the war, an Israeli brigade commander, battalion commander, and

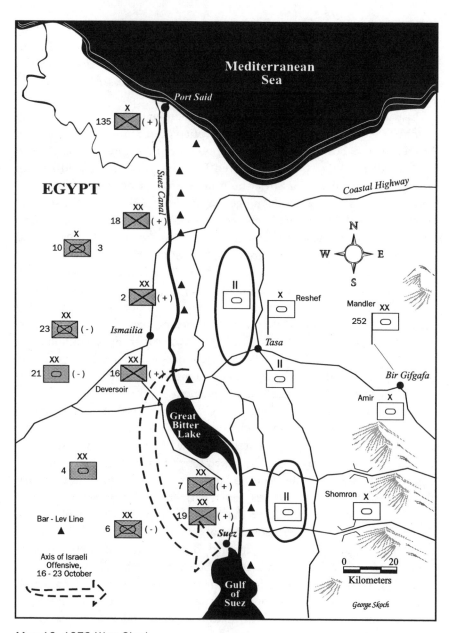

Map 12. 1973 War: Sinai

artillery commander all risked themselves to rescue personally the garrison of one of the outposts that had escaped to the east. Their involvement was in the highest traditions of Israeli leadership, showing admirable concern for the safety of their troops, but it left them unable to coordinate and control the battle.[7]

While the Israelis were frustrated in Sinai, their very survival was at risk on the Golan Heights (Map 13).[8] There, two understrength Israeli armored brigades and an infantry brigade were grossly outnumbered by the attacking Syrians. Fortunately for Israel, the Syrian offensive was a rigid caricature of Soviet doctrine, with three echelons of units moving on a fixed schedule and no one assigned to mop-up bypassed units. Although more than twenty Syrian brigades were involved, the first echelon consisted of only six brigades, seeking to rupture the Israeli defenses so that succeeding echelons could exploit. The Syrian 82d Commando Battalion seized the dominant height of Mount Hermon by an air assault. With a few such exceptions, however, the bypassed Israeli defensive outposts survived, providing intelligence and targeting data to assist the defending armor. Soviet advisers might have taught these tactics because they considered Arabs incapable of more sophisticated operations, but many Western observers wrongly concluded that the Soviet Army itself would attack in the same predictable, inflexible fashion.

Despite the numerical disparity, the Golan battle was much more conducive to Israeli tactics than was the Sinai struggle. Israeli armor fought the dense masses of Syrian armored vehicles from prepared, hull-down tank positions that exposed only their turrets, minimizing the targets they presented. The defenders changed positions between engagements instead of leaving the safety of those positions to maneuver during the battle. The overwhelming numerical superiority of the Syrian forces eventually decimated the two Israeli armored brigades, but this sacrifice bought time for other Israeli units to mobilize and reach the Golan Heights.

At first, the Israeli Air Force also sacrificed itself, conducting close support missions without first neutralizing the Syrian SAMs. Eventually, however, the Israeli fighter-bombers began to range more deeply over Syrian territory, forcing the Arabs to disperse and dilute their air defenses. Although hard-pressed by the tremendous enemy advantage in numbers and surprise, the Israelis were able to halt and counterattack the Syrians.

Thrown onto the defensive, Syrian president Hafez al-Assad asked Sadat for some Egyptian action to distract the Israelis. This appeal forced Sadat to give up his carefully designed and hard-won if shallow gains in the Sinai. By this time, nine days into the war, all surprise had been lost, and the Israeli forces in Sinai were fully mobilized and ready to fight. On 14 October

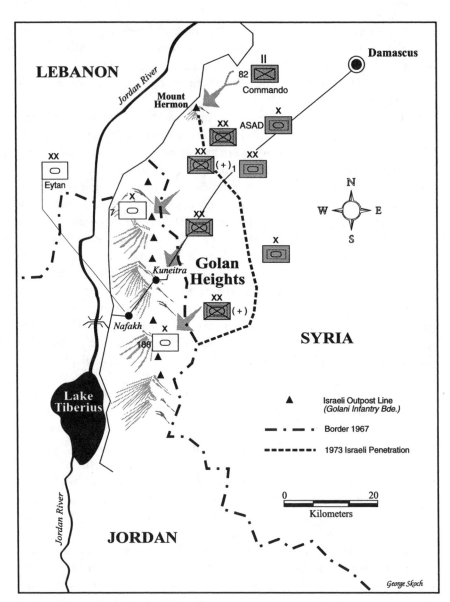

Map 13. 1973 War: Golan Heights

1973 the Egyptians attacked eastward into Sinai, away from their prepared infantry positions and air defense umbrella.[9]

In the ensuing days, the Israelis arrived at improvised solutions to their immediate problems. Because of the low regard armored commanders had for their own mechanized infantry, airborne units functioned as conventional and even armored infantry. The paratroops, however, complained of inflexibility from tank commanders; as always, tank–infantry cooperation could not be achieved without careful training and planning. After counterattacking the exposed Egyptians, the Israeli forces launched their own crossing of the Suez Canal, just north of Great Bitter Lake, on the night of 15 October. Once across, the Israeli tankers eliminated a number of Egyptian SAM sites, destroying the integrated air defense system and thereby allowing the Israeli Air Force to provide more support. By the time that fighting ceased, the Israelis had encircled an entire Egyptian field army.

Israel, Post-1973

Their ultimate success did not eliminate the serious problems that the Israelis faced as a result of the 1973 war. Although they were understandably reluctant to talk about these problems, in the ensuing years they undertook a number of measures to correct the imbalance within their forces. They showed a renewed interest in unit mortars to provide indirect-fire support and in more modern armored personnel carriers for their infantry. Israeli practice placed greater emphasis on the need for fire support and for mechanized infantry to support armor. Even if artillery fire failed to destroy antitank weapons, it could suppress them, causing the gunners of early ATGMs to lose control of their missiles. In the decade after 1973, Israel created fifteen artillery brigades, more than tripling the number of self-propelled guns, and increased its inventory of armored personnel carriers from 500 to 4,800.[10] The Israeli Air Force also began to use the package format so common in the U.S. Air Force—fighter-bombers attacked as part of a package that included remotely piloted drone aircraft (to collect targeting data), airborne control aircraft (to direct the battle), SAM suppression aircraft, air superiority fighters, and finally the actual bombers.

During the 1982 incursion into Lebanon, the Israeli Defense Forces displayed great skill in destroying Syrian air defense forces early in the campaign. Even in this struggle, however, the Israeli armor suffered significant casualties inflicted by small forces of enemy tanks and ATGMs along twisting mountain roads. The Israelis responded with a combination of air support, combat engineers, and attack helicopters to reduce losses.[11]

The Israeli advance was further hampered by Syrian attack helicopters. On 8 June 1982, as the 162d Israeli Armored Division pushed up the center of Lebanon, its leading elements were attacked by French-made Gazelle helicopters, armed with HOT ATGMs. This and subsequent attacks destroyed only seven Israeli tanks, but the HOT's four-kilometer range exceeded the range of Israeli ground-based weapons, and the leading Israeli units did not have shoulder-fired surface-to-air missiles with which to defend themselves. Thus, a few attacks by the Gazelles made the Israeli advance slow and cautious for the next several days. In turn, Israeli attack helicopters, primarily Hughes 500-MDs, used the same tactics as the Syrians, hiding behind mountainous terrain and popping up to engage them. Unlike their opponents, however, the Soviet-equipped Syrians had shoulder-fired SAMs and self-propelled antiaircraft guns; these defenses limited but could not eliminate the Israeli helicopter threat. After the 1982 campaign, both participants expanded their inventory of attack helicopters, a clear recognition of the importance of this new weapon to mechanized warfare.[12]

The Chinese Army after Mao Tse-Tung

Israel was not the only power that suffered an unexpected defeat in the 1970s. Just as in the Red Army of the 1930s, the quality of Chinese leadership suffered considerably under the pressure of the Cultural Revolution of the late 1960s. As a result, the PLA was ill-prepared for its next major military operation. Senior officers were too hidebound in their methods of operation, and junior leaders had no idea of what to do under fire.

In 1979 China again attempted to project power beyond its borders and again suffered a military shock. After years of conflict with the Vietnamese Communist regime, the Chinese decided to administer a lesson by sending 300,000 men in a limited incursion into North Vietnam. Both sides used infiltration tactics rather than frontal assaults. As in Korea, the PLA discovered that it had difficulty focusing enough combat power to mount a successful offensive against the veteran defenders. The jungle terrain neutralized China's numerical superiority, and a shortage of trucks and roads made logistical support difficult. The Vietnamese government simply used its local forces to delay the enemy until the advent of the rainy season made further attacks impossible.[13]

Despite their ideological bias, the Chinese military was shocked by the effectiveness of the Vietnamese people's war tactics. Still, the PLA's ability to dig the Vietnamese out of tunnel complexes called into question China's own plans to defend itself in a similar manner.

The embarrassing failure of 1979 accelerated previous efforts to obtain better equipment and to better integrate the different arms of the PLA. Training concentrated on combining artillery, infantry, tanks, and antitank weapons. Meanwhile, the Chinese government's unwillingness to purchase foreign weapons led the PLA's leadership to develop a mass of defense industries at home. During the 1980s China produced a new automatic rifle for its infantry as well as self-propelled artillery and improved antitank weapons. The sheer size of China's army slowed tactical and technical change, and the PLA was consistently one or more decades behind other major forces. Nevertheless, the combination of infiltration tactics, Soviet-style organization, and growing modernization of equipment made the PLA into a formidable force by the late 1990s.

The United States Army in the 1970s: Active Defense

Meanwhile, the events of 1973 affected armies far beyond Israel. In the United States, Gen. William DePuy used these events as a lever to impose his own view of the future of the U.S. Army, a doctrine known as the Active Defense.[14]

As the newly appointed commander of the army's Training and Doctrine Command (TRADOC), DePuy faced a number of critical problems. The U.S. Army had just withdrawn from Vietnam and was suffering from drug abuse, racial tension, demoralization, and the sudden shift from conscription to a volunteer force. Moreover, Depuy and other senior leaders considered the Vietnam experience of airmobility and light infantry tactics to be largely inappropriate for any future conflict in Europe. There, the heavily mechanized Soviet Army appeared poised to attack NATO. The West German government's desire to protect every kilometer of its shallow territory obligated the U.S. Army to defend along the interzonal border established in 1945, with little opportunity for delaying or maneuver tactics. At least during the first battles of such a war, the U.S. Army would have to fight outnumbered and outgunned by a superior Soviet foe.

General Depuy set out to reformulate American fighting doctrine, imposing his own view of these problems and of those posed by the 1973 war. The result was a series of "How to Fight" manuals, lavishly illustrated revisions of the army's existing field manuals (FMs) that represented the first major reorientation of doctrine since World War II. The capstone of this effort was FM 100-5, *Operations,* issued in 1976. This manual correctly identified many of the implications of the 1973 war—the increased lethality of the battlefield; the dominant role of armor; the continued value of mechanized infantry, both to accompany tanks and to operate antitank guided missiles; the growing accuracy of field artillery and air defense weapons; the mobility and armor-killing capability of helicopters; the increase in night combat made possible by night-vision devices; and the growing importance of electronic warfare for locating targets and jamming communications.[15] DePuy and his air force counterpart, Gen. Robert J. Dixon of the Tactical

Air Command, also made considerable progress in standardizing air support procedures, based on a shared view that the ground and air wars were interdependent parts of the same struggle.

Unfortunately, many soldiers viewed DePuy's solution to these issues as extremely grim and even mechanical. The How-to-Fight manuals described offensive operations as well as maneuver when on the defensive, but they seemed to emphasize a desperate struggle against overwhelming odds, much like that fought by Israeli tankers on the Golan Heights at the start of the 1973 war. Using the ROAD divisional structure, cross-attached task forces and teams of infantry and armor would fight from hull-down positions. Minefields would slow down the enemy advance, after which antitank guided missiles at long range and faster firing tank guns at short range would whittle down a huge attacking force. The defenders would rely on firepower more than on maneuver or initiative.

Commanders and staff officers learned a cold-hearted calculus of battle; army schools taught them to estimate the maximum possible Soviet force that could advance in a particular sector and then to allocate numerically inferior forces to slow down and halt that threat. When armored cavalry and other units acted as a covering force to delay the enemy advance, these units were expected to fight while outnumbered by a ratio of six enemy battalions to one American; in the main battle area, where the units had to stand and fight, the ratio of attacker to defender was to be improved to the still daunting disadvantage of three to one.[16]

DePuy's efforts were a valuable dose of reality for an army that had not fought under such disadvantages since 1950. Moreover, his arguments helped fund a number of developments, including infantry combat vehicles and attack helicopters, that came to fruition in the 1980s. However, General DePuy's procedures and solutions were too simplistic, more appropriate for teaching recruits in basic training than for equipping officers to make their own decisions based on a commonly held set of doctrinal principles.[17] As a result, the Active Defense doctrine barely survived into the 1980s.

The Soviet Army in the 1970s: Deep Operations

Preoccupied with developing an offensive mechanized army, Soviet theorists were caught off guard by the defensive success of their own ATGMs and other weapons in Egyptian hands.[18] Unlike Western tanks, Soviet combat vehicles were designed with such low silhouettes that it was difficult for them to take up a hull-down, defensive position and still depress their gun tubes to engage the enemy. Thus, imitating the defensive tactics of the Israeli and U.S. Armies was impossible for the Soviets, even if they had been inclined to reorient their doctrine.

Yet the Middle East conflict did accelerate certain preexisting trends within the Soviet Army. Even before that war, the Soviets had begun fielding their first true infantry fighting vehicle, the *Boevaya Mashina Piekhota* (BMP), or infantry combat vehicle. It included a number of firing ports, so that the eight-man infantry squad could fire their weapons from its sides and rear. More important, the BMP was designed to provide an enclosed environment, reducing the risk of radiation exposure in a nuclear war. It also carried a small turret with a short-range cannon and an ATGM launcher. It was true that the BMP was vulnerable if hit by a high-velocity antitank shell, and the commander's vision was obstructed by placing his hatch beside the gun turret. Moreover, the vehicle was too slow to keep pace with the next generation of Soviet tanks, the T-64/72. The BMP was never intended to be the sole infantry vehicle in the Soviet inventory; just as the West Germans mixed *Marders* with conventional personnel carriers, the Soviets deliberately structured high priority divisions with one regiment of BMPs and two regiments of wheeled armored personnel carriers, such as the BTR-70 introduced in 1978.[19] Nevertheless, Soviet leaders regarded the BMP as the essential tool for mounted infantry combat in high intensity conflict, and many Western observers were equally impressed.

The 1973 war encouraged other changes in the Soviet arsenal. In place of the traditional towed field guns and howitzers of the artillery, Soviet units began to acquire self-propelled guns that could keep pace with the mechanized advance. These guns could provide both indirect, long-range and direct, short-range fire to weaken antitank defenses.

Soviet analysis of the 1973 conflict and of the U.S. experience in Vietnam also expanded the role of the helicopter, moving it from simple transport to fire support. This change was symbolized by the transfer, during the mid-1970s, of most helicopters from the air transport command to the frontal aviation units supporting ground combat forces. By coincidence, in 1973 the Soviet Union had fielded the first (Hind A) version of the Mi-24 helicopter, a heavily armored aircraft that could carry eight infantrymen and significant guns and missiles. Based on the 1973 war, which had demonstrated both the power of ATGMs and the need to provide responsive support for tank units, the Mi-24 was redesigned into the Hind D, intended solely as an attack helicopter.[20]

These changes in equipment reflected a major renewal of the traditional Soviet emphasis on deep operations by large mechanized formations. In 1976 Marshal N. V. Ogarkov, Chief of the Soviet General Staff, wrote "The Deep Operation (or Battle)" for the *Soviet Military Encyclopedia.*[21] Two years later, the military theorist Gen. Makhmut A. Gareev submitted a major report on this issue to Ogarkov, which sparked both theoretical debate and practical changes.

Because all Soviet units were mechanized, the label "mobile group" no longer carried its World War II meaning of an exploitation force. In the later 1970s, Ogarkov and his assistants referred to the Operational Maneuver Group (OMG) to describe the specialized organization that would conduct deep exploitation operations. In the tradition of the tank armies and other mobile groups of World War II, the OMG would penetrate enemy rear areas to disrupt and preempt the enemy's defenses. This type of maneuver had the added advantage that it would reduce Soviet vulnerability to enemy nuclear attack. The rapid movement of the OMGs and the fact that Soviet and enemy forces would become intermingled were intended to make it impossible for the enemy to use tactical nuclear weapons, for fear of causing friendly casualties.[22]

The standard Soviet tank and motorized rifle divisions were formidable forces and became more so during the 1970s with the addition of these new weapons. However, these divisions were multipurpose forces with a traditional, rather inflexible style of tactics, command, and control. Ogarkov and other advocates of the OMG wanted to develop specialized, functional units to perform the deep operations role. Their first effort, which took shape in the later 1970s, was the addition of separate tank regiments, similar in structure to the tank brigades of 1944–1945, at the level of a combined arms or tank army. These organizations, which Western observers often misinterpreted as a new type of reserve force, were actually intended to be the nuclei for operational maneuver groups at field army level, just as the earlier separate tank battalions could become tactical maneuver groups at division level. Eventually, the Soviet Army went even further to create permanent, combined arms forces for the operational maneuver role.

Both attack helicopters and air assault infantry played major roles in such deep operations; indeed, air assault provided the vertical dimension to the traditional idea of these operations.[23] These units were intended to facilitate the long-range exploitation of mechanized units behind enemy lines; the helicopters would provide more responsive fire support than that available from fixed-wing fighter-bombers, and air assault units would seize river crossings, defiles, and other areas that might otherwise slow the advance of the ground force.

The Soviets had an opportunity to test some of these tactics in Ethiopia. In 1978 Gen. Vasili Petrov directed a major Cuban-Ethiopian-Soviet offensive to recapture the Ogadan region from Somalia. Soviet helicopters airlifted Cuban paratroops, complete with armored vehicles, to secure the flanks of an advancing mechanized force. Naturally enough, Soviet theorists viewed the resulting victory as a vindication of their concepts.[24]

To fulfill these new roles, however, the helicopter had to be integrated into army units instead of operating as a separate air force organization. During the early 1980s, the Soviets formed a number of air assault brigades. Their exact structure varied, but typically they included up to four battalions of airborne rifle troops, of which one or more might be equipped with the BMD, the air transportable counterpart of the BMP. By 1985 high priority Soviet tank and motorized rifle divisions included their own helicopter squadrons (consisting of Mi-24D attack helicopters and various transport and observation aircraft), with tank armies receiving attack helicopter regiments.[25]

Afghanistan

The Soviet Army needed these sophisticated weapons when it became involved in its first sustained combat since World War II, the 1979–1989 counterinsurgency war in Afghanistan. After a year in which Soviet advisers and helicopters assisted the disintegrating pro-Moscow government in Kabul, the Soviet Union seized control in December 1979. This coup was similar to the 1968 Warsaw Pact intervention in Czechoslovakia; airborne and special operations troops took over the capital while a large mechanized force advanced by land, occupying the highways of eastern Afghanistan that were the backbone of Soviet efforts to control the country. During the initial invasion and a subsequent counterinsurgency campaign in 1980, helicopter-borne units seized mountain passes and landed on high ground to secure the flanks of the mechanized advance.[26]

Yet the Soviet Army rapidly relearned the basics of counterinsurgency, especially the futility of using large conventional forces in an effort to bring the guerrillas to battle. The poorly trained reservists and conscripts in Soviet motorized rifle units were reluctant to leave either their vehicles or the highways, becoming passive targets for ambush. One motorized rifle company reportedly found itself isolated east of the capital in March 1980; after it had expended its large supply of ammunition by firing nervously at shadows, the company was annihilated by a night attack. Moreover, crew-served weapons, such as the standard Soviet machine gun and the AGS-17 automatic grenade launcher, were simply too heavy for dismounted troops to transport across mountainous country.[27]

Gradually, the Soviet commanders adjusted to the unusual circumstances of Afghanistan. As a diplomatic gesture of reducing their involvement, the Soviets evacuated many units that were useless in counterinsurgency—heavy artillery, air defense missiles, tank regiments, and the like. Additional motorized rifle, air assault, and special operations troops took the places of these heavier elements. When the infantry dismounted to conduct foot actions, their armored vehicles, manned only by drivers and gunners and accompanied by a few tanks, would maneuver as an independent firing element, known as a *Bronegruppa* (armored group). Beginning in 1982 and increasingly from 1985, the Soviet Army relied on elite air assault and special operations troops, operating in units of a battalion or smaller, to maintain the initiative against the Afghan resistance.[28] It also experimented with various airmobile and armored brigades. *Spetsnaz* special operations teams were highly successful at collecting intelligence and interdicting Afghani supply lines. Meanwhile, Soviet military publications and schools launched a renewed effort to encourage initiative and independence among junior leaders, especially in the elite units. By contrast, the bulk of conventional motorized rifle units were committed to securing the tenuous supply routes into the country, and the Soviets continued to substitute firepower for infantry aggressiveness.

The Afghan conflict became a series of search and destroy missions, frustrating operations that sometimes resembled those launched by the United States in Vietnam. One such operation was the Soviet sweep of Xadighar Canyon in March 1986 (Map 14).[29] Based on intelligence collected by *Spetsnaz* patrols, two Soviet motorized rifle battalions, supported by a 122mm towed artillery battalion, launched a surprise attack against a *mujahideen* force in the Xadighar Canyon region. The troops approached the valley at night, avoiding major roads because of the possibility of mines and enemy observation. In an attempt to seal off the battle area, Soviet commanders used groups of Mi-8 and Mi-24 helicopters as well as SU-17 and SU-25 fighter-bombers. Four sixteen-man teams, drawn from four separate *Spetsnaz* companies, conducted helicopter assaults onto the high ground overlooking the canyon. These teams eventually linked up with the armored vehicles of their parent battalion, vehicles that acted as a *Bronegruppa* to help the conventional motorized forces. Despite these precautions, most of the rebel forces eluded the Soviet cordon. Only platoon-sized elements of *mujahideen* remained behind to defend the two villages at the southern end of the canyon; the Soviets could claim only twenty enemy killed in the operation.

Like their American counterparts in Vietnam, the Soviet troops relied on the advantages of firepower from fighter-bombers and especially from helicopters. The SU-25 Frogfoot provided the first true Soviet close support aircraft since the 1940s, and the number of helicopters in Afghanistan grew tenfold in 1980 alone.[30] Gradually, however, the Afghan rebels found the means to reduce if not eliminate the air threat. The *mujahideen* opposition troops captured antiaircraft machine guns and SA-7 shoulder-fired air defense missiles. Then, beginning in 1986, the opposition acquired U.S. Stinger and British Blowpipe missiles, which had larger warheads and more effec-

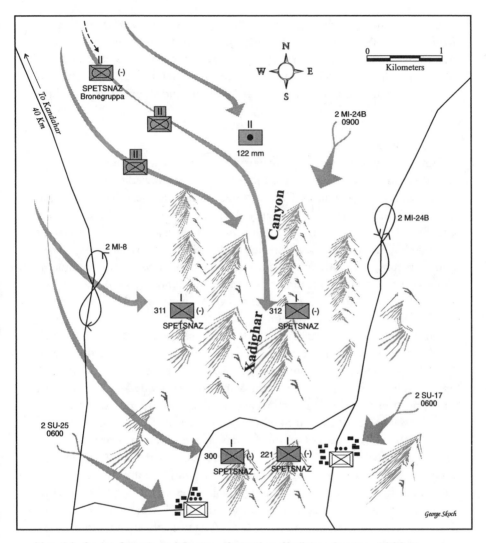

Map 14. Soviet Search and Destroy Operation, Xadighar Canyon, 1986

tive tracking systems than the SA-7. Soviet and satellite Afghani forces lost an estimated 150 to 200 aircraft to these missiles in 1987 alone, prompting one devout rebel leader to remark, "There are only two things Afghans must have—the *Koran* and Stingers."[31]

The missile threat did not eliminate Soviet air support, of course. Both helicopters and fixed-wing aircraft adopted a tactic of having one aircraft orbit at high altitude while another conducted a bombing run, allowing the first aircraft to identify and attack any missile launch team. Yet all Soviet aircraft had to fly higher, greatly reducing the accuracy of their attacks. Moreover, a pilot who was constantly looking for the telltale smoke plume of a Stinger was simply not as willing to press home the attack in support of his ground troops. Thus a few hundred air defense missiles stymied Soviet pilots almost as effectively as the Saggers had slowed Israeli tankers in 1973.

Precision Weapons

Both these missiles were part of a new generation of so-called smart weapons, weapons with improved guidance systems that allowed them to hit their targets with great regularity. For centuries, tactics had been based on firing a huge volume of shells at the enemy, even though only a small percentage ever found their targets. In the late twentieth century, however, the advent of guided bombs, cruise missiles, and individually fired missiles changed this equation fundamentally, making a few, relatively expensive, weapons far more effective than thousands of dumb bombs and shells. Although the United States led in the development of smart weapons, the other powers often produced their own versions.

Strategic air defense missiles, like the Nike series, and short-range anti-tank missiles such as the French SS-11 had been available since the 1950s. Electronic miniaturization soon allowed for much more sophisticated guidance systems, however. The Vietnam War actually provided the impetus and the opportunity to develop such weapons. The U.S. Air Force, which had traditionally used saturation bombing of an area, needed a more accurate weapon to destroy bridges and other targets while minimizing damage to nearby civilians. In 1965 air force designers learned of U.S. Army experiments in using lasers to designate targets. The air force approved two rival design efforts; each connected a laser guidance device to a set of control fins to direct existing dumb bombs onto their targets.[32] Although the air force field-tested the laser bomb over Southeast Asia in 1968, such weapons were not used widely until the waning days of the Vietnamese conflict. Television and eventually infrared guidance systems followed the laser systems; the latter permitted targeting even when the target was obscured by darkness, camouflage, or clouds. By 1985 guided bombs had become common in the USAF inventory, with kill probabilities against a Soviet T-55 tank ranging from 26 percent to 87 percent, depending on the system. Against stationary or unarmored targets, the probability was even higher.[33]

During the 1970s, small, lightweight inertial navigation packages allowed engineers to achieve the long-cherished goal of unmanned cruise missiles that could fly close to the earth and hit fixed targets with a high degree of accuracy. The miniaturization of computers permitted even more sophisticated devices, such as digitized terrain comparison and cheap, plentiful global positioning systems (GPS). An inertially guided missile could update its position by frequent input from global positioning satellites. American designers originally intended the resulting cruise missiles to deliver nuclear warheads; during the 1980s, however, similar systems using conventional warheads became common. Although the United States led the world in such technology, by the late 1980s other nations, notably France, began work on similar weapons. A decade later, the United States tried to extend the global positioning concept from cruise missiles to guided bombs and even to helicopter-launched missiles.[34]

As guided bombs and cruise missiles entered the vocabulary of world leaders, equally significant developments occurred in smart weapons for tactical ground combat. The Tactical, Optically Guided Weapon antitank system was also tested in the closing years of the Vietnam conflict. The TOW's original range of three kilometers was later boosted to close to four, and the portion of the spectrum used by its guidance system was changed to make it less vulnerable to weather and electronic interference. This ATGM and its smaller, shorter range (one kilometer) counterpart, the Dragon, were a generational improvement over earlier such weapons. For example, the Soviet Sagger that had thwarted the Israelis in 1973 had to be "flown" by an operator, who moved a control stick to guide the missile to its target. At ranges of less than one kilometer, the operator often lacked the time to get the missile under control, causing it to miss. Even at longer ranges, the Sagger would work only in the hands of a carefully trained operator.[35] By contrast, the TOW, Dragon, and similar second-generation missiles required only that the operator keep his sight locked onto the target, while the missile flew to the location indicated by the sight's crosshairs. Of course, enemy artillery fire could still wound or distract the operator, but the TOW's probability of killing the target was considerably greater than that for a Sagger.

In 1966 the United States had fielded a similar, if less reliable, ATGM, the Shillelagh, for use in a combination gun–missile system on its light cavalry tank, the XM551 Sheridan. The same gun tube could launch either a medium-velocity cannon shell or a Shillelagh missile. Beyond its minimum range of one kilometer, the Shillelagh was fairly effective, although it required considerable gunner training and frequent calibration of the sighting system.[36] In the mid-1970s American designers mounted this same system in the M60A2 variant of the main battle tank. However, the Shillelagh, like all second-generation ATGMs, suffered from the fact that the gunner had to keep the target aligned throughout a long flight time; during this period, the gunner had to remain stationary and could not engage another target. This was one consideration behind General DePuy's conception of mixing accurate but slow-firing ATGMs with more rapidly firing conventional tanks. The M60A2 proved to be a blind alley for development.

The demand for a more flexible ATGM was partially met in the 1980s by the Hellfire missile, designed for use in attack helicopters. The Hellfire had a six kilometer range. However, at least in the early 1990s, the gunner still had to remain locked on the target while the pilot either hovered or moved slowly forward. Further development was necessary to allow the Hellfire to become a fire-and-forget weapon, guided by another helicopter without constant guidance by the original gunner.[37] Another option was the Copperhead missile, fired from a conventional artillery piece and then guided onto its target by a forward observer with a laser designator.

These "smart" weapons contributed to the lethality of the battlefield in general and of air attack in particular. Given their expense and complexity of use, however, they were unable to eliminate completely the need for carefully trained combined arms forces on the ground.

Moreover, most ATGMs depended on a shaped-charge, chemical energy warhead. During the 1980s the American, British, Soviet, and other armies developed various forms of improved armor intended to attenuate the effect of a shaped charge. In addition to ceramic or composite armor, some tanks were fitted with reactive armor. Small charges, mounted outside the main armor plate, were designed to detonate when an ATGM struck the vehicle, exploding outward to neutralize the chemical blast stream of the shaped charge. Of course, no army could afford to equip all its vehicles with such advanced armor plating, and weight limitations meant that even the strongest tank had thinner armor on its sides and rear. Still, the existence of such improved armor placed some limitations on the effectiveness of ATGMs.

Attack Aviation

The advent of improved ATGMs, coupled with the perceived threat of massed Soviet armor, prompted Western armies to elevate the helicopter from a means of moving combat troops to a combat arm in its own right. The change was by no means easy, in part because ground commanders understandably feared that army aviation would go the way of fixed-wing air units, evolving into a separate force that would no longer support the ground units. This fear grew during the 1980s, when army aviators theorized about the possibility of helicopter air-to-air combat.

The evolution of attack helicopters was particularly difficult in the United States, where aviation advocates had to deal with opposition from USAF commanders, who regarded helicopters as a dissipation of airpower, and from ground commanders, who wanted to use those helicopters as a flexible tool to reinforce critical efforts and sectors. Moreover, the emphasis on attack helicopters encouraged commanders to neglect the air assault concept, which seemed to be a discredited aspect of the Vietnam War. Further, the scarcity of funds during the 1970s meant that the United States did not field a true attack helicopter, the AH-64 Apache, until 1986; before that, aviation had to make do with various low-performing substitutes.[38]

Nevertheless, senior leaders regarded the attack helicopter as essential for the defense of Europe. The 1973 war had demonstrated the effectiveness of antitank guided missiles, and helicopters were the logical means of using them. Not only could a helicopter move ATGMs rapidly from place to place, but it could also gain sufficient altitude to maximize the missiles' range, shooting over hills in the rolling terrain of Central Europe. Then-major general Donn Starry was a primary advocate of this concept; as the commander of the Armor Center and School during the mid-1970s, Starry was a proponent for air cavalry and attack helicopter doctrine. Unfortunately, field commanders tended to hold their attack helicopters in reserve until too late, in part because they worried about their survivability if exposed to the full effects of a Soviet air defense umbrella. Thus, by the time attack helicopters found themselves committed to halt a penetration, the enemy might already be unstoppable.

One of the forgotten recommendations of the 1962 Howze Board had been the creation of separate air cavalry brigades. As part of the post-Vietnam experimentation, the army finally formed the 6th Air Cavalry Combat Brigade (ACCB) at Ft. Hood, Texas, in 1975. For a decade this was the only such brigade, but it nevertheless served as a test bed for concepts and tactics. Meanwhile, in 1976 the army decided to provide each armored or mechanized infantry division with two attack and one transportation helicopter companies; the ratio was reversed in infantry divisions. Again, however, these units could be parceled out as support elements under the control of a brigade or even a battalion task force commander.

In a sense, the army of the 1970s and 1980s repeated the debates of the 1930s and 1940s, when ground commanders wanted to control fixed-wing air support and aviators deplored the resulting waste and dissipation of combat power. In 1979 aviation officers became a separate combat branch of the U.S. Army; previously, these officers had been commissioned in other branches, such as armor and transportation, and were required to maintain their proficiency in their basic branches while still qualifying as pilots. The new branch was able to concentrate on its basic mission in combat. However, many ground combat officers argued that this change would separate the aviators from their ground counterparts; by contrast, the U.S. Marine Corps continued to insist that its pilots have some experience in ground units.

During the 1980s a major review of army organization finally led to the creation of separate aviation brigades, under aviation commanders, within each division and corps. At the same time, many helicopter companies were elevated to the status of small battalions. In particular, the corps combat aviation brigade, patterned on the 6th ACCB, included up to six battalions of eighteen AH-64s each. Such a force finally gave rotary-wing aviators the tools to conduct the type of deep operations that the revised doctrine of that decade required. However, some doctrinal confusion remained because these aviation brigades could not, of themselves, occupy ground as could ground maneuver brigades.[39] The staffs of the former, particularly in divisions, had difficulty learning to control both their own aviation assets and, on occasion, ground task forces.

Much the same evolution occurred in the British Army. The Army Air Corps became an official combat arm in 1973. A decade later, each division was authorized a regiment (battalion) consisting of two squadrons of attack helicopters and one of reconnaissance. Both the Royal Air Force and the Royal Armoured Corps objected to the use of helicopters in a dedicated antitank role, but the perceived threat of Soviet armor in Central Europe was so great that the army aviators prevailed. When the British armed forces underwent major personnel cuts in 1991, the Army Air Corps emerged relatively unscathed.[40]

Despite budgetary restraints, the French Army also emphasized a semi-independent role for helicopters as a combat arm. This trend culminated in 1985 with the creation of the 4th Airmobile Division *(4e division aeromobile)*, a collection of 240 attack and transport helicopters intended for long-range operations both inside and outside Europe. By contrast, concern over Soviet armor caused the West German Army to decentralize its aviation assets, attaching antitank helicopters to ground brigades. In 1988 the Germans followed the Soviet lead by creating air-mechanized brigades, combinations of air assault infantry and antitank weapons that could function as the reserve force for a corps of several divisions.[41]

AirLand Battle: Rebirth of the U.S. Army in the 1980s

The focus on helicopters and long distance operations was in part a reflection of a larger trend throughout NATO, the trend to place the tactical, frontline battle into the larger context of the operational level of war. To some extent, this reemphasis on the operational level was a response to Soviet doctrine. More generally, however, professional soldiers in several nations began to rethink the scope and complexity of mechanized warfare. Nowhere was this reconsideration more profound than in the United States.

Beginning in the late 1970s, many senior U.S. officers began to think beyond the calculus of DePuy's Active Defense doctrine.[42] The pivotal figure was again the commander of the Training and Doctrine Command, Donn

A. Starry, by then a general. Although he had a healthy respect for technology and the sheer numbers of the Soviet adversary, Starry had always emphasized military history and psychology as the means to understand battle. After supporting DePuy in the development of Active Defense at the tactical level, Starry assumed command of a corps in Germany, a position that allowed him to think in broader, operational-level terms. It was obvious, for example, that even if Active Defense tactics defeated the first and second echelons of a Soviet attack, NATO forces would lose the war if they could not slow down and weaken other Soviet units following behind. By the time he moved to TRADOC in 1977, Starry had concluded that the United States had to integrate all its forces—chemical and nuclear weapons, airpower, and other means of attacking an opponent both on the battlefield and throughout the depth of its territories. This integration, which eventually became known as AirLand Battle, required that commanders have a larger, more coherent concept of the campaign, one driven by the traditional German doctrines of the operational level and subordinate initiative in accomplishing the commander's intent.

AirLand Battle doctrine asked commanders to prepare for not one but three simultaneous actions—the Main or Close Battle on the traditional battlefield, the Deep Battle to slow and weaken approaching enemy reserves, and the Rear Battle to counter enemy special operations, airmobile, or airborne assaults. Because it was politically impossible to give up ground in West Germany, Starry and others emphasized the Deep Battle, using fire and maneuver into the enemy rear as a means of gaining the depth necessary for a more fluid form of combat.

U.S. and NATO theorists originally intended these concepts to defeat the Soviet habit of arraying its forces in different layers or echelons, destroying them while they were still in the Soviet rear area, before they entered the battlefield. Ironically, however, AirLand Battle proved equally effective against the developing Soviet concept of the Operational Maneuver Group. These developments in Western doctrine prompted the Soviets to emphasize single-echelon attacks and early commitment of OMGs, in order to eliminate the intended targets of Deep Battle.

As described in the 1982 and 1986 manuals, AirLand Battle was in part a natural reaction against the cold calculations of the Active Defense. It reemphasized the importance of leadership and maneuver and specified that commanders of larger units must concern themselves with higher levels of the opponent's army and larger, deeper areas of the battlefield. Starry's subordinates, notably L. Donald Holder and Huba Wass de Czege at the Command and General Staff College, pushed for a thorough reexamination of warfare. For example, they encouraged the U.S. Army for the first time to differentiate between tactical and operational levels of war. This change in focus arose partly in response to OMG research conducted by John Erickson and Chris Donnelly in Great Britain, assisted by a variety of Soviet analysts in the United States.[43] AirLand Battle also involved numerous changes in technology, organization, and interservice cooperation.

After years and sometimes decades of development, a number of new weapons came to fruition in the 1980s, allowing U.S. commanders to fulfill General Starry's vision. In addition to the AH-64 attack helicopter, commanders had a variety of new missile and conventional artillery weapons to reach deeper into their opponent's territory.[44] The Multiple Launch Rocket System (MLRS) was a much more accurate and sophisticated revision of 1940s–era unguided rocket launchers. Field artillery acquired a host of new weapons to delay and "canalize" advancing Soviet forces. Both the MLRS and conventional howitzers could shoot improved conventional munitions (ICM)—hundreds of little bomblets that exploded among personnel or penetrated the thinly armored tops of enemy vehicles. The artillery could even fire scatterable mines, in essence planting a minefield around an enemy unit even as it moved down a road. Specially equipped fire support (FIST) teams consolidated both mortar and artillery forward observers to accompany company and battalion commanders, and a modified scout helicopter, the OH-58D, could use lasers for accurate ranging and designation of targets. A new generation of counterbattery radars allowed U.S. artillerymen to determine the source of enemy fire almost instantly and to fire back at it before the first enemy shells landed. The entire artillery system was connected by the automated Tactical Fire Direction System (TACFIRE). Although TACFIRE used twenty-year-old technology when it was fielded in the 1980s, it processed firing data at a speed of more than eighteen times that of its predecessor.

Global positioning systems allowed forward observers and firing units to determine their locations with unprecedented accuracy, which made the resulting artillery fires extremely precise and lethal. The same precision in determining location meant that the guns in a particular artillery battery no longer needed to park side by side, with all guns oriented on one surveyed location. Instead, beginning in 1985 most U.S. artillery battalions abandoned the traditional six-gun battery, converting to batteries of eight guns that could operate in two four-gun platoons. This change in organization and positioning reduced the artillery's vulnerability to enemy attack while increasing the firepower of an artillery battalion by one third.

Artillery was not the only area to experience massive change. In 1979 the M1 Abrams, the first completely new U.S. tank design in three decades, began production. The Abrams carried a form of composite armor that thwarted many shaped-charge, chemical energy rounds, thereby restoring the tank–antitank balance that had been upset by the development of ATGMs. The M1 also abandoned the traditional diesel engine in favor of a gas turbine power plant, solving the problem of speed for increasingly heavy vehicles. Computer-stabilized turrets and advanced suspension systems allowed the M-1 to fire its gun accurately while moving at speeds of up to forty-one miles per hour; the crew in earlier tanks were bruised and unable to hit targets at far lower speeds. Of course, any design has its drawbacks— the M1 burned fuel at an alarming rate, and infantry could no longer shelter behind the tank because of the turbine exhaust. Still, the Abrams gave American tank units far greater mobility, accuracy, and protection than they had ever had before.[45]

Nineteen seventy-nine also saw the long-delayed appearance of the Bradley Infantry Fighting Vehicle (M2) and Cavalry Fighting Vehicle (M3), the first true IFVs in the American inventory. They were almost as tall as a World War II Sherman tank, giving them a silhouette that was too prominent, especially for use by cavalry reconnaissance elements. The 1988 improved version (M2A2) of the Bradley increased its weight and armor protection but eliminated the side firing ports and periscopes that had been the hallmark of a vehicle for mounted infantry combat. Moreover, U.S. organization and doctrine had to wrestle with the problem of how to command and employ the firepower of the Bradley while its infantry was operating dismounted. Nevertheless, the M2 and M3 were more nearly compatible with the Abrams than were the few remaining M113-series vehicles, carrying the FIST teams and improved antitank missiles, which often lagged behind fast-moving task forces.[46]

Such a mobile, complex force required more effective tactical intelligence and electronic warfare. For decades, each U.S. division had been accompanied by two separate intelligence units—an intelligence company that provided analysis, prisoner interrogation, counterintelligence, and other mundane functions, and an Army Security Agency (ASA) company, whose signals intelligence functions were often so classified that tactical commanders had little access to the resulting information. During the 1970s, however, U.S. military intelligence leaders reoriented their structure and functions to make such support more readily available to commanders. The two companies merged and expanded into a Combat Electronic Warfare and Intelligence (CEWI) battalion for each division, a one-stop source of all forms of intelligence collection, including a suite of new electronic intercept, direction-finding, and jamming systems. Unfortunately, such complex systems were extremely difficult to maintain in a field environment, especially when they were scattered in small teams across the entire front of a division. CEWI battalions had some notable successes in the

1970s and 1980s, but it took years of training for the reality of intelligence support to catch up with the promise.

Having finally adjusted to the end of conscription, the U.S. Army of the 1980s was well prepared to absorb these new weapons systems. In the mid-1970s, the army had abandoned the fixed training cycle that had dominated peacetime unit activities since World War II. Instead, commanders had to identify their training deficiencies and train to overcome them. For units based in the United States, the ultimate test of training was a deployment to the National Training Center in California, a semidesert environment where a host of training devices and computers allowed units to learn from their mistakes without paying in blood.

Nevertheless, the new weapons and the new standards brought a long-standing training issue to a head. Since Vietnam if not before, company commanders and platoon leaders had been inundated with a host of different assets and communications systems, requiring a single junior officer, without any staff assistance, to make literally dozens of decisions in the first few seconds after contacting the enemy. The FIST team relieved the company commander of some of the details of fire support, but in essence leaders as well as followers were hard-pressed to integrate all the arms and services at once. Moreover, the crews of the new IFVs had to conduct vehicle maintenance as well as frequent training and testing in ATGM and automatic cannon gunnery, which inevitably complicated infantry training while separating the crews psychologically from the dismounted elements of each platoon.

Paradoxically, the most highly trained peacetime U.S. Army of the twentieth century decided that its junior leaders could not handle this complexity. The response to this issue, as reflected in the next generation of organizational designs, was to simplify small units and abandon combined arms at the lowest level, reducing most company-sized formations to a single type of weapon system. Tank platoons were reduced from five to four vehicles, and M2-equipped mechanized infantry companies lost their organic mortar and TOW antitank sections. Henceforth, tank and infantry companies would function as pure forces rather than as composite teams. In each mechanized battalion, a fifth or antiarmor company consolidated twelve improved TOW vehicles, and a tank or mechanized battalion had only a single platoon of 4.2-inch self-propelled mortars.[47] The same wave of simplification eliminated tanks from some divisional armored cavalry squadrons, leaving those units without the muscle to conduct the full spectrum of combat.

Despite such significant changes at the small-unit level, the newly designed Division 86 bore considerable resemblance to its predecessor, the modified

ROAD structure. Each division had a division base, three ground and one air combat brigade headquarters, a division artillery element, and a division support command. By the time the force designers had included the aviation brigade, the MLRS, the eight-gun artillery batteries, and a host of other new systems, the resulting division came perilously close to 20,000 soldiers, a figure that prompted much revision and often erratic personnel cuts totaling more than 3,000 spaces (Fig. 20).[48] The new heavy division, whether it was designated as armored, mechanized infantry, or cavalry, was formidable not only in its combat capability but also in the demands it made for strategic transportation and for daily resupply. It was optimized to counter the heavily armored Soviet threat but proved equally effective against Iraqi armor.

General Starry's search for methods to conduct the Deep Battle against the assumed Soviet adversary also prompted the U.S. Army and Air Force to make a major effort to resolve the recurring issue of air–ground cooperation. Indeed, as the term AirLand Battle implied, Starry regarded the two forces as completely wedded on the battlefield. The two services developed a series of agreements that became known as the 31 Initiatives.[49]

These initiatives increased cooperation on everything from air base defense to joint airlift, but not all of them were equally successful. Perhaps the most controversial issue was that of air support on the battlefield. Most army officers readily admitted that the first mission of the air force must be to achieve air superiority over the battlefield; the issue was the allocation of remaining air assets after planning for that. During the 1970s the steady growth of attack helicopters and of the Soviet armored threat had prompted the USAF to field its first custom-designed ground support aircraft, the A-10 Thunderbolt II, known colloquially as the Warthog. Although it was designed in part to destroy Soviet self-propelled air defense weapons, its 30mm automatic cannon had considerable effect against all thin-skinned armored vehicles.[50] In May 1981 the air force went further, reaching agreement with the army on a concept for Offensive Air Support. This agreement recognized that, in the European or NATO context, ground headquarters must be involved in planning not only close air support but also tactical air reconnaissance and Battlefield Air Interdiction (BAI). The latter included air attacks on deep targets that were capable of directly affecting friendly ground forces; therefore, ground commanders had a vested interest in influencing the choice and timing of air force BAI missions. Depending on his plan for Deep Battle and maneuver, the ground commander might want BAI to ignore one set of targets and to concentrate on another. The agreement explicitly recognized the possibility that air force firepower could achieve decisive results even where ground-based firepower was already employed, the very issue that had divided the two services since the 1930s. Despite the

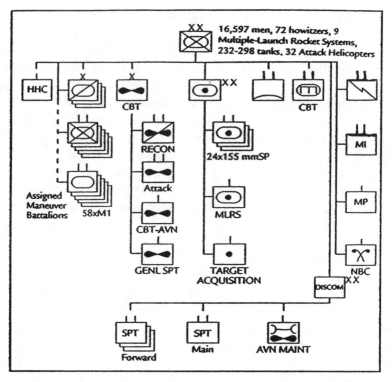

Figure 20. Type U.S. Mechanized Infantry (Heavy) Division, c. 1987

best intentions of both sides, however, the BAI issue was not resolved. Air force officers understandably resisted army efforts to define its areas of interest to ranges so deep that they restricted air force interdiction attacks. Moreover, the extremely centralized process of air force planning continued to conflict with the decentralized nature of ground operations.[51]

Light Infantry and Special Operations Forces

While much of the U.S. Army focused on the threat of a high-intensity, mechanized war in Europe or the Middle East, Western armed forces also had to confront a broad spectrum of other contingency missions throughout the world. In the U.S. Army, such missions led to renewed interest in traditional light infantry formations and in special operations forces.

The enormous amounts of equipment in mechanized and air assault units required considerable time and scarce transportation resources to deploy to areas such as Latin America, the Middle East, or Asia. Even the marine corps would be hard-pressed to conduct large-scale operations in these parts of the world, because of shortages of specialized amphibious shipping. Recognizing this problem, during the early 1980s the U.S. Army sought to create a different type of unit. In 1979 Gen. E. C. Meyer and General Starry had agreed to test a new, medium-weight unit, a form of motorized infantry division that would use new technology to perform infantry contingency missions while still being able to fight mechanized forces in Europe. Ultimately, a number of factors, in particular lack of funds to give this motorized division a light armored gun, led to its failure. When Gen. John A. Wickham Jr. became Chief of Staff of the Army in 1983, he launched a crash project to create a new light infantry division (LID). Like many officers with a light or airborne infantry background, Wickham believed that superior training and leadership could compensate for the absence of some heavy weapons. The resulting LID, numbering approximately 10,000 soldiers, was severely limited in vehicles and heavy weapons (Fig. 21). In comparison to a heavy mechanized division, the LID had 44 TOW antitank launchers instead of 162; 29 attack helicopters instead of 48; and enough trucks to move only three of the nine infantry battalions at one time.[52] Light-weight, towed 105mm guns replaced the self-propelled, heavier artillery of conventional divisions. The combat engineer battalion was less than one-third the size of its counterpart in a heavy division, with very little heavy equipment. In theory, the entire light infantry division could deploy anywhere in the world in 520 sorties of C-141B transport aircraft; in practice, more sorties were required because the aircraft would have only limited fuel capacity if they took off fully loaded.

In a throwback to General McNair's concept of streamlining and pooling, planners expected to add a number of "corps plugs" to enable the LID to fight in a highly mechanized environment such as Europe. These plugs included a variety of units, such as an attack helicopter battalion, a transport helicopter battalion, an entire field artillery brigade, an air defense battalion, and a truck company.[53] The unanswered question was how such nonorganic units would mesh into the command structure and unique operating style of an LID.

Light infantry divisions allowed the U.S. Army to create more divisions cheaply, saving procurement funds for the costly equipment of the heavy divisions. The LIDs performed remarkably well in their intended mission of low-intensity combat, operating effectively in Honduras, Somalia, and Haiti.

U.S. special operations forces also grew in the later 1970s and 1980s.[54] After a high point during the Vietnamese conflict, the special operations units of all the armed services had suffered major cuts in personnel and funding. A number of highly publicized terrorist attacks in the later 1970s helped justify an increase in such forces, as did the failed 1980 attempt to rescue

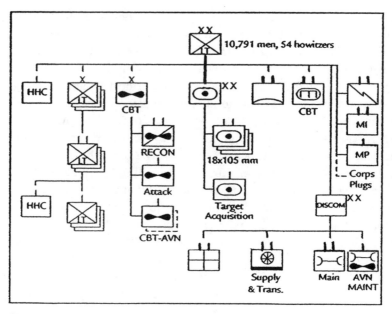

Figure 21. Type U.S. Light Infantry Division, c. 1985

the U.S. hostages held by the revolutionary government of Iran. Just as John Kennedy had campaigned on a platform of nonnuclear or flexible response in 1960, so Ronald Reagan promised to revitalize special operations forces in 1980, although his administration was slow to implement this promise. As in the case of heavy forces, however, a number of efforts begun in the later 1970s bore fruit in the following decade. For the assault or commando role, the U.S. Army trained two (later three) battalions of airborne Rangers; the Navy expanded its Sea, Air, and Land (SEAL) teams. The air force modernized and expanded its AC-series of gunships, armed transports that could provide huge volumes of fire to support special operations teams on the ground. Many of these new resources came under the control of a newly created Joint Special Operations Command, a recognition of the unique nature of their tactics.

The invasion of Grenada in October 1983 tested the revived special operations capabilities of the United States.[55] Although this operation has achieved legendary status as a comedy of military errors, most of the problems were due to the haste and excessive secrecy with which it was conducted. These limitations meant that key commanders had never even met each other, let alone coordinated their operations, and that there was insufficient time for intelligence collection and practices. Both SEALs and Special Forces teams encountered unexpected resistance and casualties in their landings. A shortage of trained air force transport crews caused the two Ranger battalions to leave many of their troops behind, and navigational problems and repeated changes in instructions led to the Rangers parachuting in broad daylight. Incompatible communications equipment and service procedures further complicated the command of this operation. Nevertheless, Grenada served as both an object lesson and a justification for the continued development of light infantry and special operations troops.

The Falklands

The U.S. Army was not the only Western force to find itself fighting an unusual war in an unexpected place. In March 1982, Argentina seized the Falklands, or Malvinas Islands, from Britain, prompting the British to recapture them. Having emphasized its NATO missions in Europe, Britain had been forced to neglect its naval air, amphibious warfare, and light infantry capabilities, all of which were in high demand in this unforeseen war.

The Falklands campaign was unusual in several ways. First, neither side was able to achieve air superiority. Except for a few light attack aircraft, the Argentine Air Force was based on the mainland rather than on the islands, forcing its pilots to operate at extreme range with little loiter time over the battlefield. The British Royal Navy and Air Force could muster only forty-two Harrier vertical takeoff and landing aircraft; these few made the invasion possible, but they could not completely dominate the air.[56]

Air defense systems downed a number of aircraft on each side, but again they could not completely protect their forces. Argentine infantry weapons and obsolescent antiaircraft guns shot down at least five Harriers, but the newer British missile systems had difficulties. This was particularly true when they were mounted on the cliffs overlooking the landing site at Port San Carlos; frequently, the Argentine aircraft were flying over the water below the firing batteries, making the attackers difficult to engage except in a manual, optical mode. Still, Rapier surface-to-air missiles claimed to have killed at least fourteen Argentine aircraft, and the British government ordered additional Rapier units after the war.[57]

The initial Argentine seizure of the islands had been a near-textbook example of U.S. Marine tactics using amphibious armored vehicles; the light antitank rockets available to the few British defenders proved inaccurate or ineffective against such vehicles.[58] Once the Argentine Marines secured the main island, however, they turned it over to an army force that included a number of partially-trained conscripts. As a result, the Argentine defense proved at best uneven in its effectiveness.

The difficult terrain and remote location of the Falklands made the ground war a dismounted infantry contest. Only a few specialized tractors could move across this terrain. Early in the campaign, the Argentine Air Force sank a British container ship that was carrying fourteen helicopters, including all but one of the larger Chinook transports. The remaining British helicopters were insufficient even to move the available artillery pieces and critical supplies, let alone to transport large troop units. Instead, the British light infantry units, especially the Royal Marines, two battalions of the Parachute Regiment, and a Gurkha battalion, conducted the campaign on foot.

The most famous battle of the Falklands War was the British attack on Goose Green on 29 May 1982 (Map 15).[59] The Second Battalion, the Parachute Regiment, made an exhausting cross-country advance to Goose Green on 28 May, only to lose the element of surprise when a British news service announced the movement. Because of the lack of transportation, the battalion carried only two of its 81mm mortars and was supported by only three 105mm howitzers and a single 4.5-inch naval gun, which unfortunately jammed at the critical time.

Not only did the Argentines outnumber the British by two to one at Goose Green, but they were also defending a narrow neck of land that made flanking attacks impossible. Yet the defenders were an ad hoc mixture of units that included soldiers of varying levels of training; some of the defensive positions were well prepared and defended, but others failed on both counts. The 12th Argentine Regiment was also missing a number of its mortars, recoilless rifles, and other heavy weapons.

After British engineers had searched for minefields, the initial night attack by the parachute battalion made some progress, pushing back the reinforced Company A, 12th Argentine Regiment, that provided outpost security. By daylight, the British advance had reached the Argentine main defenses. There, the defenders pinned many of the British down with machine-gun fire; only the personal sacrifice of the parachute battalion commander restored confidence. Even then, Argentine 35mm antiaircraft guns, firing in a ground role, delayed the British advance, as did air strikes by several light attack aircraft. The officer who succeeded to command of the British troops finally used 66mm MILAN antitank missiles to knock out Argentine positions; this example, and the belated arrival of Harrier air support, led eventually to the Argentine surrender. Despite the lack of fire support and coordination, the British attackers had succeeded, losing 14 dead but killing more than 55 and capturing over 1,200.

This victory, in turn, gave the British psychological ascendancy for the balance of their campaign to recapture the Falklands. It also reinforced the

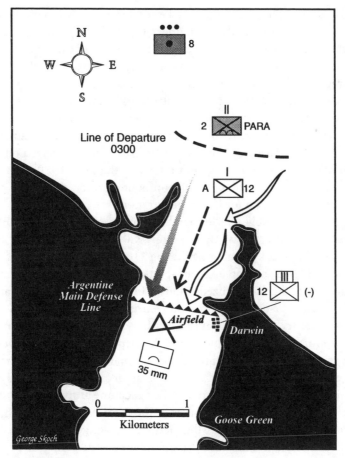

Map 15. Battle of Goose Green, 1982

time-tested concept that combat units must have their own, organic fire support and air defense weapons so that they can operate if necessary without outside aid.

European Armies in the 1980s and 1990s

Although the Falklands campaign restored luster to British arms, the nation also had to deal with terrorism in Northern Ireland and the ever-present threat of Soviet attack in Germany. For the latter role, the British Army of the Rhine went through a dizzying series of reorganizations. In 1975 this field army found itself reduced to four armored divisions of two brigades each. By 1983 this had become three larger divisions, each typically including three armored brigades, an engineer battalion, a helicopter battalion, and three artillery battalions; reconnaissance and larger caliber artillery were relegated to the corps level. As an experiment in increased mobility, the 3d Armoured Division substituted an airmobile brigade for one of its heavier counterparts. Many lower priority infantry units had to make do with wheeled instead of tracked personnel carriers. Despite such budgetary concerns, the British were able to improve their tank force just as their American counterparts were doing. Existing Chieftain main battle tanks received additional compound armor and laser range finders; in the later 1980s, the British fielded the Challenger tank with full compound armor, digital fire control, and a 120mm main gun for better penetration.[60]

The Federal Republic of Germany continued to field *Marder* IFVs and improved Leopard II tanks. By the late 1980s, a typical armored brigade consisted of three tank battalions (totaling 110 tanks), a tank destroyer company with ATGMs, and an eighteen-gun self-propelled artillery battalion; its counterpart, the mechanized infantry brigade, included one tank battalion, two *Marder*-equipped mechanized infantry battalions, an M113–mounted infantry battalion, and an artillery battalion. The German Air Force supported such brigades with the Alphajet, a light, low-altitude ground attack aircraft that had to depend on maneuverability for survival.[61]

The end of the Cold War and the reunification of Germany caused a major reorganization of the German Army. The peacetime army needed a smaller, more balanced force. Once the forces of the former Soviet Union began to withdraw from Eastern Europe, the Germans no longer felt the need for an armor-heavy force and instead began a major restructuring in 1992. The most common type of unit in the entire army became a square panzer grenadier brigade (Fig. 22). Two battalions of *Marder*-equipped infantry and two battalions of Leopard or Leopard II tanks were supported by a self-propelled artillery battalion and by companies for armored reconnaissance and engineers. Even the remaining airborne brigades were equipped with lighter armored personnel carriers, making the German Army more uniform in its capabilities and tactics.

France continued to maintain a multitiered army, with mechanized units in and near Germany, conventional infantry divisions for home defense, and a variety of specialized units for contingency operations, especially in the former French colonies of Africa. The mechanized force received a limited number of LeClerc tanks, modern, fifty-two-ton vehicles intended to replace the much lighter AMX-30. Because of the continuing French emphasis on tactical nuclear missiles, the artillery elements and tactical

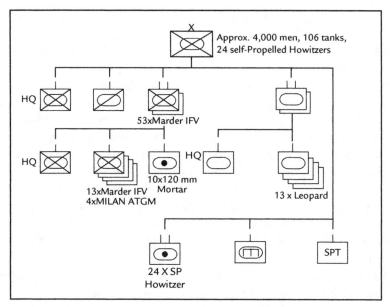

Figure 22. Type German Mechanized Brigade, c. 1992

air support of the mechanized force were still small by comparison to their NATO counterparts.[62]

In 1984 France created the Force d'Action Rapide (FAR), a corps-sized headquarters for contingency missions. Having reduced its divisions to the size of large brigades, the French Army grouped different capabilities under the FAR headquarters. In a design similar to that of the larger U.S. XVIII Airborne Corps, the FAR could choose from five different divisions—a light armored division (which represented France in the 1991 Gulf War), a mountain division, a marine infantry division, an airborne division, and the airmobile division.[63] Many of these units were Foreign Legion or other volunteer formations, although for legal reasons the FAR sometimes had to leave national service conscripts behind when operating outside Europe. Even this issue disappeared in the mid-1990s, when France greatly reduced its dependence on conscription for manpower.

Soviet Army to Russian Army: Continuity in Change

Military planners in Moscow remained dedicated to the concept of deep operations throughout the last three decades of the twentieth century. However, a number of technical and political issues drove them to modify their force structures and tactics for achieving it.

Since the depths of World War II, Soviet commanders had arrayed their units in multiple layers or echelons, each echelon pushing forward on the battlefield to achieve breakthrough and exploitation along a relatively narrow front. An echelon was not any particular-sized unit but a group of forces with a specific timetable and mission. Divisions, armies, and even *fronts* deployed in two or more such echelons; second or third echelons might well include operational maneuver groups. The first echelon at each level of command would engage the enemy, and successive echelons would reinforce this battle or (preferably) pass through the units in front so as to exploit into the rear. This method allowed Soviet commanders to conduct sustained operations without regard for the heavy casualties that any individual unit might suffer. The Soviet Army envisioned a single-echelon formation only when the enemy was extremely weak.

Beginning in the late 1950s, Soviet planners had to modify this method of employment slightly, seeking to close rapidly with the enemy so that the rearward echelons did not become lucrative nuclear targets. During the 1980s, however, the United States and NATO developments in precision-guided weapons and deep strike operations threatened the entire basis of Soviet echeloned deployments. As later demonstrated in the 1991 Gulf War, fighter-bombers and helicopters armed with smart weapons could devastate troop and supply concentrations behind the front lines. As a result, second-echelon organizations and OMGs could no longer wait for frontline units to create a penetration.

Rather than abandon their entire premise of echelonment and deep operations, Soviet and later Russian military planners decided to avoid the conventional, linear nature of initial battles, seeking instead to achieve an intermingling of forces from the first skirmishes of a war. Instead of concentrating OMGs and other follow-on forces in their own rear areas, where they would be subject to enemy air attack, the Soviets wanted to project such exploitation units rapidly into their opponent's rear areas. Just as the Chinese had used infiltration and hugging tactics to offset the firepower of U.S. forces in Korea, so the Soviet Army planned to break into the NATO defenses as rapidly as possible, becoming so intermixed with their opponents that precision-guided munitions and deep attack concepts would be stymied for fear of friendly casualties.[64]

In order to do this, the Soviet Army planned to conduct a surprise attack at the very start of hostilities, seeking to penetrate NATO defenses before they were completely ready. All Soviet units had to be capable of deep operations without an extensive mobilization period prior to war. Although the leading divisions might be deployed in several tactical echelons, larger headquarters, such as a combined arms army, or front, would attack while arrayed in a single, operational-level echelon.

In turn, this led Soviet planners to go beyond the concept of specialized units that were organized to function as OMGs. Instead, the Soviets envisioned converting all or virtually all of their conventional, motorized rifle and tank units into combined arms structures capable of deep maneuver at the tactical or operational level. Traditionally, the Soviet military labeled such unusual units as "brigades" and "corps" rather than as "regiments" and "divisions." This practice is still in use.

Beginning about 1985, the Soviet Union reorganized its ground forces, creating more flexible, combined arms organizations for the nonlinear battlefield (Fig. 23). Each motorized rifle or tank brigade could function as a forward detachment for a corps, which in turn would in theory have all the combat arms and logistics necessary for independent operations. An air assault infantry battalion and, in some cases, attack and transport helicopter units would give the corps the means to seize key locations ahead of its forward detachment. Fire support came from a family of self-propelled artillery weapons, such as the 152mm gun designated as the 2S5. This reorganization meant that three levels of command (front, corps, and brigade) would replace the four levels (front, army, division, regiment) that had become traditional in the Soviet structure. The changes also meant abandoning the fixed structure of the motorized rifle and tank divisions and regiments. Regardless of how the structures appeared in the abstract, the new brigades and corps were supposed to cross-attach and tailor combined arms task forces in much the same manner as the U.S. ROAD division.

To achieve such flexibility, Soviet theorists emphasized a concept that in the West became known as "information warfare." In modern warfare, they argued, a commander and staff might be inundated with enormous amounts of information, most of which was not essential to understand and direct the battle. The key to success was therefore to use sophisticated communications and data analysis systems to sort through the various reports, seeking the key pieces of information necessary to manage the battle.

Yet Soviet theory was always more advanced than Soviet practice. Even while the Soviet Union survived, commanders found it expensive and difficult to obtain the type of communications and computer resources necessary for the new units to function as designed. Despite the best Soviet efforts to encourage initiative by junior leaders in Afghanistan and elsewhere, it was time-consuming to develop the type of independent judgment necessary to

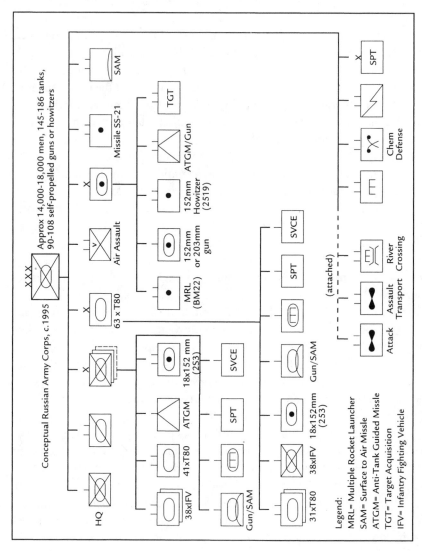

Figure 23. Ideal Russian Army Corps, c. 1995

Conceptual Russian Army Corps, c.1995 Approx 14,000-18,000 men, 145-186 tanks, 90-108 self-propelled guns or howitzers

XXX

HQ

SAM

Missile SS-21

Air Assault

63 x T80

38xIFV   41xT80   ATGM   18x152 mm (2S3)

Gun/SAM   SPT   SVCE

31xT80   38xIFV   18x152mm (2S3)   Gun/SAM

SPT   SVCE

MRL (BM22)   152mm or 203mm gun   152mm Howitzer (2S19)   ATGM/Gun   TGT

(attached)

Attack   Assault Transport   River Crossing

SPT   Chem Defense

Legend:
MRL= Multiple Rocket Launcher
SAM= Surface to Air Missile
ATGM= Anti-Tank Guided Missle
TGT= Target Acquisition
IFV= Infantry Fighting Vehicle

command corps and brigades that were to conduct high-speed exploitation operations in the enemy rear. During the later 1980s, Mikhail Gorbachev's efforts to reduce tensions with the West prompted Soviet publications and organizations to pay lip service to a lighter, more defensive form of warfare, delaying but not ending the effort to reform the military.

Once the Soviet Union expired, reality made it even harder to match theory. The army of the Russian Federation lost many of its newest and most effective weapons, units, and facilities during the collapse of the USSR; thereafter, the Russian Army was hard-pressed to man and equip even part of its huge force structure, let alone to develop the sophisticated structures necessary for nonlinear and information warfare. As a result, Russia ended the century with its reorganization in limbo: the force structure included a hodgepodge of armies, divisions, corps, and brigades.[65]

The Soviet Union had encouraged its Warsaw Pact allies to restructure their armies in the same manner, but this process also stopped short of completion after both the former and the latter imploded in 1989–1992. Some armies, notably the Hungarians, implemented the new structure, but others did not. In Poland, for example, the older Soviet-style tank divisions, each with three tank and one motor rifle regiments, were replaced by divisions in which each regiment contained two tank and two mechanized infantry battalions. Such regiments closely resembled the panzer grenadier brigades of Germany. The Poles even experimented with an air assault division, which, given the Polish military tradition, was designated as a cavalry formation.[66]

The Iran–Iraq War

During the 1980s, the most intense sustained conflict in the world was that between Iran and Iraq. In the fall of 1980 Iraqi dictator Saddam Hussein invaded Iran, apparently calculating that the revolutionary upheaval there would make the Iranians easy prey. After initial Iraqi successes, however, the conflict settled into a positional stalemate not unlike that of World War I. The Iranians had lost many of their professional officers and suffered great difficulties in getting parts and ammunition for their advanced, U.S.–designed weapons systems. Like the Russians in World War I, the Iranians compensated for these deficiencies by using large masses of poorly trained, poorly armed infantrymen in an attempt to overwhelm their foes. The effort often failed, but it was sufficient to prevent a clear-cut Iraqi victory. Iraqi armor often proved vulnerable to Iranian light infantry attacks.

Although equipped with Soviet-made weapons, the Iraqi Army clearly showed its heritage as a former British colony and protectorate. Division organization and staff procedures closely followed those of the pre-1943

British Army. Like the Egyptian Army of the 1950s and 1960s, the Iraqi Army tended to plan set-piece, centralized battles and could not react quickly to unexpected opportunities or threats. In some instances, an attack plan went all the way to Baghdad for approval, making any subsequent changes politically unhealthy for the field commander. Iraqi artillery followed the same tendencies, favoring strict planning and large volumes of fire over flexibility and accuracy. During the long war, Iraqi artillerymen generally fell behind world trends in the increasing accuracy and responsiveness of fire support. They were, however, able to integrate poison gas shells into their artillery preparations.[67]

When Saddam continued his aggression by invading Kuwait in August 1990, he understandably believed that his veteran armed forces could perform well against any threat. Unfortunately, the Iraqi Army had fallen into the mental habits of a World War I force and was completely unprepared for the speed, lethality, and flexibility of a late twentieth-century army.

CHAPTER 9

· · · · · · · · · · · · · · ·

War in the 1990s

Air Operations, 1991

The most striking aspect of the 1991 Gulf War was the enormous effect of Coalition airpower on the Iraqi forces. For the first time, airpower at least approached the effectiveness that its advocates had preached for generations. Yet many civilian observers, including political leaders, received an exaggerated idea of what it could accomplish, leading to future policy errors. In fact, the air triumph of 1991 was due in large measure to a unique combination of circumstances that are unlikely to recur in future conflicts.[1]

First, the Iraqi dictator allowed the U.S.–led Coalition five months to build up an enormous air and ground force on his doorstep. Had Saddam Hussein invaded Saudi Arabia and the Gulf states in the early fall of 1990, the Coalition air forces would have had neither the planning time nor the initiative to conduct a classic so-called "offensive air campaign." Second, Saudi Arabia had spent years constructing a magnificent network of military bases, unlikely to be equaled in future theaters of war. Third, the flat, open terrain and favorable winter weather of the Middle East made flight plans and target identification unusually simple. Particularly outside urban areas, the Iraqi forces were so isolated that Coalition pilots could attack any suspected target with little fear of causing casualties among the innocent. By contrast, later NATO air attacks in various parts of former Yugoslavia encountered much greater challenges of terrain, weather, and possible civilian casualties.

Once the initial air and cruise missile attacks dismembered Iraqi command, control, and air defense systems, the air operation became a traditional battle of attrition. By attacking Iraqi factories, utilities, and lines of communication, the Coalition air strikes inflicted a frightening toll on the enemy. Within the region of Kuwait, however, the results, though impressive, were somewhat different. Airpower advocates had correctly argued that high performance aircraft have difficulty detecting and engaging enemy vehicles that are dispersed and protected by earthen revetments. Those army and air force aircraft that flew at sufficiently low speeds and altitudes to hit dispersed vehicles were exposed to ground fire, suffering much higher casualty rates than did other airframes. A-10 Warthogs armed with Maverick precision bombs had an enviable kill record, but the same could not be said for other aircraft using more conventional munitions. For example, an F-16 fighter had a bombing sight, a "pip" in the pilot's computer-driven visual display, that, at the altitudes normally flown in the Gulf, covered a ground area of more than ten square meters; such a pip was so large that it would completely obscure an individual target vehicle, and in any event conventional bombs might fall anywhere within a forty meter circle.[2] There were not enough smart munitions in the world to destroy the entire Iraqi armored force from the sky. As in Kosovo, airpower found it much easier to destroy command, control, and logistical targets than to expel an enemy ground force from its area of operations.

This does not deny the importance of airpower or belittle the tactical success of Coalition pilots. Certainly the relentless air attacks had a phenomenal effect on the already poor morale of Iraqi forces. Airpower in the Gulf War, as in previous conflicts, was absolutely indispensable for victory but could not, by itself, achieve it within reasonable parameters of time or expenditure. Despite the enormous damage inflicted on the Iraqi economy and on Baghdad's command and control structure, the Iraqi regime's control over its people proved remarkably resilient. Coalition ground forces still had to fight their way into Kuwait to enforce the eviction notice served by the air attacks.

Moreover, the centralized air planning system that proved so effective in orchestrating an elaborate series of air operations was, of necessity, slow to respond to unexpected emergencies. The most notable example was the so-called battle of Khafji, 29–31 January 1991, when elements of three Iraqi divisions attacked the Saudi border city of Al Khafji. Advanced airborne intelligence platforms, including the experimental Joint Surveillance and Target Acquisition System (J-STARS),[3] allowed Coalition commanders to observe the Iraqi advance literally as it happened. Despite this advantage, the Tactical Air Control Center directing Coalition air operations took some four hours to recognize the magnitude of the threat. In the interim, the U.S. Marines in the Khafji area received some air support, but the commander of Arab troops had to threaten to pull Saudi air elements out of the Coalition if the U.S. planners would not provide sufficient aircraft.[4] Once diverted, air elements did a superb job of destroying the Iraqis. In one instance, a brigade of the Iraqi 5th Mechanized Division was trapped while traversing a narrow gap through a minefield, and Coalition aircraft virtually destroyed the entire unit. Nevertheless, the delay in diverting airpower from the "air campaign" to missions that would assist the ground forces might have had serious consequences.

This issue reappeared several weeks later, as the air phase of Desert Storm reached its height and Coalition ground forces prepared to attack. Ground commanders expressed concern that their targeting requests were ignored by the air planners. To some extent, this concern was a matter of perception, because the ground headquarters did not receive timely feedback concerning targets that were, in fact, engaged by the air elements. Nevertheless, these commanders wondered whether air support would be available in sufficient quantities once their offensive began. Lt. Gen. Charles Horner, who commanded the Coalition air forces, repeated the reasonable, long-standing contention that airpower should be sent on specific missions instead of being held in reserve for contingencies. Yet by mid-February 1991, large numbers of aircraft were being assigned as "killer scouts," patrolling specified boxes of enemy airspace while looking for targets of opportunity. The purpose was to prevent any Iraqi movement in daylight, but the air planners apparently preferred such potentially wasteful, independent operations to providing direct support to the ground forces.[5]

One should note, of course, that the ground forces were sometimes equally reluctant to work closely with their air counterparts. Once the ground offensive began, XVIII Airborne Corps placed the coordination line for air–ground operations well forward of its own units, in order to give attack helicopters freedom of maneuver without having to coordinate with the Coalition air forces. This action obviously restricted the ability of airpower to support the ground advance.[6]

As in previous wars, professional soldiers of the air and ground components worked together to overcome issues and achieve the common goal. Their differences were not due to selfish emotions but to different priorities and sincere beliefs on both sides. Nevertheless, the much-heralded success of airpower in Desert Storm inevitably increased single-service parochialism rather than improving the combination of the air–ground team. It also gave political leaders unreasonable expectations about what could be accomplished by airpower alone.

SCUD Busting

Although tactical ballistic missiles had appeared in several previous wars in the Middle East, Desert Storm represented the first occasion in which they had a significant political, if not military, effect on a war. Given the recent proliferation of ballistic missiles, particularly in southern and eastern Asia, the Desert Storm example may be a harbinger of future warfare.

As originally designed by the Soviet Union, the SCUD-B tactical missile had a range of 300 kilometers. Iraqi modifications produced theoretical ranges of 600 kilometers (Al-Husayn) and 750 kilometers (Al-Hijirah), but at such extreme distances, the notoriously inaccurate SCUD missile might land almost anywhere.[7]

Coalition air planners neutralized the fixed launching sites for these missiles at the beginning of the war, but the mobile launcher trucks proved remarkably elusive. Reportedly, a launcher could fold up and evacuate its launch site within six minutes of firing the missile, long before it could be targeted in response. Missile launches against Saudi Arabia were primarily a propaganda tool for Saddam Hussein, but the forty-one missiles fired at Israel might well have brought that nation into the war, causing major political problems for the anti–Iraqi Coalition. Thus the Coalition had to commit large numbers of air sorties and special operations reconnaissance teams in a generally vain effort to locate and destroy the mobile launchers.

Much has been written about the role of the U.S. Army's Patriot air defense system in countering the Iraqi missiles, but most of the criticism is based on misunderstandings about its purpose. The United States originally designed the Patriot to shoot down high performance aircraft; a subsequent modification called PAC-2 gave the Patriot a limited capability to intercept ballistic missiles. The PAC-2 Patriot's intended role was to protect a single, extremely valuable target such as a headquarters or a port facility. It functioned superbly in this mission but was never intended nor able to provide complete protection for a large area such as Israel or the northeastern coast of Saudi Arabia. Moreover, the improvised Al-Husayn and Al-Hijirah missiles were so fragile that, even when a Patriot intercepted one, the warhead might break free, inflicting casualties and damage on a nearby area.

Under these circumstances, the Patriot system and its operators performed admirably in the Gulf War. Both its success and its limitations argue strongly for further development of air defense systems to protect future armies from ballistic missiles, cruise missiles, and aircraft.

The Ground Conflict

From the moment it invaded Kuwait in August 1990, the Iraqi Army was overextended logistically, leaving it vulnerable to air interdiction. Airpower also smashed Iraq's only offensive effort, the attack on Khafji, and denied any Iraqi aerial reconnaissance of Coalition forces. Based on their experiences of the trench warfare in the Iran-Iraq War, however, Iraqi leaders still expected to inflict unacceptable casualties when the Coalition finally attacked on the ground. Events proved them wrong.

Technology was part of the reason for Coalition success in defeating the Iraqi defenses. Counterbattery radars and TACFIRE communications links enabled U.S. artillerymen to determine the source of any shells fired by their opponents and to fire back at the Iraqi guns literally before the Iraqi barrage impacted on the ground. A variety of engineer devices, such as a system to project a string of charges across a minefield and detonate the mines, neutralized many of the Iraqi obstacles. Of even greater significance was the proliferation of Global Positioning System receivers. They enabled units to maneuver with confidence even at night in a featureless desert and allowed U.S. commanders to have unprecedented accuracy in knowing the locations of their subordinates.[8]

However, the Coalition's success on the ground was not just a product of airpower and technology. To a considerable extent, the ground war was a result of the carefully developed doctrine and training of the U.S. Army and Marine Corps.

At the operational level, the Coalition dislocated Iraqi plans by shifting forces far to the west and then advancing in a clockwise sweep, enveloping the entire Iraqi Army in Kuwait. The Iraqi command, control, and communications system had difficulty reacting in a timely manner to this threat to its right flank; and when Iraqi units did begin to move, their redeployment was hamstrung by continuing air attacks.

At the tactical level, U.S. and other Coalition forces demonstrated a remarkable ability to maneuver and to respond to changing situations, an ability that far exceeded that of their Iraqi opponents. Nowhere were this speed and flexibility more evident than at the so-called battle of 73 Easting.[9] For want of any distinguishing geographic feature, this encounter was named for its location on military maps, where the line of contact for some time was the north-south grid line labeled 73.

The first and more publicized part of this battle occurred on the afternoon of 26 February 1991. By this point in the war, the U.S. VII Corps had completed its swing west of Kuwait and was advancing almost due eastward to intercept the Iraqi Republican Guards divisions, the most potent enemy force in the region. The VII Corps attacked with three subordinate units abreast—from north to the south, the 1st and 3d Armored Divisions and the 2d Armored Cavalry Regiment (ACR). Unlike divisional cavalry units, which had lost their tanks during the Army of Excellence modifications of the 1980s, the 2d ACR remained a fully capable combined arms force, with M1A1 tanks, M3 cavalry versions of the Bradley fighting vehicle, and a host of specialized weapons integrated down to the troop (i.e., company) level.

The Second Squadron (i.e., battalion), 2d ACR, rolled eastward with two troops forward, followed by another cavalry troop and the squadron's tank company (see Map 16, left side). Ordinarily, each squadron had its own battery of self-propelled artillery, but for this operation the entire 6th Battalion, 41st Field Artillery, was in direct support of 2d Squadron. Shortly after 1600 on 26 January 1991, the two leading cavalry troops crossed a low rise at the 70 Easting grid line and encountered a large body of enemy troops. Subsequent analysis indicated that this was a battalion-sized task force acting as a combat outpost for the Tawalkalna Republican Guard Mechanized Division. This task force was equipped with T-72 tanks and BMP infantry fighting vehicles and had assumed hasty defensive positions, but M1A1 tanks and other weapons destroyed most of the battalion in less than six minutes of firing. The Iraqis' tank guns were adjusted to fire at ranges of 1,500 meters, which meant that their shots fell short when the U.S. tanks engaged at 2,400 meters. Moreover, the Iraqis assumed that the American tanks, like all previous armored vehicles, halted in order to fire their guns accurately. In fact, however, the stabilized, computerized gun of the M1A1 allowed the cavalry to continue its advance without halting. The cavalry squadron moved forward so rapidly, firing on the move, that its fire support teams had to cancel two successive artillery missions for fear of hitting U.S. troops.

Although 2d Squadron was supposed to halt at the point where it first encountered the Iraqis, the momentum of the advance carried it eastward for another three kilometers, to the 73 Easting grid line. This brought the cavalry into contact with the main defenses of the Tawalkalna Division and its neighbor to the south, the 12th Iraqi Armored Division. During the advance, supporting artillery used proximity-fused airbursts to halt Iraqi dismounted infantry and then switched to improved conventional munitions when an Iraqi BMP company vainly attempted to outflank the American squadron. For the next six hours, the armored cavalry fought off a series of Iraqi counterattacks. Finally, at 2240 that night, the supporting artillery (including MLRS) fired a synchronized, time-on-target barrage that destroyed a number of Iraqi bunkers and vehicles.[10]

After this barrage, the 1st Brigade, 1st U.S. Infantry Division, passed through the 2d Squadron to conduct the main attack. Because the Iraqis

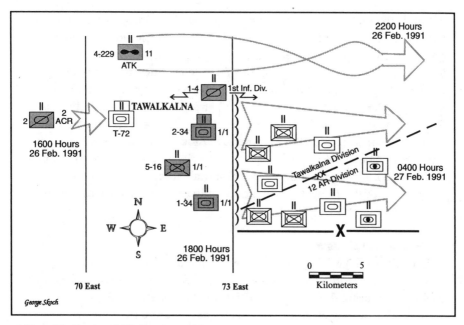

Map 16. Battle of 73 Easting, 1991

were oriented to defend to the southwest rather than directly to the west, the 1st Brigade cut diagonally across the divisional boundary between the Tawalkalna and the 12th Armored, a situation that further complicated the Iraqi defense. The 12th, being part of the Iraqi regular army instead of the Republican Guard, was equipped with aging T-55 tanks. Still, the U.S. attack was by no means flawless or bloodless. Task Force 2d Battalion, 34th Armor, on the 1st Brigade's left, temporarily lost contact with two of its companies that crossed over a low rise; star clusters fired to guide them back to their parent unit alerted the Iraqi defenders in the area. To the south, Task Force 1st Battalion, 34th Armor, began the advance with its scout (reconnaissance) platoon in the lead, rather than tanks. The scout platoon lost two Bradleys, one soldier killed, and four wounded when it first encountered the Iraqi defenses. For the rest of the night, the U.S. mechanized forces moved eastward, fighting a series of short, intense encounter battles with various Iraqi companies and battalions. By morning, two Iraqi brigades were destroyed, and the Americans were moving through the Iraqi rear areas. After a brief halt for refueling, the 1st Infantry Division continued the advance for two more days. As one Iraqi prisoner remarked of another U.S. attack, "You were like the wind. You come, blow and go away. You cannot shoot the wind."[11]

While the night advance continued, the VII U.S. Corps had put the doctrine of deep attack into practice. The Iraqi 10th Armored Division was in assembly areas approximately eighty kilometers east of the battle front at 73 Easting. The task of attacking this force fell to the 4th Battalion, 229th Aviation Regiment, of the 11th Aviation Brigade. Last-minute communication problems delayed the attack until after 2200 hours on 26 February. Still, in two separate strikes, the eighteen AH-64s of 4th Battalion did considerable damage to the 10th Armored Division. Unfortunately, the ground advance had been too rapid, going so far that it came up against the coordination line between Coalition ground and air forces. The AH-64s had to allow many of the Iraqis to escape rather than risk fratricide with the high-performance aircraft.[12]

Organizational Issues

The 1991 Gulf War revived the same issue that had plagued France before 1914—the readiness of army reserve units to fight as soon as they were mobilized. During the 1970s and 1980s the U.S. Defense Department had come to rely heavily on reserve and National Guard units to provide capabilities that the active armed forces could not afford to maintain in peacetime. The U.S. Marine Corps accomplished this by putting active duty officers in command of its reserve units, at least at higher levels. In the Army Reserve and National Guard, however, the American tradition of citizen-leaders and units, not just individual citizen-soldiers, meant that reserve component officers provided the commanders and staff for their own battalions and brigades. Hundreds of these reserve elements, from special forces and field artillery to maintenance and supply units, proved themselves during Desert Shield and Desert Storm. Yet one group of National Guard infantry and armored units had a different experience.

For reasons of economy, by 1990 each of three active duty U.S. Army divisions had only six of its ten maneuver battalions and two of its three brigade headquarters—National Guard "roundout" units provided the remaining four battalions and one brigade in each division. To ensure that these units were ready to deploy, they had received new equipment and training even before their parent active-duty units. National Guard leaders and staff officers regularly trained with their parent headquarters. Contingency plans called for the roundout units to deploy immediately after their active duty colleagues, which in practice meant that the National Guard troops would have one or more months to mobilize and complete any necessary training. Yet when the Kuwait crisis came, these roundout units were left behind. One explanation was that the reserve components were initially mobilized for a period of only three to six months, and the Defense Department saw little purpose in using scarce shipping space to move units that could not remain in the Middle East for an undetermined period.[13] When the roundout units were finally mobilized, they remained in the United States for extensive periods of training. National Guard leaders understandably concluded that the active army doubted the ability of these units and their leaders to perform.[14] Considering how much the United States has relied on reserve components for various missions since the Gulf War, this issue had serious implications for the future.

Structurally, Desert Storm accelerated one change that was long over-due. At over 900 men, the combat engineer battalion was the largest in a U.S. mechanized or armored division. Moreover, the great demand for combat engineers to build and breach obstacles meant that, doctrinally, this organic battalion was always reinforced by a second, nondivisional or "corps combat" engineer battalion. When hostilities commenced, the U.S. Army was already moving to change this, forming a divisional engineer brigade with three smaller, 433-man battalions, one to support each maneuver brigade. During Desert Storm, four of the five U.S. heavy divisions implemented this change on an ad hoc basis; and immediately after the war, the U.S. Army began to formally restructure its engineers in this manner. Unfortunately, the engineer brigade headquarters fell victim to the next round of cost-cutting, leaving the engineers with inadequate central control.[15]

In the later 1990s the U.S. Army was pulled in two directions. On the one hand, laptop computers linked tactical units together, instantaneously transmitting orders and depicting the locations of enemy and friendly forces down to the level of individual vehicles. Such experiments had the potential to give commanders unprecedented clarity and accuracy in understanding the battlefield they faced.

On the other hand, the post–Cold War U.S. Army found itself with vast new missions while its size and resources declined by at least one-third. The organizational response to these challenges was evolutionary rather than revolutionary. The new design, known as Division XXI, sought to retain most of the existing capabilities of a heavy division while reducing its strength by approximately 1,000 soldiers to 15,700 (Fig. 24).[16] This relatively slight reduction resulted in a 24 percent decrease in the number of actual combat systems in the division. Tank and mechanized infantry battalions were reduced from four to three line companies. Company-level mortars, already deleted from many units under Division 86, disappeared permanently, and the battalion mortar platoon consisted of only four 120mm tubes. The new division also lost its chemical defense company and the engineer brigade headquarters.

The new division did give each brigade its own reconnaissance troop, which potentially offered the brigade commander more ability to see and

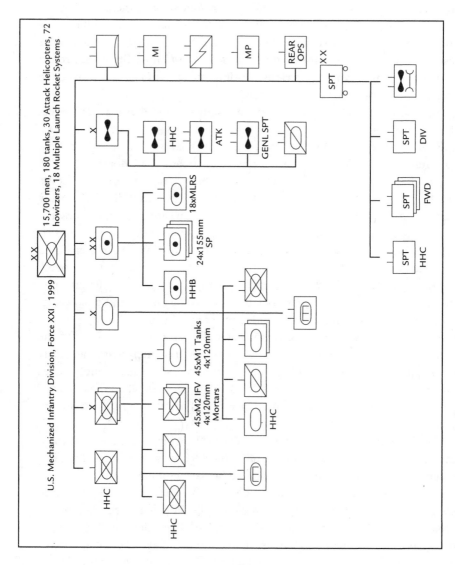

U.S. Mechanized Infantry Division, Force XXI , 1999

15,700 men, 180 tanks, 30 Attack Helicopters, 72 howitzers, 18 Multiple Launch Rocket Systems

HHC

HHC

45xM2 IFV 45xM1 Tanks
4x120mm 4x120mm
Mortars

HHC

18xMLRS

24x155mm
SP

HHB

HHC

ATK

GENL SPT

MI

MP

REAR OPS

SPT

SPT
DIV

SPT
FWD

SPT
HHC

Fig. 24. U.S. Division XXI Design, 1999

secure the battlefield. Portable computers and other electronic equipment would extend the capabilities of this troop, but its primary vehicle was a lightly armored version of the High-Mobility Multipurpose Wheeled Vehicle (HMMWV). Such equipment permitted rapid deployment to trouble spots worldwide and sufficed in most low-intensity operations. However, this development only extended the declining ability of reconnaissance units to fight for information when necessary.[17] Even a cavalry regiment was equipped with HMMWVs for low-intensity contingency missions, and by 2000 the army was revisiting the concept of light, wheeled armored vehicles to increase strategic mobility to the Third World.

Wars and Rumors of War

Indeed, after Desert Storm, most military operations of the 1990s fell in the category that the U.S. Army labeled "operations other than war"—disaster relief, peacekeeping, refugee assistance, and the like.[18] American, French, Russian, and other national leaders came to rely on the discipline, organization, and communications of military forces to resolve almost any world crisis. The end of the Cold War reduced but did not eliminate the risk of wider war inherent in such interventions. To some extent, such missions were a throwback to the late nineteenth and early twentieth century, when American and European troops found themselves interfering in the affairs of numerous Third World nations or colonies. Unfortunately, military organizations have always existed to apply maximum force against a clearly defined enemy, not to use minimum force in a quasi-police role. Placing ground forces, whether American or other, in such ambiguous situations inevitably led to problems. The most prominent example of this was the 1992–1993 intervention in the lawless environment of Somalia. U.S. special operations forces, sent to abduct a prominent terrorist leader, lost the advantage of surprise and were mauled by the local populace. These troops might still have prevailed had they had access to armored support or to the aerial component of the special operations team, the AC-130 gunship. Because of the political perception that such gunships caused unacceptable civilian damage, the AC-130s had left Somalia before the fatal Ranger operation.[19]

Doctrinally, the U.S. Army attempted to respond to these new missions by increasing its strategic perspective, connecting tactical operations and capabilities to the national political objectives. In the process, however, the operational level of war, the focus that had aided army commanders in conducting the 1991 war, received a lower emphasis. When a new edition of *Field Manual 100-5* was published in 1993, it moved the definitions of the three levels of war from the second chapter, which discussed basic concepts, to the sixth. The authors also eliminated most of the historical case studies that had illuminated the previous editions, substituting examples drawn almost entirely from U.S. operations in Panama (1989) and the Gulf War (1991).[20] In short, a well-intentioned effort to adjust to the strategic realities of the era caused the army to neglect much of the intellectual excellence of the 1980s, including the operational level of war and Deep Battle.

Chechnya

Like the U.S. Army, the armed forces of the Russian Federation became involved in numerous operations other than war. In addition to sending peacekeepers to Bosnia and Kosovo, the Russian Army was caught in the middle of secession struggles in the former Soviet republics of Georgia, Moldova, and Azerbaijan. The most serious and embarrassing conflict occurred in Chechnya, a rebellious area along the southern border of the Russian Federation.[21] When Chechnya's dictator, Dzhokhar Dudayev, broke away from Russia, he took with him a supply of Soviet-made modern weapons, including 50 tanks, 100 armored personnel carriers, and numerous artillery pieces. The resulting conflict was far more than a police action.

Unable to resolve the issue politically, the Russian leadership ordered its army and paramilitary interior troops to invade Chechnya in December 1994. The crisis came on New Year's Eve, when three brigade-sized columns of Russian troops attempted to seize the capital city, Grozny, by a sudden coup. Unfortunately for the Russian troops, their army was in no condition to conduct such an attack. The Federation's army was unpaid, unfunded, undermanned, and almost untrained. Certainly the troops involved had no special preparation for fighting in urban areas; some of the soldiers thought they were to maintain order in a city that had already surrendered. Rules of engagement instructed the Russian troops to fire their weapons only if the Chechens fired first. For matters of prestige, the Russians had maintained a large number of corps and divisions even though they had neither the manpower nor the funds to make them effective. The force structure was so understrength that all the units sent to Chechnya were provisional, with companies and even tank crews composed of strangers from a variety of different parent battalions.

Ignoring the example of Desert Storm, where Coalition aircraft destroyed carefully selected targets to dislocate the enemy's command and control structure, Russian pilots simply bombed residential areas of Grozny at random. When the ground attack was supposed to begin, only one of the three Russian columns actually attacked into the city. The other two commanders, hamstrung by poor communications and logistics, falsely reported imaginary advances. This permitted the Chechen forces to concentrate their efforts against the third Russian attack. This column, built around the 131st Motorized Rifle Brigade, lost 20 out of 26 tanks, 102 out of 120 infantry fighting vehicles, and its 6 self-propelled air defense weapons in one night.[22] The battle for Grozny dissolved into a savage struggle of individual weapons in which the venerable Soviet rocket propelled grenade (RPG) became the favored weapon for both antitank defense and general fire support. The Chechen leadership employed cellular telephones and other commercial communications links to direct their defense, while cleverly using intimidation and propaganda to discredit and demoralize the attackers.

By bringing in *Spetsnaz* and other elite forces, the Moscow government succeeded in pacifying Grozny after a month of fighting. This debacle caused Russian professional soldiers to reevaluate many of their basic assumptions about warfare. When the Russian government again attacked Chechnya in the fall of 1999, it used a prolonged aerial and artillery bombardment in a brutal if effective effort to weaken the defenders before any ground offensive.

Ultimately, however, Chechnya did not necessarily disprove the relevance of mechanized, combined arms warfare. It simply reinforced the principles of command, control, training, and combat in urban areas, where armored units are particularly vulnerable to short-range ambush.

PART III CONCLUSION

The Russian neglect of first principles in Chechnya is a fitting end point for this book, because it brings us back to the trends or concepts that have recurred throughout the century. Some of these are so self-evident that soldiers rarely discuss them, yet because these ideas have survived the test of different technologies and armies over most of a century, they merit some attention.

First, individual weapons, however powerful, do not win wars. Heavy machine guns, tanks, fighter-bombers, tank destroyers, and attack helicopters, at one time or another, have been segregated into separate units and touted as invincible. Once the opponent has time to develop his defenses, however, these weapons lose much of their initial effectiveness.

The second trend, then, is for major armies to integrate more and more arms and services at progressively lower levels of organization, in order to combine different capabilities of mobility, protection, and firepower while posing more complicated threats to enemy units. Even the "simplified" U.S. companies of the 1980s and 1990s have a host of weapons and capabilities undreamed of as late as 1939.

This integration does not necessarily mean combining individual weapons or even companies of different arms together in a permanent organization in garrison; indeed, such a rigid structure would be almost as ineffective tactically as the current garrison structure of "pure" infantry and tank battalions, because any fixed battalion structure could not adjust the balance of weapons in response to changes in terrain, enemy, and mission. To be effective the different arms and services must train together at all times, changing task organization frequently.

When making such changes in temporary organization, however, it is more effective to begin with a large combined arms unit, such as a division or a fixed brigade, and select elements of it to form a specific task force, rather than to start with a smaller organization and attach nondivisional elements to the formation. In the former case, all elements of the resulting task force are accustomed to working together and have a common sense of unit identity that can overcome many misunderstandings. In the latter case, confusion and delay may occur until the nondivisional additions adjust to their new command relationships and the gaining headquarters learns the capabilities, limitations, and personalities of these attachments. Frequent changes in the partnership of units, especially changes that are not practiced in peacetime, will produce inefficiency, misunderstanding, and confusion. Only the need to adjust the proportion of arms to different tactical situations limits the degree to which those arms can be grouped together permanently.

Implicit in this discussion is the third principle, the enduring value of a division command structure that can plan, identify critical information, and react quickly to changing situations. Since the Cold War, critics have frequently charged that the U.S. Army and other major armies are hanging onto familiar but obsolete force structures. Yet the complexity of military operations, whether in war or not, is possible only when those operations are orchestrated by a tightly knit group of commanders and staff planners, supported by adequate communications and transportation. Whether it is called a division, a corps, or a strike force, this command and control organization is largely responsible for the phenomenal ability of armies to accomplish the broad spectrum of missions found at the turn of the twenty-first century.

A well-trained division or higher headquarters cannot win a battle by itself, however. All elements must be capable of working toward a common goal with minimum guidance; excessive centralization and micromanagement can make any military organization too rigid to respond to a flexible, fast-moving opponent. The British and French Armies between 1915 and 1942, like the Egyptian, Syrian, and Iraqi Armies in more recent conflicts, suffered repeated defeats in part because their command, control, and communications were too centralized and slow-paced.

In the 1990s, this requirement to respond quickly to changes in battle has given rise to the concept, in the United States, China, and the former Soviet Union, of information warfare.[23] Taken to an extreme, "information warfare" could describe terrorist attacks against the hardware or software of the entire Internet. More narrowly defined, this term includes a number of technologies that can accelerate or retard the decision-making cycle of a commander and staff. GPS units can locate friendly units and guide smart weapons, highly sensitive radar systems provide an instantaneous view of the battlefield, and microcomputers linked together offer unparalleled opportunities for the commander to coordinate operations and logistics. In fact, the two greatest dangers of such technology are that the commander and staff may become overwhelmed with information or that the enemy will take actions to jam or impede this flow of information.

Even if a headquarters uses such technology to combine the different arms and services in a flexible manner, two related requirements must be met to ensure effective combinations on the battlefield. First, all arms and services need the same mobility and almost the same degree of armor protection as the units they support. Not only infantry, engineers, field artillery, and air defense but also logistics units need to be able to go where the tanks go in order to support one another. During Desert Storm, a number of these support elements, including TOW antitank launchers, engineer mine-laying equipment, and signals intercept teams, were mounted in 1960s–era tracked vehicles that could not keep pace with the M1 tanks and M2/M3 IFVs in cross-country movement.

Another corollary is that the arms must be balanced within an organization, grouped together to perform according to a particular doctrine. Units above battalion level in which one arm outnumbers the others may be useful in certain circumstances but lack flexibility. Similarly, specialized arms and elites of all kinds, like the tank destroyers of World War II, have special capabilities that must be balanced against their vulnerability when not supported by other arms. This conclusion is reflected in the trend, during the 1990s, to more balanced unit organizations in the British, Polish, Russian, and other armed forces.

Special operations forces may appear to be an exception to this generalization about balance, with such forces having grown in prominence during the last two decades of the century. Guerrilla forces may operate in conjunction with, or independent of, the rest of the combined arms team. Ranger or commando-style assault and reconnaissance units appear to be pure light infantry, yet they function best as part of their own specialized form of combined arms, including intelligence, helicopter transport, long-range communications, and aerial gunship support.

Perhaps the greatest test of the ability of an army to orchestrate the different arms and services is the problem of defense against enemy penetration and exploitation. Part of this involves the necessity for a flexible command structure that functions on the basis of mission-type orders. Yet even with such an effective command structure, troops must be prepared psychologically and technically to continue to fight when penetrated and bypassed by enemy forces. A coordinated defense against an enemy who holds the initiative is a true measure of training and combat effectiveness.

More generally, a high degree of training remains fundamental to combat success in all situations. Throughout the twentieth century, professional soldiers repeatedly neglected or distrusted the training level of their country's reserve forces because the professionals expected a short war that would be fought between the opposing regular armies. Thus the French Army underestimated its own reservists and Germany's in 1914. In 1940 the French suffered an even greater disaster in part because they had failed to train and motivate the next generation of reserves. The Red Army survived the disasters of 1941 largely because it had a huge pool of trained manpower on call.

Today many observers again expect short conflicts and therefore reject the necessity for a system to mobilize manpower and industry. Yet no nation can afford to maintain in peacetime the full force structure it might need for war. Moreover, as demonstrated by the U.S. Defense Department's reliance on reservists in the former Yugoslavia, prolonged peacekeeping missions still require reserve forces, if only to give the active army a respite from constant operational deployments.

Artillery and infantry learned to function together in World War I; and with some difficulty tanks, antitank weapons, engineers, and antiaircraft artillery joined that team during and after World War II. Yet after nearly a century of effort, the aircraft is still not integrated fully into the combined arms team. In four wars since 1941, the U.S. Army and U.S. Air Force have had to develop ad hoc compromises and procedures for air–ground cooperation because they sincerely differed on the basic function of airpower and therefore frequently neglected air–ground cooperation in training and doctrine. During the 1980s intensive joint training attempted to close this gap, but in 1991 and perhaps in 1999 air force planners remained dedicated to the doctrine that an air campaign could win a war not only without a ground attack but also without sufficient consideration to the legitimate concerns of ground commanders who must complete the battle and occupy the terrain. No one disputes the indispensable value of air superiority and air-delivered munitions; the issue is what to do once air superiority is achieved and the air forces turn to attacking enemy ground units.

To some extent, the development of airmobility and attack aviation represents a concerted effort by most major armies to acquire their own air support when it receives a low priority from their air force counterparts.

Properly employed, the helicopter is integrated physically and tactically into the ground battle, providing a third dimension to the combined arms team. The deep attack at the battle of 73 Easting is a superb example of how aviators can use their own capabilities and tactics for independent operations that nevertheless support the overall goals of the force.

In this regard, the dispute between air and ground commanders focused on the difficult link, not only between two aspects of a single campaign but also between the tactical and operational levels of war. As delineated in the AirLand Battle doctrine of the 1980s, the two services must look beyond individual air strikes and ground actions to see how their operations combine to accomplish higher level goals. If a nation attempts to conduct an offensive campaign solely by airpower, it may be so focused on the operational or even strategic-political goal of the campaign that it overlooks the practical reality of how individual battles, both air and ground, can accomplish that overall goal with minimum loss of time and life.

One final issue concerns the technical and tactical trade-offs involved in projecting land- and airpower over long distances. Sophistication and complexity in military units are worthless if they cannot deploy quickly to a theater of war and sustain their operations once they arrive. With the end of the Cold War, the United States retained the need to project power in many parts of the world but lost many of the advanced bases and prepositioned weapons stockpiles that made such power projection possible. Like the British in the Falkland Islands campaign, or the French in their 1994 peacekeeping mission in central Africa, the U.S. armed forces must overcome enormous distances with inadequate means of air and sea transportation.

One obvious solution to this dilemma, as illustrated by the light infantry divisions of the 1980s, is simply to dismantle the combined arms team, to return to a simpler, infantry-oriented form of warfare. Such a force gains speed and mobility by leaving behind most of the large and clumsy tracked vehicles and helicopters of mechanized warfare. That solution has worked well in various low-intensity or noncombat operations during the 1990s. Unfortunately, however, regional powers such as Iraq and North Korea are unlikely to reduce their own armed forces to the relatively primitive level of the light infantry division. Moving a lightly equipped force rapidly may accomplish the political goal of deterring attack; but if deterrence fails, such a force will suffer the fate of Task Force Smith, overrun by an opponent who has shorter supply lines, better weaponry, and greater familiarity with the terrain. Even with wheeled light tanks, a light infantry force, projected thousands of miles from its home, simplifies the tactical problem presented to an opponent. Just as the Egyptians in 1973 found a technical solution to reduce the effectiveness of Israeli tanks and aircraft, so a future enemy may find a response—hugging tactics, chemical weapons, or some other means—that will neutralize the few effective weapons of a light infantry force.

As a minimum, the prospect of light units being forced to defend against heavily armed opponents argues for a much closer and more harmonious relationship between air and ground components. Yet even resolving this issue will not completely eliminate the problem. Throughout the twentieth century, critics consistently argued that U.S. ground forces placed excessive reliance on firepower, preferring to stand on the defensive and blast an enemy instead of maneuvering aggressively to accomplish decisive results. Faced with a well-equipped opponent on some future battlefield, however, a lightly equipped American force may have no choice but to adopt such tactics.

Any nation whose vital interests require the ability to project air- and land-power over long distances must face this dilemma, balancing the needs of strategic mobility against the tactical effectiveness of the force it deploys. For the foreseeable future, therefore, some portion of every major army must remain equipped and trained to conduct the type of integrated, mobile mechanized warfare that has dominated European and Middle Eastern battlefields since 1939.

APPENDIX:
TERMS AND ACRONYMS

• • • • • • • • • • • • • • •

AAF—[U.S.] Army Air Forces, 1942–1947.

ACR—Armored Cavalry Regiment. A U.S. Army combined arms organization of 4,000 to 6,000 men that is capable of reconnaissance, security, and combat missions.

Air assault or airmobility—Concept of using helicopters to move ground combat forces rapidly over long distances, usually behind the forward defenses of the opponent.

Airborne—As a unit designation, indicates that the majority of soldiers in a unit are capable of assault landing by parachute or glider.

APC—Armored Personnel Carrier. A lightly armored vehicle from which infantry dismounts to fight. See IFV.

Armored Infantry—Infantry mounted in lightly armored vehicles whose function is to provide security and support to tank units.

ARVN—Army of the Republic of [South] Vietnam.

Assault gun—Self-propelled armored gun, usually without a turret, which is intended to provide direct-fire support for infantry attacks.

ATGM—Anti-Tank Guided Missile.

BAI—Battlefield Air Interdiction. Air attacks on deep targets that are capable of directly affecting friendly ground forces; ground commanders have a major interest in influencing the choice and timing of BAI targets.

Battalion—A permanent military organization of approximately 300 to 1,000 soldiers (depending on the nation and the purpose of the unit), consisting of two or

---

Note: None of the definitions is official; they are provided here to assist the reader who is unfamiliar with military terminology.

more subordinate companies and normally commanded by a major or lieutenant colonel. The battalion is usually the highest level of organization in which one weapon or combat arm (infantry, artillery, tanks, and so forth) predominates; it is also the standard measure of computing and maneuvering combat power at a tactical level.

BMP—*Boevaya Mashina Piekhota.* Soviet infantry combat vehicle.

Brigade—A military organization of approximately 1,500 to 7,000 soldiers, consisting of two or more subordinate battalions and normally commanded by a colonel or brigadier general. The structure of a brigade may be either fixed or adjustable, depending on the type of brigade and the doctrine of that army. Although some brigades consist almost entirely of one arm or service (field artillery, engineer, military police, and so forth), most are combined arms organizations.

Bronegruppa—Armored group. Soviet term for a group of tanks and infantry fighting vehicles that maneuver separately from their dismounted infantry.

BTR—*Bronetransportr.* Soviet term for an armored personnel carrier.

Caliber—Method of describing the size of a weapon by measuring the width of its projectile in terms of inches or fractions thereof. For example, the bullet on a caliber .45 pistol is forty-five one-hundredths of an inch in diameter.

CAS—Close Air Support. Aerial attacks on ground targets in close proximity to friendly ground forces in order to support them. CAS requires the greatest degree of coordination to avoid friendly casualties and involves the greatest risk to the aircraft conducting the attack.

Cavalry—Military force organized to capitalize on its mobility, especially to search for enemy units and to prevent them from gathering information about friendly units. Historically, cavalry was mounted on horses, but the United States and other armies have used the term to describe armored or helicopter units with similar capabilities. See ACR.

Company team—A combined arms force, temporarily built around a company headquarters but including elements of more than one combat arm.

Corps—A military organization of 20,000 to 200,000 soldiers, typically commanded by a lieutenant general or an equivalent. In addition to controlling two or more subordinate combined arms divisions, a corps typically also has smaller organizations (brigades, regiments, or battalions) of various arms and services, including field artillery, engineers, air defense, military intelligence, and so forth. In Soviet or Russian parlance, a corps may also be a specialized, experimental organization that is somewhat smaller in size.

Covering force—In the defense, an organization whose mission is to force the enemy to slow down, deploy, and expend time and resources before reaching the defender's main defenses.

Defilade—Terrain, such as a hill, that protects against direct fire.

Deep Battle—Operations intended to influence the enemy beyond the range of conventional artillery, using airpower, helicopters, long-range missiles, or ground units in the enemy rear areas.

Direct fire—Any firearm that shoots directly at (or slightly above, to compensate for gravity) its target. Such weapons have a relatively flat trajectory, as opposed to the high-angle, curved trajectory of indirect-fire weapons such as howitzers, mortars, and grenade launchers.

Division—A military organization composed of 7,000 to 20,000 troops. The division is typically the lowest level that includes all combat and combat support arms and is therefore the organizational reflection of an army's concept of how to employ them.

Doctrine—The generally understood concepts by which a particular army performs particular tasks. Contrary to its civilian meaning, "doctrine" does not imply rigid obedience but a flexible application of principles to different circumstances.

Exploitation—Phase of offensive operation after the attacker has penetrated the defender's prepared positions and is able to move rapidly, disrupting the latter's logistics and communications.

FAC—Forward Air Controller. A person or team authorized to direct air strikes at ground targets.

FDC—Artillery fire-direction center.

Field army—A military organization of 50,000 to 400,000 soldiers, consisting (except in the Soviet Army) of two or more subordinate corps. The adjective "field" is used to denote that this organization maneuvers forces in battle. By contrast, "army" may mean all the ground combat forces of a nation, and "theater army" is the administrative organization that controls all soldiers of a nation in a particular region or theater of war.

FIST—U.S. fire support/forward observer team.

FM—Field Manual. A U.S. Army doctrinal publication.

Front—Soviet term for a group of field armies, numbering in the aggregate more than 400,000 troops.

GPS—Global Positioning System. An electronic device that determines locations based on signals from a combination of artificial satellites.

Half-track—A lightly armored vehicle that has a pair of front wheels for steering and a continuous track to support the rest of the vehicle.

Heavy—When used with the name of a particular weapon or arm (for example, "tank heavy") the term implies a force in which that weapon or arm predominates. When used to describe a division or other unit, heavy indicates that it is equipped with armored vehicles, as opposed to lighter, dismounted infantry units.

Howitzer—An artillery piece designed to throw its shell at a relatively low velocity over a relatively steep arc, as opposed to a field gun, which has a higher velocity and flatter trajectory.

Hull down—A position that a tank or other armored vehicle takes so that most of it is protected from enemy fire, usually by a natural or man-made ridge of earth. Only the turret is exposed, permitting the vehicle to fire on the enemy while its own vulnerability is reduced.

ICM—Improved Conventional Munitions. A U.S. artillery shell or bomb that scatters tiny shaped-charge bomblets on an enemy force.

IFV—Infantry Fighting Vehicle. A lightly armored vehicle that carries antitank weapons and permits its infantry to fight either mounted or dismounted, depending on circumstances. See APC.

Indirect fire—The process of engaging a target by firing an explosive shell in a curving trajectory instead of aiming directly at the target. The firing unit is usually behind a hill or otherwise not in direct line of sight with its target. Although known for centuries, indirect fire became standard procedure for artillery in World War I, revolutionizing the battlefield.

Light—Lightly equipped, especially foot-mobile infantry with portable weapons and few or no vehicles.

LZ—Landing Zone. Usually for helicopter assault.

Mechanized—A military unit in which the majority of soldiers operate in some form of armored vehicle that provides them with both transportation and some measure of protection from enemy fire. See Motorized.

MLRS—Multiple Launch Rocket System. U.S. improved, long-range rocket launcher.

Mobile group—Soviet term for a combined arms force with superior mobility and striking power that seeks to disrupt and exploit the enemy's rear areas as part of Deep Battle or deep operations. See OMG.

Mortar—A simple form of indirect-fire weapon that launches explosive shells at an extremely high angle of trajectory. Infantry and armor units often have mortars to provide dedicated fire support.

Motorized—A military unit in which the majority of soldiers operate in some form of wheeled motor vehicle that allows them to move quickly between battles but that does not protect them during battle. See Mechanized.

NATO—North Atlantic Treaty Organization.

NKPA—North Korean People's Army.

OMG—Operational Maneuver Group. Soviet/Russian term (since late 1970s) for an organization designed to perform deep exploitation missions. See Mobile group.

Organic—Adjective indicating that a weapon or element is part of the permanent, official structure of a particular organization. For example, a mortar platoon is frequently organic to an infantry company or battalion. By contrast, tanks, engineers, and other arms are usually not organic to infantry units but must be attached temporarily.

OSS—U.S. Office of Strategic Services. World War II predecessor of Central Intelligence Agency.

Panzer—German term for armored or tank, originally from *Panzerkampfwagen*.

Panzer grenadier—German infantry specialized for support of armored forces.

PLA—Chinese People's Liberation Army.

RAF—[British] Royal Air Force.

RCT—Regimental Combat Team. U.S. term (c. 1940–1957) for an infantry regiment with its supporting artillery, medical, engineer, and other units.

ROAD—Reorganization Objective Army Division. The U.S. division design from the early 1960s to the mid-1980s.

SAM—Surface-to-Air Missile. An air defense weapon.

SIGINT—Signals Intercept. The process of intercepting enemy radio signals for intelligence purposes.

Signal—Traditional term for communications. In the twentieth century, signals usually involve radio-electronic messages.

Special operations—Unconventional military forces, either guerrillas or specially trained assault troops, which use light infantry tactics and the element of surprise to operate against stronger enemy forces. Such operations are usually conducted in enemy rear areas.

Spetsnaz—Soviet special operations troops.

SS—(1) *Shutztaffel.* A politically oriented organization of Nazi Germany that had many branches. In the context of this study, SS units were part of the combat or *Waffen* SS, an army separate from the German Army. *Waffen* SS units usually had high priority for men and equipment. (2) A surface-to-surface missile.

Strong point—A defensive position, including various field fortifications, designed to be defended from all directions, even if bypassed or surrounded.

Suppress—To temporarily disrupt, but not necessarily destroy, an enemy unit. Most commonly, suppression means intensive artillery or other firing that disturbs the enemy soldiers' aim or causes them to seek protective cover, allowing the suppressing force to maneuver more safely.

Task force—An organization, usually of battalion size or larger, that temporarily includes elements of multiple combat and support arms in order to best perform a specified tactical mission. The organization may be designated by a code word, by the name of its commander, or by the numerical designation of the headquarters of that unit (for example, TF 1-34 means a task force built around the 1st battalion, 34th regiment of some arm).

Task organization—The process or result of combining different units and weapons systems to provide a temporary grouping (company team or task force) that sees to accomplishing a specific tactical mission.

TOW—Tactical, Optically Guided Weapon. U.S. antitank guided missile.

TRADOC—U.S. Army Training and Doctrine Command.

NOTES

• • • • • • • • • • • • • •

Introduction

1. Gerald Gilbert, *The Evolution of Tactics* (London, 1907), 183–84.

2. John F. C. Fuller, *The Foundations of the Science of War* (London, 1925), 148.

3. U.S. Department of the Army, Field Manual 100-5: *Operations* (Washington, DC, 1982), 7-4.

4. See, for example, the problems experienced by the British Army in dealing with new weapons in World War I, as brilliantly analyzed by Tim Travers, *The Killing Ground: The British Army, the Western Front and the Emergence of Modern Warfare, 1900–1918* (London, 1987), 69–76.

5. Slim's speech to the 10th Indian Division, 1941, quoted in John Masters, *The Road Past Mandalay* (London, 1961), 45.

Chapter 1. Prologue to 1914

1. Bernard Brodie and Fawn M. Brodie, *From Crossbow to H-Bomb*, rev. ed. (Bloomington, IN, 1973), 82.

2. Ibid., 132.

3. Dennis E. Showalter, *Railroads and Rifles: Soldiers, Technology, and the Unification of Germany* (Hamden, CT, 1976), 75–139.

4. On the development of quick-firing artillery, see Bruce I. Gudmundsson, *On Artillery* (Westport, CT, 1993), 7–8.

5. David Woodward, *Armies of the World, 1815–1914* (New York, 1978), 30, 46, 74. The numerical strength of maneuver divisions was about half the size of these totals.

6. Martin Samuels, *Command or Control? Command, Training and Tactics in the British and German Armies, 1888–1918* (London, 1995), 79.

7. For a discussion of the problems of the British officer corps, see Correlli Barnett, *Britain and Her Army, 1509–1870* (New York, 1970), 333–46. See also Hubert C. Johnson, *Breakthrough! Tactics, Technology, and the Search for Victory on the Western Front in World War I* (Novato, CA, 1994), 33.

8. Samuels, *Command or Control?* 10–58.

9. David G. Chandler, *The Campaigns of Napoleon* (New York, 1966), 351–63.

10. The division structures discussed here are derived primarily from James E. Edmonds and Archibald F. Becke, *History of the Great War: Military Operations France and Belgium, 1914,* 3d ed. (London, 1933), 1:6–7, 490–97. For Russian division organization, see A. A. Strokov, *Vooruzhennye sily i voennoe iskusstvo v Pervoi Mirovoi Voine* [The armed forces and military art in the First World War] (Moscow, 1974), 142, 144–47, 588–89. On machine gun organization, see Wilhelm von Balck, *Development of Tactics—World War* (Ft. Leavenworth, KS, 1922), 176.

11. Pascal Lucas, *Evolution des Idées Tactiques en France et en Allemagne Pendant la Guerre de 1914–1918* [Evolution of Tactical Ideas in France and Germany during the 1914–1918 War], 3d ed. (Paris, 1932), 29.

12. Michael Howard, *The Franco-Prussian War: The German Invasion of France, 1870–1871* (New York, 1969), 156–57.

13. Lucas, *Idées tactiques,* 6; H. Burgess, *Duties of Engineer Troops in a General Engagement of a Mixed Force,* Occasional Paper no. 32, U.S. Army Engineer School (Washington, D.C., 1908) 20, 28–32.

14. Samuels, *Command or Control?* 83; Travers, *The Killing Ground,* 64–65.

15. Howard, *Franco-Prussian War,* 85–117.

16. Jonathan M. House, "The Decisive Attack: A New Look at French Infantry Tactics on the Eve of World War I," *Military Affairs* 40 (December 1976): 164–65.

17. Balck, *Tactics,* 31.

18. See, for example, S. L. A. Marshall, *World War I* (New York, 1971), 76–88, 100–105. On relative mobility, see Jack Snyder, *The Ideology of the Offensive: Military Decision Making and the Disasters of 1914* (Ithaca, NY, 1984), 21–22.

19. Travers, *The Killing Ground,* especially 37–50, 254; House, "The Decisive Attack," 164–67; Samuels, *Command or Control?* 71–83.

20. Henri Bonnal, *La première bataille: le service de deux ans; Du caractère chez les chefs; Discipline—Armée nationale; Cavalerie* [The First Battle: Two-year service; the Character of commanders; Discipline—National Army; Cavalry] (Paris, 1908), 59–60.

21. House, "Decisive Attack," 165–67. See also Ronald H. Cole, "'Forward with the Bayonet!' The French Army Prepares for Offensive Warfare, 1911–1914" (Ph.D. diss., University of Maryland, 1975), esp. 148–50.

22. Woodward, *Armies,* 30–33; Lucas, *Idées tactiques,* 31–34. Snyder, *The Ideology of the Offensive,* 16, contends that some French professional soldiers used the difficulty of the offensive as a reason to demand longer active duty training and to discard reservists as good only for the defensive.

· ·

23. Wilhelm von Leeb, *Defense,* trans. Stefan T. Possany and Daniel Vilfroy (Harrisburg, PA, 1943), 62; Lucas, *Idées tactiques,* 6.

24. Shelford Bidwell and Dominick Graham, *Fire-Power: British Army Weapons and Theories of War, 1904–1945* (Boston, 1982), 14.

25. Robert H. Scales, "Artillery in Small Wars: The Evolution of British Artillery Doctrine, 1860–1914" (Ph.D. diss., Duke University, 1976), 308–17.

26. Bidwell and Graham, *Fire-Power,* 10–11. On the early development of indirect fire techniques, see Boyd L. Dastrup, *King of Battle: A Branch History of the U.S. Army's Field Artillery* (Ft. Monroe, VA, 1992), 127–50; Jonathan Bailey, "The First World War and the Birth of the Modern Style of Warfare," Occasional Paper no. 22, (Camberley, England, 1996); and Albert P. Palazzo, "The British Army's Counter-Battery Staff Office and Control of the Enemy in World War I," *Journal of Military History* 63:1 (January 1999): 55–74.

27. Gudmundsson, *On Artillery,* 20, 32.

28. Ibid., 21–22, 31; Bidwell and Graham, *Fire-Power,* 17; Lucas, *Idées tactiques,* 37; Marshall, *World War I,* 44. See also Robert M. Ripperger, "The Development of the French Artillery for the Offensive, 1890–1914," *Journal of Military History* 59:4 (October 1995): 599–618.

29. Samuels, *Command or Control?* 84–85; Gudmundsson, *On Artillery,* 29.

30. Brodie and Brodie, *Crossbow to H-Bomb,* 126.

Chapter 2. World War I

1. Ferdinand Foch, *The Memoirs of Marshal Foch,* trans. T. Bentley Mott (London, 1931), 16–17. On rifle fire and officer control, see Bruce I. Gudmundsson, *Stormtroop Tactics: Innovation in the German Army, 1914–1918* (Westport, CT, 1989), 24.

2. Gudmundsson, *Stormtroop Tactics,* 80, 83.

3. Lucas, *Idées tactiques,* 115. On Haig, see Travers, *The Killing Ground,* 94.

4. Lucas, *Idées tactiques,* 55–58.

5. Gudmundsson, *On Artillery,* 50–53.

6. Ibid., 57–58, 103n; Bidwell and Graham, *Fire-Power,* 101–7.

7. See, for example, Travers, *The Killing Ground,* 114, 161.

8. Balck, *Tactics,* 244.

9. Gudmundsson, *Stormtroop Tactics,* 175.

10. Samuels, *Command or Control?* 149–50.

11. Graeme C. Wynne, *If Germany Attacks: The Battle in Depth in the West* (London, 1940), 26, 30.

12. John Monash, *The Australian Victories in France in 1918* (New York, n.d.), 124–25, 171–72.

13. Robert A. Doughty, *The Breaking Point: Sedan and the Fall of France, 1940* (Hamden, CT, 1990), 27–30 and passim.

14. Ian V. Hogg, *Barrage: The Guns in Action* (New York, 1970), 28.

15. Wynne, *If Germany Attacks,* 19; Lucas, *Idées tactiques,* 75.

16. The best account of this German defensive system remains Wynne, *If Germany Attacks,* 126–29, 202–12. See also Wynne's "The Development of the German Defensive Battle in 1917 and Its Influence on British Defence Tactics," *Army Quarterly* 24 (1937): 15–32, 248–66; Balck, *Tactics,* 151–68; Samuels, *Command or Control?* 171–91; Gudmundsson, *On Artillery,* 70.

17. Samuels, *Command or Control?* 181. For the spread of machine guns, see Gudmundsson, *Stormtroop Tactics,* 93–100.

18. Samuels's study is a brilliant comparison of the German and British approaches to battlefield change.

19. Wynne, "The Development of the German Defensive Battle," 22–27; Samuels, *Command or Control?* 194–97.

20. Samuels, *Command or Control?* 198–228; Lucas, *Idées tactiques,* 75–76; George C. Marshall, *Memoirs of My Services in the World War, 1917–1918* (Boston, 1976), 61–63.

21. Alan Clark, *The Donkeys* (New York, 1962), 77–82, 84–85, 145–49.

22. This discussion of close air support is based on Peter C. Smith, *Close Air Support: An Illustrated History, 1914 to the Present* (New York, 1990), 5–13; B. Frank Cooling, ed., *Case Studies in the Development of Close Air Support* (Washington, DC, 1990), 17–25; Gary C. Cox, "Beyond the Battle Line: U.S. Air Attack Theory and Doctrine, 1919–1941" (Maxwell Air Force Base, AL, 1996), 5–9; Brereton Greenhous, "Evolution of a Close Ground-Support Role for Aircraft in World War I," *Military Affairs* 39 (February 1975): 22–28; and James S. Corum, *The Luftwaffe: Creating the Operational Air War 1918–1940* (Lawrence, KS, 1997), 26–48.

23. Smith, *Close Air Support,* 8.

24. J. P. Harris, *Men, Ideas, and Tanks: British Military Thought and Armoured Forces, 1903–1939* (Manchester, England, 1995), 57–63.

25. Ibid., 71–74.

26. Bryan Cooper, *The Battle of Cambrai* (New York, 1968), 107–20; Greenhous, "Evolution," 25.

27. George H. Raney, "Tank and Anti-Tank Activities of the German Army," *Infantry Journal* 31 (1927): 151–58.

28. John Wheldon, *Machine Age Armies* (London, 1968), 24–26. See also Samuel D. Rockenbach, "Tanks" (Ft. Meade, MD, 1922), 17, and Balck, *Tactics,* 130–31.

29. Johnson, *Breakthrough!* 171–73; Dale E. Wilson, *Treat 'Em Rough! The Birth of American Armor, 1917–20* (Novato, CA, 1989), 10, 41.

30. Gudmundsson, *On Artillery,* 74–77.

31. Ibid., 74–75; Ian V. Hogg, *The Guns, 1914–1918* (New York, 1971), 72–76.

32. Balck, *Tactics,* 39–40, 181.

33. Wynne, *If Germany Attacks,* 53–58; Lucas, *Idées tactiques,* 109. See also France, Ministère de la Guerre, *Instructions for the Offensive Combat of Small Units,* dated 2 January 1918, translated and printed by Headquarters, American Expeditionary Force, March 1918.

34. Paddy Griffith, *Battle Tactics of the Western Front: The British Army's Art of Attack, 1916–18* (New Haven, CT, 1994), 77–78.

35. This discussion is based largely on Samuels, *Command or Control?* 87–93.

36. John A. English, *A Perspective on Infantry* (New York, 1981), 24–26; Lucas, *Idées tactiques,* 230–31. See also Lazlo M. Alfoldi, "The Hutier Legend," *Parameters* 5 (June 1976): 69–74.

37. Gudmundsson, *On Artillery,* 88–97; Samuels, *Command or Control?* 238.

38. Georg Bruckmüller, *The German Artillery in the Break-Through Battles of the World War,* 2d ed., trans. J. H. Wallace and H. D. Kehm (Berlin, 1922), 41, 44–46; Gudmundsson, *On Artillery,* 89.

39. Bruckmüller, *The German Artillery in the Break-Through Battles,* 49–50, 65–70, 73; S. L. A. Marshall, *World War I,* 346–47; Gudmundsson, *On Artillery,* 91–93; Balck, *Tactics,* 186.

40. U.S. War Department, General Staff, Historical Branch [of] War Plans Division, *A Survey of German Tactics, 1918* (Washington, DC, 1918), 12. See also Bruckmüller, *The German Artillery in the Break-Through Battles,* 54, 73–74; Robert R. McCormick, *The Army of 1918* (New York, 1920), 171.

41. Erwin Rommel, *Infantry Attacks,* trans. G. E. Kidde (Washington, DC, 1944), 177–204.

42. John F. C. Fuller, *Memoirs of an Unconventional Soldier* (London, 1936), 253.

43. Gudmundsson, *On Artillery,* 97–100.

44. John Terraine, *To Win a War—1918, the Year of Victory* (Garden City, NY, 1981), 45, 55–57.

45. This account of Second Armageddon is based primarily on Cyril Falls, *Armageddon, 1918* (Philadelphia, 1964), and on Archibald P. Wavell, *The Palestine Campaigns,* 2d ed. (London, 1929).

46. Smith, *Close Air Support,* 14–15.

47. Hugh Skillen, *Spies of the Airwaves: A History of Y Sections During the Second World War* (Pinner, Middlesex, UK, 1989), 10–11; John Ferris, "The British Army and Signals Intelligence in the Field During the First World War," *Intelligence and National Security* 3:4 (October 1988): 23–48; A. M. Nikolaieff, "Secret Causes of German Successes on the Eastern Front," *Coast Artillery Journal* (September–October 1935): 373–77.

48. See, for example, the role of the British Expeditionary Force's intelligence chief, Brig. John Charteris, as described in Travers, *The Killing Ground,* 115 ff.; Kenneth Strong, *Men of Intelligence: A Study of the Roles and Decisions of Chiefs of Intelligence from World War I to the Present Day* (London, 1970), 28–34, is more sympathetic.

49. Gudmundsson, *On Artillery,* 71–73.

50. Balck, *Tactics,* 188.

51. Ibid., 259.

52. Ibid., 37; Lucas, *Idées tactiques,* 299–300; René Altmayer, *Étude de tactique générale,* 2d ed. (Paris, 1937), 20. For the organizational turmoil caused by the British change to a triangular structure, see Samuels, *Command or Control?* 221, 249.

53. U.S. War Department, General Staff, War College Division, *Order of Battle of the United States Land Forces in the World War: American Expeditionary Forces,*

Divisions (Washington, DC, 1931), 446–47; John B. Wilson, *Maneuver and Fire-power: The Evolution of Divisions and Separate Brigades* (Washington, DC, 1998), 53–54; George C. Marshall, *Memoirs*, 25, 61–63; John J. Pershing, *My Experiences in the World War* (New York, 1931), 1:101n.

54. For the fate of Pershing's tactical ideas, see Henry J. Osterhoudt, "The Evolution of U.S. Army Assault Tactics, 1778–1919: The Search for Sound Doctrine" (Ph.D. diss., Duke University, 1986), 173–92.

Chapter 3. The Interwar Period

1. Constance M. Green, Harry C. Thomson, and Peter C. Roots, *The Ordnance Department: Planning Munitions for War,* U.S. Army in World War II series (Washington, 1955), 205.

2. Edward J. Drea, *Nomonhan: Japanese-Soviet Tactical Combat, 1939,* Leavenworth Paper no. 2 (Ft. Leavenworth, KS, 1981), 17–20; John J. T. Sweet, *Iron Arm: The Mechanization of Mussolini's Army, 1920–1940* (Westport, CT, 1980), 49, 73, 184.

3. C. Serré, *Rapport fait au nom de la commission chargée d'enquêté sur les événements survenues en France de 1933 à 1945* (Paris, 1947–1951), 2:298–99, 313.

4. Sweet, *Iron Arm,* 17; Harold Winton, "General Sir John Burnett-Stuart and British Military Reform 1927–1938" (Ph.D. diss., Stanford University, 1977), 2–3.

5. See, for example, the British cavalry influence described in Harris, *Men, Ideas, and Tanks,* 253–61, 186.

6. This section is based on Cooling, ed., *Case Studies in the Development of Close Air Support,* 28–79; Smith, *Close Air Support,* 16–50; and Cox, *Beyond the Battle Line,* 6–23.

7. The comment "hasten slowly" by Gen. Sir George Milne, chief of the Imperial General Staff, on a 1926 proposal for mechanized formations; quoted in Kenneth Macksey, *The Tank Pioneers* (London, 1981), 69.

8. Fuller, *Memoirs,* 318–41.

9. Harris, *Men, Ideas, and Tanks,* 165–69.

10. Winton, "Burnett-Stuart," 95, 198.

11. Great Britain, War Office, *Field Service Regulations,* vol. 2, *Operations (1924)* (London, 1924), 25; Great Britain, War Office, *Field Service Regulations,* vol. 2, *Operations–General (1935)* (London, 1935), 16; Bidwell and Graham, *Fire-Power,* 261–74.

12. Giffard LeQ. Martel, *In the Wake of the Tank,* 2d ed. (London, 1935), 243, 251–55; Winton, "Burnett-Stuart," 4.

13. Great Britain, War Office, *Field Service Regulations,* vol. 2, *Operations (1924),* 122–23; Great Britain, War Office, *Field Service Regulations,* vol. 2, *Operations (1929)* (London, 1929), 117.

14. Quoted in Harris, *Men, Ideas, and Tanks,* 225–26.

15. Martel, *In the Wake,* 147–48; Winton, "Burnett-Stuart," 198; Macksey, *Tank Pioneers,* 81.

16. Martel, *In the Wake*, 227–28.

17. Ibid., 120–22; Macksey, *Tank Pioneers*, 99.

18. Macksey, *Tank Pioneers*, 132–34; Winton, "Burnett-Stuart," 341–54; Harris, *Men, Ideas, and Tanks*, 251.

19. Winton, "Burnett-Stuart," 509 and passim; Macksey, *Tank Pioneers*, 139, 180–81; Harris, *Men, Ideas, and Tanks*, 283–97.

20. Macksey, *Tank Pioneers*, 86–87. On the problems of British radio communications, see the brilliant essay by John Ferris, "The British Army, Signals, and Security in the Desert Campaign, 1940–42," *Intelligence and National Security* 5:2 (April 1990): 255–91.

21. G. MacLeod Ross, *The Business of Tanks, 1933 to 1945* (Ilfracombe, UK, 1976), 81–137; Harris, *Men, Ideas, and Tanks*, 275–79.

22. On antitank expectations, see Helmut Klotz, *Les Leçons militaires de la Guerre Civile en Espagne*, 2d ed. (Paris, 1937), 61–62, 85–93; Ross, *Business of Tanks*, 130–39; Macksey, *Tank Pioneers*, 142, 160–62.

23. English, *Perspective*, 74–78; Bidwell and Graham, *Fire-Power*, 194–95.

24. Bidwell and Graham, *Fire-Power*, 164–65, 198.

25. "Strike concentrated, not dispersed," by Heinz Guderian, 1940, quoted in Guderian, *Panzer Leader* (New York, 1952), 105.

26. James S. Corum, *The Roots of Blitzkrieg: Hans von Seeckt and German Military Reform* (Lawrence, KS, 1992), 29–31.

27. Germany, [Weimar] Ministry of National Defense, *Command and Control of the Combined Arms (1921–23)*, trans. U.S. Army General Service Schools (Ft. Leavenworth, KS, 1925), 107.

28. Ibid., 23–24, 91, 116, 118.

29. Corum, *The Roots of Blitzkrieg*, 107–8.

30. Ibid., 111–43; Richard L. DiNardo, *Germany's Panzer Arm* (Westport, CT, 1997), 73–77.

31. Kenneth Macksey, *Guderian: Creator of the Blitzkrieg* (New York, 1976), 6–10. Corum is much more critical of Guderian's originality in *The Roots of Blitzkrieg*, 137–43.

32. Guderian, *Panzer Leader*, 24.

33. Charles Messenger, *The Blitzkrieg Story* (New York, 1976), 77–79.

34. Macksey, *Guderian*, 65.

35. Robert M. Kennedy, *The German Campaign in Poland (1939)*, Dept. of the Army Pamphlet 20-255 (Washington, DC, 1956), 28–30; Matthew Cooper and James Lucas, *Panzer: The Armoured Force of the Third Reich* (New York, 1976), 22–23; DiNardo, *Germany's Panzer Arm*, 96–97.

36. Kennedy, *Poland*, 28.

37. Corum, *Luftwaffe*, 202–4, 221–23, 271.

38. Ibid., 266–69: Corum argues that Udet's preoccupation with dive-bombers resulted in major errors in the development of Germany's other aircraft; Smith, *Close Air Support*, 31–33, 48–50; Cooling, ed., *Case Studies in Close Air Support*, 81.

39. Paul Deichmann, *German Air Force Operations in Support of the Army,* USAF Historical Studies no. 13 (Maxwell Air Force Base, AL, 1962), 33–36, 131–33; Cooling, ed., *Case Studies in Close Air Support,* 82–90.

40. Cooper and Lucas, *Panzer,* 23.

41. France, Ministère de la Guerre, *Instruction sur l'emploi tactique des grandes unités, 1936* (Paris, 1940), 16; Doughty, *The Breaking Point,* 8–11.

42. France, Ministère de la Guerre, *Instruction provisoire sur l'emploi tactique des grandes unités, 1921* (Paris, 1922), 23.

43. Ibid., 24.

44. Jeffrey J. Clarke, "Military Technology in Republican France: The Evolution of the French Armored Force, 1917–1940" (Ph.D. diss., Duke University, 1969), 94–97.

45. Jean-Baptiste Estienne, "Étude sur les missions des chars blindées en campagne," 25 May 1919, quoted in Georges Ferré, *Le Défaut de l'armure* (Paris, 1948), 34–47; Estienne, *Conférence faite le 15 Fevrier 1920 sur les chars d'assaut* (Paris, 1920), 37–38, 42.

46. Messenger, *Blitzkrieg Story,* 89; Robert A. Doughty, "The Enigma of French Military Armored Doctrine, 1940," *Armor* 83 (September–October 1974): 41; on light cavalry division composition, see Doughty, *The Breaking Point,* 81.

47. Charles de Gaulle, *Vers l'armée du métier* (Paris, 1963 [1934]), 97.

48. Serré, *Rapport,* 2: 377; Clarke, "Military Technology," 152; Robert Jacomet, *L'Armement de la France, 1936–1939* (Paris, 1945), 123–26.

49. Ferré, *Le Défaut,* 74–75.

50. Ministère de la Guerre, *Instruction (1936),* 16, 18, 21, 23, 44, 46, 111, 154; Clarke, "Military Technology," 82.

51. Smith, *Close Air Support,* 18–20, 38–39; Cooling, ed., *Case Studies in Air Support,* 28–29, 31.

52. Doughty, *The Breaking Point,* 105–19.

53. Albert Seaton, *Stalin as Military Commander* (New York, 1976), 62–63.

54. The Soviet concept of Deep Battle is summarized in David M. Glantz, "Soviet Operational Formation for Battle: A Perspective," *Military Review* 63 (February 1983): 4. See also G. S. Isserson, "Operational Prospects for the Future," a 1938 article translated by Harold S. Orenstein in *The Evolution of Soviet Operational Art, 1927–1991: The Documentary Basis* (London, 1995), 1:78–90.

55. USSR, Commissariat of Defense, General Staff, *Field Regulations, 1936 (Tentative),* trans. Charles Berman (Washington, DC, typescript, 1937), 4–6, 83–84.

56. A. Ryazanskiy, "The Creation and Development of Tank Troop Tactics in the Pre-War Period," *Voyenni vestnik* 11 (1966): 25, 32; George F. Hofman, "Doctrine, Tank Technology, and Execution: I. A. Khalepskii and the Red Army's Fulfillment of Deep Offensive Operations," *Journal of Slavic Military Studies* 9:2 (June 1996): 283–334.

57. A. Yekimovskiy and N. Makarov, "The Tactics of the Red Army in the 1920s and 1930s," *Voyenni vestnik* 3 (1967): 11.

58. John Erickson, *The Road to Stalingrad: Stalin's War with Germany* (New York, 1975), 1:6, 19–20. For a more recent Russian account, see O. F. Suvenirov, "Vsearmeiskaia tragediia" [An army-wide tragedy], *Voyenno-istoricheskii zhurnal* 3 (March, 1989): 42.

59. Messenger, *Blitzkrieg Story*, 100. See also David M. Glantz, "Observing the Soviets: U.S. Military Attachés in Eastern Europe During the 1930s," *Journal of Military History* 5:2 (April 1991): 153–83; John L. S. Daley, "The Theory and Practice of Armored Warfare in Spain" (part 1) *Armor* 108:2 (March–April 1999): 30, 39–43. For a detailed discussion of the Soviet experience and lessons learned, see Steven J. Zaloga, "Soviet Tank Operations in the Spanish Civil War," *Journal of Slavic Military Studies* 12:3 (September 1999): 134–62.

60. Erickson, *The Road to Stalingrad*, 18, 32.

61. This account of Khalkin-Gol is based primarily on Drea, *Nomonhan*, 23–86.

62. U.S. War Department, General Staff, War College, *Provisional Manual of Tactics for Large Units* (Washington, DC, 1922–1923). William O. Odom argues in *After the Trenches: The Transformation of U.S. Army Doctrine, 1918–1939* (College Station, TX, 1999) that this pro-French manual was in fundamental conflict with the offensively oriented American doctrine of other interwar publications.

63. U.S. War Department, General Staff, *Field Service Regulations, U.S. Army, 1923* (Washington, DC, 1924), 11.

64. John J. Pershing, Indorsement (forwarding report of the AEF Superior Board on Organization and Tactics) to secretary of war, General Headquarters, AEF, 16 June 1920; Fox Conner, letter to Maj. Malin Craig (concerning division structure), 24 April 1920, copy in U.S. Army Military History Institute.

65. The primary sources for this discussion of the triangular division's development are Kent R. Greenfield, Robert R. Palmer, and Bell I. Wiley, *The Army Ground Forces: The Organization of Ground Combat Troops*, U.S. Army in World War II series (Washington, DC, 1947), 271–78; Harry C. Ingles, "The New Division," *Infantry Journal* 46 (November–December 1939): 521–29; Janice E. McKenney, "Field Artillery Army Lineage Series," typescript (Washington, DC, 1980), 255–61; Wilson, *Maneuver and Firepower*, 126–45; U.S. Army, Headquarters Second Division, "Special Report Based on Field Service Test of the Provisional 2d Division, Conducted by the 2d Division, U.S. Army, 1939," National Archives File 52-83.

66. Ernest F. Fisher Jr., "Weapons and Equipment Evolution and its Influence upon Organization and Tactics in the American Army from 1775–1963" (Washington, DC, n.d.), 61–67, 77.

67. Blanche D. Coll, Jean E. Keith, and Herbert H. Rosenthal, *The Technical Services: The Corps of Engineers: Troops and Equipment*, U.S. Army in World War II series (Washington, DC, 1958), 10–20; Paul W. Thompson, *Engineers in Battle* (Harrisburg, PA, 1942); on the German engineer role in 1940, see Doughty, *The Breaking Point*, 57, 142, 155–57.

68. On the development of fire-direction centers, see McKenney, "Field Artillery Army Lineage," 266–73; Riley Sunderland, "Massed Fires and the F.D.C." *Army* 8 (1958): 56–59.

69. The primary sources for this discussion of U.S. armor between the wars are Timothy K. Nenninger, "The Development of American Armor, 1917–1940" (master's thesis, University of Wisconsin, 1968), 55–188; James M. Snyder, ed., *History of the Armored Force, Command, and Center,* Army Ground Forces Study no. 27 (Washington, DC, 1946), 1–17; and Mildred H. Gillie, *Forging the Thunderbolt: A History of the Development of the Armored Force* (Harrisburg, PA, 1947), 37–178.

70. Coll et al., *The Corps of Engineers,* 17.

71. Cox, "Beyond the Battle Line," 9–33; Cooling, ed., *Case Studies in Close Air Support,* 41–60; Frank D. Lackland, "Attack Aviation," student paper written at the Command and General Staff College (Ft. Leavenworth, KS, 1931); U.S. War Department, General Staff, FM 100–5: *Tentative Field Service Regulations—Operations* (Washington, DC, 1939), 22.

Chapter 4. The Axis Advance, 1939–1942

1. This critique of the Polish campaign is based on Guderian, *Panzer Leader,* 65–72; Kennedy, *Poland,* 130–35; and Cooper and Lucas, *Panzer,* 25–28. See also Klaus A. Maier et al., *Germany's Initial Conquests in Europe,* vol. 2, *Germany and the Second World War,* trans. Dean S. McMurry and Ewald Osers (Oxford, England, 1991), 98–106.

2. Kennedy, *Poland,* 61–62. The Polish campaign did include at least one encirclement battle, as described by Rolf Elble, *Die Schlacht an der Bzura im September 1939 aus deutscher und polnischer Sicht* (Freiburg, FRG, 1975).

3. Cooper and Lucas, *Panzer,* 24, 27, 29.

4. Kennedy, *Poland,* 28–30 and 133; Guderian, *Panzer Leader,* 89.

5. Cooper and Lucas, *Panzer,* 27.

6. J. L. Moulton, *A Study of Warfare in Three Dimensions: The Norwegian Campaign of 1940* (Athens, OH, 1967), especially 61–63; Gerhard L. Weinberg, *A World at Arms: A Global History of World War II* (Cambridge, England, 1994), 113–19.

7. Hanson W. Baldwin, *The Crucial Years, 1939–1941* (New York, 1976), 93.

8. Maier et al., *Germany's Initial Conquests,* 281–82; Doughty, *The Breaking Point,* 45, 53–54.

9. James E. Mrazek, *The Fall of Eben Emael* (Washington, DC, 1971); see also James Lucas, *Kommando: German Special Forces of World War II* (New York, 1985), 53–69.

10. R. H. S. Stolfi, "Equipment for Victory in France in 1940," *History: The Journal of the Historical Association* 55 (February 1970): 1–20.

11. Ibid.; Ferré, *Le Défaut,* 185–95.

12. On Gort, see Brian Bond, *Britain, France, and Belgium, 1939–1940,* 2d ed. (London, 1990), 62.

13. Weinberg, *A World at Arms,* 129.

14. On the problems of building French units, see the classic study of the 55th Infantry Division in Doughty's *The Breaking Point*, especially 111–28.

15. Macksey, *Tank Pioneers*, 160–62; Bryan Perrett, *Knights of the Black Cross: Hitler's Panzerwaffe and Its Leaders* (New York, 1986), 77.

16. For the air–ground coordination problem, see Guderian, *Panzer Leader*, 97–98, 101–2; Doughty, *The Breaking Point*, 136–37.

17. Cooling, ed., *Close Air Support*, 92–94; Robert Jackson, *Air War over France, May-June, 1940* (London, 1974), 50–62.

18. I. S. O. Playfair, F. C. Flynn, C. J. C. Molony, and S. E. Toomer, *The Mediterranean and the Middle East*, vol. 2, *The Germans Come to the Help of Their Allies (1941)*, History of the Second World War (London, 1956), 175.

19. Ibid.; Ross, *The Business of Tanks*, 151.

20. This discussion of British developments after Dunkirk is derived largely from Nigel Hamilton, *Monty: The Making of a General, 1887–1942* (New York, 1981), 394–545; Martel, *Our Armoured Forces* (London, 1945), 75–181; and English, *Perspective*, 155–63, 199–201; and on artillery, Bidwell and Graham, *Fire-power*, 230–34.

21. See the unfortunate Canadian experience of battle drills and fire support in John A. English, *The Canadian Army and the Normandy Campaign* (New York, 1991), 111–19.

22. Hamilton, *Monty*, 459–60, 533–55.

23. Martel, *Our Armoured Forces*, 159–61.

24. Ibid., 379–80. The detailed organization is included in Great Britain, War Office letter number 20/GEN/6059 (S.D. 1), dated 1 October 1942, CARL N-6136.

25. Playfair et al., *The Mediterranean and the Middle East*, vol. 1, *The Early Successes Against Italy (to May 1941)* (London, 1954), 261–68. Barrie Pitt, *The Crucible of War: Western Desert, 1941* (London, 1980), 85–117.

26. Michael Carver, *Dilemmas of the Desert War: A New Look at the Libyan Campaign, 1940–1942* (Bloomington, IN, 1986), 15.

27. Ferris, "The British Army, Signals and Security in the Desert," 255–91.

28. On equipment in North Africa, see Carver, *Dilemmas of the Desert War*, esp. 25–52; English, *The Canadian Army*, 163–72; Playfair et al., *The Mediterranean and Middle East*, 2:13–14, 173–75, 341–45, and *British Fortunes Reach Their Lowest Ebb* (London, 1960), 3:27–28; U.S. War Department, General Staff, Military Intelligence Service, *Artillery in the Desert*, Special Series no. 6, 25 November 1942 (Washington, DC, 1942), 39–41; J. A. I. Agar-Hamilton and L. C. F. Turner, *The Sidi Rezegh Battles, 1941* (Cape Town, SA, 1957), 33–55.

29. Carver, *Dilemmas of the Desert War*, 69.

30. Ibid.

31. The account of the 16th RTR, which other sources attribute to the 6th RTR, is from U.S. War Department, General Staff, Military Intelligence Service, Special Bulletin no. 36, *The Battle of Salum, June 15–17, 1941* (Washington, DC, 1941), 26. See also Pitt, *The Crucible of War*, 300–301.

32. Great Britain, War Office Middle East Command, "Notes on Main Lessons of Recent Operations in the Western Desert," dated 10 August 1942, typescript copy by U.S. War Department, General Staff, Military Intelligence Service, in CARL, N-3915. Playfair et al., *The Mediterranean and the Middle East,* 3:213–14, 223–24, 254, 287.

33. Hamilton, *Monty,* 653–54, 680–81.

34. Ibid., 732–844.

35. Andrei Eremenko, *The Arduous Beginning* (Moscow, 1966), 12–22; U.S. War Department, General Staff, Military Intelligence Division, Special Bulletin no. 2: "Soviet-Finnish War: Operations from November 30, 1939 to January 7, 1940," dated 10 January 1940, G2 document 2657-D-1054; Malcolm McIntosh, *Juggernaut: A History of the Soviet Armed Forces* (New York, [1968]), 113–16.

36. On Soviet reforms in 1940 and 1941, see David M. Glantz and Jonathan M. House, *When Titans Clashed: How the Red Army Stopped Hitler* (Lawrence, KS, 1995), 23–27, 33–41; Erickson, *Road to Stalingrad,* 13–49; A. Yekimovskiy, "Tactics of the Soviet Army During the Great Patriotic War," *Voyenni vestnik* 4 (1967): 12.

37. Erickson, *Road to Stalingrad,* 32–34. For the evolution of the T-34, see Steven Zaloga and Peter Sarson, *T-34/76 Medium Tank, 1941–1945* (London, 1994). On the effect of such weapons in battle, see Steven Zaloga, "Technological Surprise and the Initial Period of War: The Case of the T-34 Tank," *Journal of Slavic Military Studies* 6:4 (December 1993): 634–46.

38. Guderian, *Panzer Leader,* 138; Cooper and Lucas, *Panzer,* 40–42; Perrett, *Knights of the Black Cross,* 77–78.

39. Martin Van Crevald, *Supplying War: Logistics from Wallenstein to Patton* (Cambridge and New York, 1977), 160–80; Daniel K. Beaver, "Politics and Policy: The War Department Motorization and Standardization Program for Wheeled Transport Vehicles, 1920–1940," *Military Affairs* 47 (October, 1983): 101; Guderian, *Panzer Leader,* 190. For a good summary of German strengths and weaknesses in 1941, see Kenneth Macksey, "The German Army in 1941," in *The Initial Period of War on the Eastern Front, 22 June–August 1941,* ed. David M. Glantz (London, 1993), 55–65.

40. The Soviet explanation is summarized in N. Kobrin, "Encirclement Operations," *Soviet Military Review* 8 (1981): 36–39. In *Stalingrad: Anatomy of an Agony* (London, 1992), V. E. Tarrant has argued that this emphasis on encirclement is not, strictly speaking, a part of blitzkrieg but reflects the nineteenth-century German concept of *Vernichtungsgedanke,* centrally controlled battles to destroy the opponent. This apparent contradiction is explained by reference to the levels of war I have described in the introduction. Tarrant is correct that, at the *tactical* level, blitzkrieg envisaged decentralized execution, aiming to disrupt and disorganize rather than to destroy the enemy. At the *operational* and *strategic* levels, however, the outcome of these disruptive attacks had to be the capture or destruction of Soviet units so that they could not reorganize themselves and fight again. Indeed, as described in this book, one major cause of German failure in 1941 to 1942 lay in the failure to eliminate the Red Army before pushing deep for political and economic objectives.

41. On the Soviet response, see Glantz and House, *When Titans Clashed,* 62–72.

42. Erickson, *Road to Stalingrad,* 173; V. D. Sokolovsky, ed., *Soviet Military Strategy,* 3d ed., ed. and trans. Harriet Fast Scott (New York, 1975), 163.

43. Glantz and House, *When Titans Clashed,* 99–103; A. Ryazanskiy, "Tactics of Tank Forces During the Great Patriotic War," *Voyenni vestnik* 5 (1967): 13–20; Sokolovsky, *Soviet Military Strategy,* 165.

44. Stalin's Order no. 306, 8 October 1942, captured and translated by the German Army and retranslated by U.S. Army, copy in CARL N-16582.256. E. Bolton, "Talks on the Soviet Art of War: 2, Strategic Defence, the Struggle to Capture the Initiative," *Soviet Military Review* 6 (1967): 45.

45. Glantz and House, *When Titans Clashed,* 143–47.

Chapter 5. *Allied Response and Armored Clashes*

1. This discussion of McNair's concepts is based on Greenfield, Palmer, and Wiley, *Organization of Ground Combat Troops,* 271, 273, 185.

2. Ibid., 356–59. The only nondivisional U.S. regiments for much of the war were infantry and dismounted cavalry, many of which became separate units when the National Guard divisions were reduced from square to triangular formation, losing their fourth regiments. These regiments proved useful for everything from protecting the Pacific Coast to assaulting islands too small for a full division; their function was retained even after the war, as evidenced by modern separate brigades. See the wry commentary by Melvin Walthall, *We Can't All Be Heroes* (Hicksville, NY, 1975). For a more serious example of their role, see Edward J. Drea, *Defending the Driniumor: Covering Force Operations in New Guinea, 1944,* Leavenworth Paper no. 9 (Ft. Leavenworth, KS, 1984), concerning the 112th Cavalry Regimental Combat Team.

3. On the combined arms effort to break through the hedgerows, see Michael D. Doubler, *Closing with the Enemy: How the GIs Fought in Europe, 1944–1945* (Lawrence, KS, 1994), 33–62.

4. Greenfield, Palmer, and Wiley, *Organization of Ground Combat Forces,* 307.

5. Ibid., 322–35; Snyder, *History of the Armored Force,* 29–43.

6. U.S. Army, European Theater of Operations, General Board, "Organization, Equipment, and Tactical Employment of the Armored Division," Study no. 48 (Washington, DC, n.d.), appendix 1.

7. On armored infantry, see W. Blair Haworth Jr., "The Bradley and How It Got That Way: Mechanized Infantry, Organization, and Equipment in the United States Army, 1936–1996" (1999 typescript, based on Ph.D. diss., Duke University, 1995), 22–26. Early in the war, the United States met the sudden demand for armored, self-propelled artillery by mounting a 105mm howitzer on the same chassis as a Sherman tank. The resulting weapon was the M7, called the Priest because of the pulpit-shaped machine gun position located at the front of the vehicle. See, for example, Dastrup, *King of Battle,* 205.

8. For an excellent nontechnical explanation of antitank design factors, see John Weeks, *Men Against Tanks: A History of Anti-Tank Warfare* (New York, 1975), 12–16.

• •

9. Ibid., 100–102, 67–69; Green, Thomson, and Roots, *Planning Munitions,* 355–61.

10. This dual track continued even in the Canadian Army of 1944, where brigades equipped with heavy Churchill tanks trained for infantry support; those brigades equipped with American-built Shermans trained for independent maneuver (English, *The Canadian Army,* 170).

11. On the institutional development of the tank destroyer, see Christopher R. Gabel, *Seek, Strike, and Destroy: U.S. Army Tank Destroyer Doctrine in World War II,* Leavenworth Paper no. 12 (Ft. Leavenworth, KS, 1985), 5–18.

12. Ibid., 17–25; Green, Thomson, and Roots, *Planning Munitions,* 388–90, 402–4; Charles M. Baily, "Faint Praise: The Development of American Tanks and Tank Destroyers During World War II" (Ph.D. diss., Duke University, 1977), 12–38.

13. U.S. Army, European Theater of Operations, General Board, "Organization, Equipment, and Tactical Employment of Tank Destroyer Units," Study no. 60 (Washington, DC, n.d.), 1–2, 10; Gabel, *Seek, Strike, and Destroy,* 58.

14. Pavel Rotmistrov, "Cooperation of Self-Propelled Artillery with Tanks and Infantry," *Zhurnal bronetankovykh i mekhanizirovamykh voisk* 7 (1945): 8–13, trans. U.S. Department of the Army, G2, n.d.; K. Novitskiy, "Coordination Between Medium and Heavy Tanks in Offensive Combat," *Tankist,* no. 9 (1947): 40–43, trans. U.S. Department of the Army, G2, n.d.

15. Green, Thomson, and Roots, *Planning Munitions,* 246–56, 283–86; Cooper and Lucas, *Panzer,* 54ff.

16. Perrett, *Knights of the Black Cross,* 103–5; DiNardo, *Germany's Panzer Arm,* 18–19.

17. Guderian, *Panzer Leader,* 299, 311; Perrett, *Knights of the Black Cross,* 104–5. For a summary of these German vehicles in the context of Kursk, see Jonathan M. House, "Waiting for the Panther: Kursk, 1943," in *The Limits of Technology in Modern Warfare,* ed. Andrew J. Bacevich and Brian R. Sullivan (Cambridge, England, forthcoming).

18. This discussion of Kursk is based on House, "Waiting for the Panther"; David M. Glantz and Jonathan M. House, *The Battle of Kursk* (Lawrence, KS, 1999); Janusz Piekalkiewicz, *Operation "Citadel": Kursk and Orel, the Greatest Tank Battle of the Second World War* (Novato, CA, c. 1987); and Walter S. Dunn, *Kursk: Hitler's Gamble, 1943* (Westport, CT, 1997).

19. Steven J. Zaloga and James Grandsen, *Soviet Tanks and Combat Vehicles of World War II* (London, 1984), 156–66.

20. Pavel A. Rotmistrov, *Stal'naia gvardiia* [Steel guards] (Moscow, 1988), 180. See also Richard N. Armstrong, *Red Army Tank Commanders: The Armored Guards* (Atglen, PA, 1994), 346–53.

21. Richard M. Ogorkiewicz, *Design and Development of Fighting Vehicles* (Garden City, NY, 1968), 36.

22. DiNardo, *Germany's Panzer Arm,* 15–20.

23. Army Regulation 850-15, cited in Green, Thomson, and Roots, *Planning Munitions,* 278.

24. On the M4 Sherman, see ibid., 282–87 and 302–4; Baily, "Faint Praise," iii, 7, 41–48. The M26, equipped with the same 90mm gun as the M36 tank destroyer but carrying more armor, made the latter largely redundant. See Gabel, *Seek, Strike, and Destroy,* 64.

25. Ross, *Business of Tanks,* 213, 263–300.

Chapter 6. The Complexity of Modern Warfare, 1943–1945

1. On ULTRA and its limitations, see Ralph Bennett, *Ultra in the West: The Normandy Campaign, 1944–45* (New York, 1979), 13–20, and Patrick Beesly, *Very Special Intelligence: The Story of the Admiralty's Operational Intelligence Centre, 1939–1945* (London, 1977), 63–65, 110–11. For the Pacific, see Edward Drea, *MacArthur's Ultra: Codebreaking and the War Against Japan, 1942–1945* (Lawrence, KS, 1992).

2. Skillen, *Spies of the Airwaves,* 206. Although somewhat disorganized, Skillen's massive study is one of the few effective histories of tactical-level SIGINT in the two world wars.

3. F. H. Hinsley, E. E. Thomas, C. F. G. Ransom, and R. C. Knight, *British Intelligence in the Second World War: Its Influence on Strategy and Operations* (London, 1979), 1:144.

4. Ibid. (London, 1981), 2:582–87, 739–46, and 757–63. Skillen, *Spies of the Airwaves,* 272–90.

5. Anthony Cave Brown, *Bodyguard of Lies* (New York, 1975), 102–4; for the structure and accomplishments of Unit 621, see Skillen, *Spies of the Airwaves,* 88–202.

6. Skillen, *Spies of the Airwaves,* 149–50, 171.

7. Ibid., 56–58, 123, 169–87; Hinsley et al., *British Intelligence,* 1:148; Hamilton, *Monty,* 1:777. See also Aileen Clayton, *The Enemy Is Listening* (London, 1980); H. W. Everett, "The Secret War in the Desert," *British Army Review* 60 (December 1978): 66–68.

8. George R. Thompson and Dixie R. Harris, *The Signal Corps: The Outcome (Mid-1943 Through 1945),* U.S. Army in World War II series (Washington, DC, 1966), 301–50; Skillen, *Spies of the Airwaves,* 241–55, 456–90; J. P. Finnegan, "U.S. Army Signals Intelligence in World War II," paper presented at Society for Military History Convention, Washington, DC, March 1990.

9. Dulany Terrett, *The Signal Corps: The Emergency (to December 1941),* U.S. Army in World War II series (Washington, DC, 1956), 118–20, 141–47, 178–85.

10. Eisenhower letter to the General Board, 4 October 1945, reproduced in U.S. Army, European Theater of Operations, the General Board, "Army Tactical Information Service," Study no. 18 (Washington, DC, n.d.), appendix.

11. Glantz and House, *When Titans Clashed,* 182 and passim; David M. Glantz, *Soviet Military Deception in the Second World War* (London, 1989).

12. Vasili I. Chuikov, *The Fall of Berlin,* trans. Ruth Kisch (New York, 1968), 30–33; Glantz and House, *When Titans Clashed,* 182.

13. See, for example, S. Alferov, "Wartime Experience: Breakthrough of Enemy Defenses by a Rifle Corps," *Voyenno-istoricheskiy zhurnal* [Military History Journal] 3 (1983): 53–56.

14. David M. Glantz, *Autumn Storm*, Leavenworth Paper no. 7 (Ft. Leavenworth, KS, 1984).

15. On the development of forward detachments, see ibid., and N. Kireyev and N. Dovbenko, "From the Experience of the Employment of Forward Detachments of Tank (Mechanized) Corps," *Voyenno-istoricheskiy zhurnal* 9 (1982): 20–27.

16. For the German shortages in 1942, see Earl F. Ziemke, *Stalingrad to Berlin: The German Defeat in the East* (Washington, DC, 1968), 17, and George E. Blau, *The German Campaign in Russia: Planning and Operations (1940–1942)* (Washington, DC, 1955), 128–30; for a detailed discussion of German personnel problems in 1943, see Dunn, *Kursk: Hitler's Gamble*, 38–50.

17. Timothy A. Wray, "Standing Fast: German Defensive Doctrine on the Russian Front During the Second World War" (master's thesis, Ft. Leavenworth, KS, 1983), 162, 221–35. Regrettably, only the first part of Wray's brilliant work has been published as *Standing Fast: German Defensive Doctrine on the Russian Front During World War II: Prewar to March 1943*, Research Survey no. 5 (Ft. Leavenworth, KS, 1986).

18. Wray, "Standing Fast," 227, 273; U.S. War Department, Technical Manual E 30-451, *Handbook on German Military Forces, 15 Mar 1945* (Washington, DC, 1945), 162, 221–35. On German artillery command and control issues, see Gudmundsson, *On Artillery*, 126–29.

19. Wray, "Standing Fast," 275, 298–303.

20. Ibid.; Wray, *Standing Fast*, 167–68; Guderian, *Panzer Leader*, 290–91, and 249–51.

21. Richard E. Simpkin, *Mechanized Infantry* (Oxford, England, 1980), 22.

22. TM E 30-451, II-46 to II-51, IV-9 to IV-13.

23. David M. Hazen, "Role of the Field Artillery in the Battle of Kasserine Pass" (master's thesis, Ft. Leavenworth, KS, 1973), 38–42, 77–79, 147, 174–75, 187; Snyder, *History of the Armored Force*, 17; Dastrup, *King of Battle*, 210.

24. Samuel Milner, *Victory in Papua*, U.S. Army in World War II series (Washington, DC, 1957), 92–95, 135, 375–76; see also Jay Luvaas, "Buna: 19 November 1942–2 January 1943: A 'Leavenworth Nightmare,'" in *America's First Battles, 1776–1965*, Charles E. Heller and William A. Stofft, eds. (Lawrence, KS, 1986), 186–225.

25. This account of the battle of Ormoc Valley is based on a number of sources, including M. Hamlin Cannon, *Leyte: The Return to the Philippines*, U.S. Army in World War II series (Washington, DC, 1954); Hideto Kida, "Imperial Japanese Army's Operations in New Guinea and the Decisive Battle at Leyte (Shoichigo Operation)," paper presented at the U.S.–Japanese Military History Exchange Conference, Tokyo, 1984; Headquarters, 32d Infantry Division, "Report After Action of 32d Infantry Division in Leyte Engagement or K-2 Operation for the Period 16 November–25 December 1944"; U.S. War Dept., Headquarters, 128th Infantry Regiment, "Historical Record, 128th Infantry, from 14 November 1944 to 25 November 1944."

26. Martin Blumenson, *Breakout and Pursuit,* U.S. Army in World War II series (Washington, DC, 1961), 41–43, 208.

27. Ian Gooderson, *Air Power at the Battlefront: Allied Close Air Support in Europe 1943–45* (London, 1998), 85–94.

28. U.S. Army Armor School, "Armor in the Exploitation," student project (typescript, 1949), 60–61, CARL N-2146.74.

29. The following discussion of close air support is based on numerous sources, including Gooderson, *Air Power at the Battlefront;* Thomas A. Hughes, *Over Lord: General Pete Quesada and the Triumph of Tactical Air Power in World War II* (New York, 1995); Cooling, ed., *Case Studies in the Development of Close Air Support;* Kent R. Greenfield, *Army Ground Forces and the Air–Ground Battle Team, Including Organic Light Aviation,* Army Ground Forces Study no. 35 (Washington, DC, 1948); and Smith, *Close Air Support.* See also Richard P. Hallion, *Strike from the Sky: The History of Battlefield Air Attack, 1911–1945* (Washington, DC, 1989).

30. Gooderson, *Air Power at the Battlefront,* 23.

31. Ibid., 24–26; W. A. Jacobs, "Air Support for the British Army, 1939–1943," *Military Affairs* 46 (December, 1982): 174–82.

32. Greenfield, *Army Ground Forces and the Air–Ground Battle Team,* 47.

33. Edward Mark, *Aerial Interdiction: Air Power and the Land Battle in Three American Wars* (Washington, DC, 1994), 107, 129–30, 236–49.

34. Smith, *Close Air Support,* 81–82, 96.

35. Greenfield, *Army Ground Forces and the Air–Ground Battle Team,* 30, 43, 74.

36. Ibid., 49, 53. See also Daniel R. Mortensen, *A Pattern for Joint Operations: World War II Close Air Support, North Africa* (Washington, DC, 1987).

37. Gooderson, *Air Power at the Battlefront,* 27–28.

38. Ibid., 35.

39. Hughes, *Over Lord,* esp. 186–88.

40. Gooderson, *Air Power at the Battlefront,* 41, 203.

41. Ibid., 48, 32. As a final commentary on the lack of air–ground coordination, Leslie McNair himself was a victim of this lack of communication, killed by bombs that fell on friendly troops while he was observing an attempt to use heavy bombers to break the stalemate in Normandy.

42. Playfair et al., *The Mediterranean and Middle East,* 3:71, 241.

43. Deichmann, *German Air Force Operations in Support,* 37–48.

44. Headquarters, U.S. Army Forces, Far East, Japanese Research Division, *Japanese Night Combat* (Tokyo, 1955), 1:8–15.

45. William Slim, *Defeat into Victory* (New York, 1961), 260, 337–38, 346.

46. Ibid., 249, 260–62, 321, 379, 409; for a vivid account of the Fourteenth Army's advance, see Masters, *The Road Past Mandalay,* 292–320.

47. Philip A. Crowl, "Command Relationships in Amphibious War" (typescript, 1953, on file in CARL); Albert N. Garland and Howard M. Smith, *Sicily and the Surrender of Italy,* U.S. Army in World War II series (Washington, DC, 1965), 175–82.

48. Gooderson, *Air Power at the Battlefront,* 95–97.

49. James M. Gavin, *Airborne Warfare* (Washington, DC, 1947), 81.

50. U.S. Army, European Theater of Operations, General Board, "Organization, Equipment, and Tactical Employment of the Airborne Division," Study no. 16 (Washington, DC, 1946), 20, 29–30.

51. David M. Glantz, *The Soviet Airborne Experience,* Research Survey no. 4 (Ft. Leavenworth, KS, 1984), esp. 109–11.

52. Crowl, "Command Relationships," 17–22; on the failure of air–ground coordination on the Normandy beaches, see Gordon A. Harrison, *Cross-Channel Attack,* U.S. Army in World War II series (Washington, DC, 1951), 300–301.

53. Jeter A. Isely and Philip A. Crowl, *The U.S. Marines and Amphibious War: Its Theory, and Its Practice in the Pacific* (Princeton, NJ, 1951), 3–4, 36–39, 59, 233, 334. On JASCOs and related problems of amphibious fire support, see Thompson and Harris, *The Signal Corps: The Outcome,* 229–33.

54. See, for example, James F. Gebhardt, *The Petsamo-Kirkenes Operation: Soviet Breakthrough and Pursuit in the Arctic, October 1944,* Leavenworth Paper no. 17 (Ft. Leavenworth, KS, 1989), 99–114.

55. Michael J. King, *Rangers: Selected Combat Operations in World War II,* Leavenworth Paper no. 11 (Ft. Leavenworth, KS, 1985).

56. Slim, *Defeat into Victory,* 133–35, 220–43. Masters (*Road Past Mandalay,* 111–282) describes his experiences commanding a Chindit unit, the 111th Brigade. Robin Neillands, *In the Combat Zone: Special Forces Since 1945* (New York, 1998), 20–40, gives a good summary of British and U.S. special operations forces (SOF) in World War II, although his definition of SOF is so broad that it includes any airborne or "elite" unit. James Lucas's *Kommando: German Special Forces of World War II* (New York, 1985), provides a systematic account of all German SOF.

57. Two recent German reviews of Stalingrad are Paul Carell [Paul Schmidt], *Stalingrad: The Defeat of the German 6th Army,* trans. David Johnston (Atglen, PA, c. 1993), and Joachim Wieder and Henrich von Einsiedel, *Stalingrad: Memories and Reassessments,* trans. Helmut Bogler (London, 1995). For a brief summary of the Soviet side, see Glantz and House, *When Titans Clashed,* 120–23, 132–36.

58. On Metz, see Hugh M. Cole, *The Lorraine Campaign,* U.S. Army in World War II series (Washington, DC, 1950), 372–449, and the largely derivative study by Anthony Kemp, *The Unknown Battle: Metz, 1944* (Briarcliff Manor, NY, 1981).

Chapter 7. "Limited" Warfare, 1945–1973

1. A. Ryazanskiy, "Land Forces *Podrazdeleniye* Tactics in the Postwar Period," *Voyenni vestnik* 8 (1967): 15; David Glantz, "Soviet Offensive Military Doctrine Since 1945," *Air University Review* 34 (March–April 1983): 25–34.

2. John Erickson, "The Ground Forces in Soviet Military Policy," *Strategic Review* 4 (April 1976): 65.

3. Simpkin, *Mechanized Infantry,* 31; M. V. Zakharov, ed., *50 let vooruzhennykh sil SSSR* [50 years of the Soviet Armed Forces] (Moscow, 1968), 483. On the limi-

tations of the BTRs, see David C. Isby, *Weapons and Tactics of the Soviet Army,* 3d ed. (London, 1988), 167–75.

4. The Soviet "Revolution in Military Affairs" should not be confused with the same label used by Western analysts in the 1980s. On Sokolovsky's role, see Harriet Fast Scott and William F. Scott, *Soviet Military Doctrine: Continuity, Formulation, and Dissemination* (Boulder, CO, 1988), 22–25.

5. Erickson, "The Ground Forces," 66; Glantz, "Soviet Offensive Military Doctrine Since 1945," 27.

6. Matthew Allen, *Military Helicopter Doctrines of the Major Powers, 1945–1992: Making Decisions About Air–Land Warfare,* Contributions in Military Studies no. 127 (Westport, CT, 1993), 71–80.

7. U.S. Department of the Army, Assistant Chief of Staff for Intelligence, Intelligence Research Project no. A-1729, "Soviet Tank and Motorized Rifle Division" (Washington, DC, 1958).

8. Nikita Khrushchev's destruction of the traditional Soviet Army angered many senior officers and was undoubtedly one reason for his eventual overthrow. On Khrushchev and the nuclear issue, see Vladislav Zubok and Constantine Pleshakov, *Inside the Kremlin's Cold War: From Stalin to Khrushchev* (Cambridge, MA, 1996), esp. 184–265, and Steven J. Zaloga, *Target America: The Soviet Union and the Strategic Arms Race, 1945–1964* (Novato, CA, 1993).

9. S. Shtrik, "The Encirclement and Destruction of the Enemy During Combat Operations Not Involving the Use of Nuclear Weapons," *Voyennaya mysl'* 1 (1968): 279.

10. Glantz, "Soviet Offensive Military Doctrine," 29; Kireyev and Dovbenko, "Forward Detachments," 20–27; R. Y. Malinovskiy and O. Losik, "Wartime Operations: Maneuver of Armored and Mechanized Troops," *Voyenno istoricheskiy zhurnal* 9 (1980): 18–25.

11. Quoted in Robert A. Doughty, *The Evolution of U.S. Army Tactical Doctrine, 1945–76,* Leavenworth Paper no. 1 (Ft. Leavenworth, KS, 1979), 4; see also U.S. Army, European Theater of Operations, General Board, "Organization, Equipment, and Tactical Employment of Tank Destroyer Units," Study no. 60 (Washington, DC, n.d.), 27–29; "Types of Divisions—Postwar Army," Study no. 17 (Washington, DC, n.d.), 8.

12. "The New Infantry, Armored, and A/B [Airborne] Divisions," typed explanation and tables issued by Headquarters, U.S. Army Ground Forces, 24 January 1947, copy in CARL.

13. George S. Patton Jr., quoted in Virgil Ney, *Evolution of the U.S. Army Division 1939–1968* (Ft. Belvoir, VA, 1969), 114.

14. These figures are based on a comparison between Staff Officers' Field Manual 101-10 (Tentative), U.S. Army Command and General Staff College, with changes to 1943; "Armored Division" table in "The New Infantry, Armored, and A/B Divisions"; and Greenfield et al., *Organization of Ground Combat Troops,* 320–21. See also U.S., Department of the Army, the Armored School, "Armored Division Or-

ganizational and Manning Charts, TO&E 17N," Instructional Pamphlet no. CS-2 (Ft. Knox, KY, 1949), 1–2.

15. James F. Schnabel, *Policy and Direction: The First Year,* U.S. Army in the Korean War (Washington, DC, 1972), 54; William W. Epley, "America's First Cold War Army, 1945–1950," Land Warfare Papers no. 15, Association of the United States Army (Arlington, VA, 1993), esp. 17–21.

16. For an evocative account of the Eighth Army's troubles in Korea, see Blair, *The Forgotten War,* esp. 46–159.

17. Ibid., 98; Fehrenbach, *This Kind of War,* 126–51.

18. S. L. A. Marshall, *Commentary on Infantry Operations and Weapons Usage in Korea, Winter of 1950–51,* Operations Research Office Study ORO-R-13 (Chevy Chase, MD, 1951), 6–7.

19. Ibid., 128–31; Marshall, "CCF in the Attack (Part II)," typescript, Operations Research Office Study ORO-S-34 (EUSAK) (Tokyo, 1951), 7–18.

20. John A. English and Bruce I. Gudmundsson, *On Infantry,* rev. ed. (Westport, CT, 1994), 145–48. For a detailed description of an actual Chinese assault, see Anthony Farrar-Hockley, *The Edge of the Sword* (London, 1954), 42–44.

21. S. L. A. Marshall, *Operation Punch and the Capture of Hill 440, Suwon, Korea, February 1951,* Technical Memorandum ORO-T-190 (Chevy Chase, MD, 1952), 9–10, 43–44, 69.

22. Walter G. Hermes, *Truce Tent and Fighting Front,* U.S. Army in the Korean War (Washington, DC, 1966), 119–21.

23. S. L. A. Marshall, *Pork Chop Hill: The American Fighting Man in Action; Korea, Spring, 1953* (New York, 1956), 47, 196; Hermes, *Truce Tent,* 370–72.

24. Maj. Gen. Otto P. Weyland, quoted in Mark, *Aerial Interdiction,* 274.

25. Harold K. (Johnny) Johnson, commander, 8th Cavalry Regiment, 1st Cavalry Division, quoted in Blair, *The Forgotten War,* 577n.

26. James F. Schnabel, "History of the Korean War, Part Two, Volume I, General Headquarters Support and Participation (25 June 1950–30 April 1951)," typescript (n.d.) in U.S. Army Center of Military History, 9–10. This decision was not a normal procedure of joint operations at the time; in fact, MacArthur overruled Lt. Gen. George Stratemeyer, commander of Far East Air Forces (FEAF), who protested that air superiority and attacks on airfields were a FEAF responsibility.

27. Bergeron, *The Army Gets an Air Force,* 52; Hermes, *Truce Tent,* 325–28. The strong opinions of senior commanders are evident in letters such as that of 2 January 1952 from Lt. Gen. Edward Almond (U.S. Army War College) to General Van Fleet, commander, Eighth Army, and Van Fleet's reply, 22 January 1952 (copies in author's files). One USMC forward controller echoed the concerns of these senior officers, describing how the 5th Air Force's Joint Operations Center "Allocates all A/C [aircraft] to all corps sectors along the front, a request has to be made through a long, tiresome channel prior to the approval, and it could be you wouldn't even be given the A/C you asked for in the end anyway. What a business!" (Bernard W. Peterson, *Short Straw: Memoirs of Korea by a Fighter Pilot/Forward Air Controller* [Scottsdale, AZ, 1996], 228). The debate on controlling airpower was also played

out in professional journals. See, for example, William R. Kintner, "Who Should Command Tactical Air Forces?" *Combat Forces Journal* 1:4 (November 1950): 34–37, and Elmer G. Owens and Wallace F. Veaudry, "Control of Tactical Air Power in Korea," *Combat Forces Journal* 1:9 (April 1951): 19–21.

28. Charles D. Hightower, "The History of the United States Air Force Airborne Forward Air Controller in World War II, the Korean War, and the Vietnam Conflict" (master's thesis, Ft. Leavenworth, KS, 1984, 63–64).

29. John R. Galvin, *Air Assault: The Development of Airmobile Warfare* (New York, 1969), 254–56.

30. On special operations in Korea, see Ed Evanhoe, *Darkmoon: Eighth Army Special Operations in the Korean War* (Annapolis, 1995); Lynn Montross and Nicholas A. Canzona, *U.S. Marine Operations in Korea, 1950–1953,* vol. 2, *The Inchon-Seoul Operation* (Washington, DC, 1955); and Shelby L. Stanton, *America's Tenth Legion: X Corps in Korea, 1950* (Novato, CA, 1989), 54, 66.

31. For the early history of Special Forces troops, see Andrew F. Krepinevich Jr., *The Army and Vietnam* (Baltimore, 1986), 100–106, and Aaron Bank, *From OSS to Green Berets: The Birth of Special Forces* (Novato, CA, 1986), 155–70.

32. Joseph J. Bermudez Jr. *North Korean Special Forces,* 2d ed. (Annapolis, MD, 1998), esp. 1–89; Daniel P. Bolger, *Scenes from an Unfinished War: Low Intensity Conflict in Korea, 1966–1969,* Leavenworth Paper no. 19 (Ft. Leavenworth, KS, 1991).

33. This discussion of the People's Liberation Army is based primarily on Alexander L. George, *The Chinese Communist Army in Action; The Korean War and Its Aftermath* (New York and London, 1967), esp. 197–225; Michael Carver, *Twentieth Century Warriors: The Development of the Armed Forces of the Major Military Nations in the Twentieth Century* (New York and London, 1987), 419–24; Ellis Joffre, *The Chinese Army After Mao* (Cambridge, MA, 1987); Curt Bartholomew, "China's People's Liberation Army: Basic Doctrine and Infantry Tactics," in U.S. Army Intelligence and Threat Analysis Center, *How They Fight: Armies of the World* (Washington, DC, 1992), 19–23; and Department of Defense, Defense Intelligence Agency, *Handbook on the Chinese Armed Forces* (Washington, DC, 1976).

34. This Chinese infantry organization is from the Defense Intelligence Agency's *Handbook on the Chinese Armed Forces,* appendix A.

35. Doughty, *Evolution,* 23; Wilson, *Maneuver and Firepower,* 244–53.

36. This analysis of the Pentomic Division is derived from Theodore C. Mataxis and Seymour L. Goldberg, *Nuclear Tactics, Weapons, and Firepower in the Pentomic Division, Battle Group, and Company* (Harrisburg, PA, 1958), 103–12; John H. Cushman, "Pentomic Infantry Division in Combat," *Military Review* 37 (January 1958): 19–30; Letter, U.S. Continental Army Command, 8 January 1959, Subject: Changes in ROCID TOE (U), with supporting CGSC documentation (CARL N-17935.62–U). See also Donald A. Carter, "From G.I. to Atomic Soldier: The Development of U.S. Army Tactical Doctrine, 1945–1956" (Ph.D. diss., Ohio State University, 1987).

37. Haworth, "The Bradley and How It Got That Way," chap. 3.

38. Letter, U.S. Continental Army Command, 13 January 1959, Subject: The Armored Cavalry Regiment (U); and Letter of Instruction, U.S. Army Command and General Staff College, 14 January 1959, Subject: DA approved Divisional changes, with accompanying charts (CARL N-17935.62–U).

39. Doughty, *Evolution*, 19–25; Ney, *Evolution*, 75. For a detailed exposition of how task organization was supposed to work, see U.S. Department of the Army, Field Manual 71-2, *The Tank and Mechanized Infantry Battalion Task Force* (Washington, DC, 1977), 3-4 to 3-11.

40. Malcolm Postgate, *Operation Firedog: Air Support in the Malayan Emergency 1948–1960* (London, 1992), 99–108; Allen, *Military Helicopter Doctrines*, 129–30, 191–92.

41. Mataxis and Goldberg, *Nuclear Tactics*, 122; Galvin, *Air Assault*, 264.

42. Bergeron, *The Army Gets an Air Force*, 80–106; Hamilton H. Howze, "The Howze Board," part 1, *Army* (February 1974): 12–14.

43. Howze, "The Howze Board," part 2, *Army* (March 1974): 18–24, and part 3, *Army* (April 1974): 18–24; Howze, "Tactical Employment of the Air Assault Division," *Army* (September 1963): 44–45, 52.

44. Galvin, *Air Assault*, 280–87, 293; John J. Tolson, *Airmobility, 1961–1971* (Washington, DC, 1973), 51–59.

45. Tolson, *Airmobility*, 25–28.

46. Ibid., 32, 43–44; Bergeron, *The Army Gets an Air Force*, 119. The 10 percent CAS figure is from Krepinevich, *The Army and Vietnam*, 200.

47. Dave R. Palmer, *Summons of the Trumpet: U.S.–Vietnam in Perspective* (San Rafael, CA, 1978), 98–102; Galvin, *Air Assault*, 194–95; for a full account of the initial battle, see Heller and Stofft, *America's First Battles*, 300–326.

48. Tolson, *Airmobility*, 88, 102–4, 195, 201–2.

49. Ibid., 120.

50. Krepinevich, *The Army and Vietnam*, 11 and 125.

51. English and Gudmundsson, *On Infantry*, 156–60, Robert H. Scales Jr., *Firepower in Limited War*, rev. ed. (Novato, CA, 1995), 140–41, 151.

52. Scales, *Firepower in Limited War*, 141–53; Krepinevich, *The Army and Vietnam*, 198–200.

53. This account of Lam Son 719 is based on U.S. Army, Headquarters, 101st Airborne Division (Airmobile), "Final Report—Airmobile Operations in Support of Operation Lam Son 719" (24 April 1971), vol. 1 (CARL N-18430.56-A); Tolson, *Air Assault*, 236–52; Palmer, *Summons*, 238–43; Scales, *Firepower in Limited War*, 149–51.

54. 101st Airborne Division, "Final Report," 1:12, 1:15.

55. Ibid., 1:52; Allen, *Military Helicopter Doctrines*, 24.

56. 101st Airborne Division, "Final Report," 1:42 to 1:43; Tolson, *Airmobility*, 251–52.

57. Simpkin, *Mechanized Infantry*, 25; Jean Marzloff, "The French Mechanized Brigade and Its Foreign Counterparts," *International Defense Review* 6 (April 1973): 178.

58. France, Ministère de la Guerre, École Superieure de Guerre, *Études operations 1er cycle, 1953–1954* (Paris, 1953), Book 1, Annex 3; Marzloff, "French Mechanized Brigade," 176–78; International Institute for Strategic Studies, "The Military Balance 1977–78," *Air Force* 60 (December 1977): 80–81; Philippe C. Peress, "The Combined Arms Battalion Concept in the French Army," student study project, U.S. Army Command and General Staff College, 1977, 1–41 (CARL N-13423.472); France, École Superieure de Guerre, "Memorandum on Nuclear Weapons," undated, 1976 translation no. K-6556 by U.S. Department of the Army, Assistant Chief of Staff for Intelligence, 17-8.

59. Peress, "Combined Arms Battalion," 1–41.

60. W. Blair Haworth Jr., "The Bradley and How It Got That Way," 50–54.

61. Ibid., 67–69; Marzloff, "French Mechanized Brigade," 177–78; Edgar D. Arendt, "Comparative Analysis of Contemporary Non–U.S. Army Small Infantry Unit Organizations" (Washington, DC, 1967), 17–22.

62. Moshe Dayan, *Moshe Dayan: Story of My Life* (New York, 1976), 45–49, 97–98, 100–120.

63. Moshe Dayan, *Diary of the Sinai Campaign* (New York, 1966), 34–35.

64. This account of Abu Agheila in 1956 is derived from S. L. A. Marshall, *Sinai Victory* (New York, 1967), 94–140; Edward Luttwak and Dan Horowitz, *The Israeli Army* (New York, 1975), 148–49, 151; and George W. Gawrych, *Key to the Sinai: The Battles for Abu Agheila in the 1956 and 1967 Arab-Israeli Wars*, Research Survey no. 7 (Ft. Leavenworth, KS, c. 1990). For the organization of the Egyptian and Israeli forces at Abu Agheila, see Gawrych, 14–25.

65. Gawrych, *Key to the Sinai*, 59–65.

66. Avraham Adan, *On the Banks of the Suez* (San Rafael, CA, 1980), 207–13.

Chapter 8. Combined Arms, 1973–1990

1. Trevor N. Dupuy, *Elusive Victory: The Arab-Israeli Wars, 1947–1974* (New York, 1978), 231, 612–13; Edgar O'Ballance, *No Victor, No Vanquished: The Yom Kippur War* (San Rafael, CA, 1978), 55.

2. Adan, *Banks of the Suez*, 57; Jac Weller, "Armor and Infantry in Israel," *Military Review* 55 (April 1975): 3–11.

3. Adan, *Banks of the Suez*, 491–92.

4. For the Egyptian side of the 1973 conflict, see Mohammed Heikal, *The Road to Ramadan* (London and New York, 1975), 14–43, 208, 240–41; Charles Wakebridge, "The Egyptian Staff Solution," *Military Review* 55 (March 1975): 3–11; Chaim Herzog, *The War of Atonement: October 1973* (Boston, 1975), 273; O'Ballance, *No Victor,* 115. The fullest recent discussion of the Egyptian side of the 1973 war is by George W. Gawrych, *The 1973 Arab-Israeli War: The Albatross of Decisive Victory,* Leavenworth Paper no. 21 (Ft. Leavenworth, KS, 1996), 8–27.

5. *Times* (London), *The Yom Kippur War* (Garden City, 1974), 189.

6. Gawrych, *The 1973 Arab-Israeli War,* 26–27.

7. Ibid., 36; Adan, *Banks of the Suez,* 141; Herzog, *War of Atonement,* 194.

8. *Times* (London), *The Yom Kippur War,* 157–82; Luttwak and Horowitz, *The Israeli Army,* 372–76. Although several evocative memoirs have described the desperate fighting on the Golan, the Syrian aspects of the 1973 war have not been as thoroughly analyzed as the Egyptian front.

9. Herzog, *War of Atonement,* 84–113, 205; Gawrych, *The 1973 Arab-Israeli War,* 55–57.

10. Richard A. Gabriel, *Operation Peace for Galilee: The Israeli–PLO War in Lebanon* (New York, 1984), 21.

11. Ibid., 97–98, 74. See also F. H. Toase, "The Israeli Experience of Armoured Warfare," in *Armoured Warfare,* ed. J. P. Harris and Toase (New York, 1990), esp. 181–85.

12. George W. Gawrych, "Attack Helicopters in Lebanon, 1982," in *Combined Arms in Battle Since 1939,* ed. Roger J. Spiller (Ft. Leavenworth, KS, 1992), 35–41.

13. This section is based on Joffre, *The Chinese Army After Mao,* especially 80–180, and on Harlan W. Jencks, "Lessons of a 'Lesson': China–Vietnam, 1979," in *The Lessons of Recent Wars in the Third World,* ed. Robert E. Harkavy and Stephanie G. Neuman (Lexington, MA, 1985), 1:139–60.

14. Paul H. Herbert, *Deciding What Has to Be Done: General William E. DePuy and the 1976 Edition of FM 100-5, Operations,* Leavenworth Paper no. 16 (Ft. Leavenworth, KS, 1988), 36. This discussion of the Active Defense is based heavily on Herbert's brilliant analysis.

15. U.S. Department of the Army, Field Manual 100-5, *Operations* (Washington, DC, 1 July 1976), esp. chaps. 2 and 3.

16. The author was personally taught and tested on this "calculus of battle" in various army service schools between 1978 and 1982.

17. Herbert, *Deciding What Has to Be Done,* 100–105.

18. Allen, *Military Helicopter Doctrines,* 86–89.

19. Haworth, "The Bradley and How It Got That Way," chap. 5. On the BMP–BTR mix, see Donald L. Madill, "The Continuing Evolution of the Soviet Ground Forces," *Military Review* 62:8 (August 1982): 52–58.

20. Allen, *Military Helicopter Doctrines,* 81–89.

21. N. V. Ogarkov, "Glubokaya operatsiya (boy)" in *Soverskaya voyennaya entsiklopediya* (1976), 2:574–78, trans. and ed., Harold S. Orenstein, *The Evolution of Soviet Operational Art, 1928–1991: The Documentary Basis,* vol. 2, *Operational Art, 1965–1991* (London, 1995), 191–99.

22. The author gratefully acknowledges the assistance of David M. Glantz on the evolution of Soviet doctrine. This discussion of the OMG is based on personal conversations with Colonel Glantz.

23. Allen, *Military Helicopter Doctrines,* 95–97.

24. On the Ogadan, see Isby, *Weapons and Tactics of the Soviet Army,* 90. The development of Soviet operational concepts exceeds the scope of this present work. For further information, see Orenstein, ed., *Evolution of Soviet Operational Art,* vol. 2.

25. Allen, *Military Helicopter Doctrines,* 97–100.

26. Robert F. Baumann, *Russian-Soviet Unconventional Wars in the Caucasus, Central Asia, and Afghanistan,* Leavenworth Paper no. 20 (Ft. Leavenworth, KS, 1993), 138–39.

27. Ibid., 141–42; David C. Isby, *War in a Distant Country—Afghanistan: Invasion and Resistance* (London, 1989), 24–25; Lester W. Grau and Mohammad Y. Nawroz, "The Soviet Experience in Afghanistan," *Military Review* 75:5 (September–October 1995): 17–27.

28. Isby, *War in a Distant Country,* 28, 37; Baumann, *Russian-Soviet Unconventional Wars,* 142–45. For the development of Soviet fire support in Afghanistan, see Scales, *Firepower in Limited War,* 155–97.

29. This description and the accompanying map of Xadighar Canyon are based on the Soviet account in Lester W. Grau, ed., *The Bear Went over the Mountain: Soviet Combat Tactics in Afghanistan* (Washington, DC, c. 1996), 56–59.

30. Baumann, *Russian-Soviet Unconventional Wars,* 153.

31. Isby, *War in a Distant Country,* 38, 44, 114. On the *mujahideen* side of the Afghan conflict, see Ahmad Jalili and Lester W. Grau, *The Other Side of the Mountain: Mujahideen Tactics in the Soviet-Afghan War* (Quantico, VA, 1995).

32. Peter de Leon, "The Laser-Guided Bomb: Case History of a Development," RAND Report no. R-1312-1-PR (Santa Monica, CA, 1974).

33. David C. Isby and Charles Kamps Jr., *Armies of NATO's Central Front* (London and New York, 1985), 52.

34. W. Seth Carus, *Cruise Missile Proliferation in the 1990s,* Washington Paper no. 156, Center for Strategic and International Studies (Westport, CT, and London, 1992), 16–60.

35. Department of the Army, Field Manual 71-2, 1-5.

36. Haworth, "The Bradley and How It Got That Way," chap. 6.

37. James W. Bradin, *From Hot Air to Hellfire: The History of Army Attack Aviation* (Novato, CA, 1994), 29, 145.

38. Allen, *Military Helicopter Doctrines,* 16–29. This discussion of the evolution of attack aviation is based primarily on Allen.

39. Ibid., 35, 45.

40. Anthony Farrar-Hockley, *The Army in the Air: The History of the Army Air Corps* (Dover, NH, and Stroud, Gloucestershire, 1994), esp. 211–29; Allen, *Military Helicopter Doctrines,* 135–51.

41. Allen, *Military Helicopter Doctrines,* 182–296.

42. In addition to the author's own experiences, this discussion of AirLand Battle is based on John L. Romjue, *From Active Defense to AirLand Battle: The Development of Army Doctrine 1973–1982* (Ft. Monroe, VA, 1984). See also Robert H. Scales, *Certain Victory: The U.S. Army in the Gulf War* (Washington, DC, 1993), 12–36. For a comparison of AirLand Battle and Active Defense, see Huba Wass de Czege and L. Donald Holder, "The New FM 100-5," *Military Review* 62:7 (July 1982): 53–70.

43. AirLand Battle was first described in the 1982 edition of Field Manual 100-5, *Operations;* it was fully developed in a 1986 edition, which received only limited

distribution within the army, and is not readily available. This discussion of the two editions is based on Wass de Czege and Holder, "The New FM 100-5," and L. Donald Holder, "Education and Training for Theater Warfare," in *On Operational Art*, ed. Clayton R. Newell and Michael D. Krause (Washington, DC, 1994), 171–87; and Holder, "Maneuver in Deep Battle," *Military Review* 62:5 (May 1982): 54–61.

44. The changes in U.S. artillery during the 1980s are summarized in Dastrup, *King of Battle*, 290–310.

45. R. P. Hunnicutt, *Abrams: A History of the American Main Battle Tank* (Novato, CA, 1990), 2:174–202. For the long history of tank design prior to the M1, see Robert S. Cameron, "American Tank Development During the Cold War," *Armor* 107:4 (July–August 1998): 30–38.

46. Haworth, "The Bradley and How It Got That Way," chaps. 7 and 8.

47. Lawrence M. Jackson II, "Force Modernization—Doctrine, Organization, and Equipment," *Military Review* 62:12 (December 1982): 2–12.

48. On the iterations of Division 86, see Glen R. Hawkins, *United States Army Force Structure and Force Design Initiatives, 1939–1989* (Washington, D.C., n.d.), especially 63–71, 81–83, and Wilson, *Maneuver and Firepower*, 369–401.

49. Richard G. Davis, *The Thirty-one Initiatives: A Study in Air Force–Army Co-operation* (Washington, DC, 1987).

50. Charles E. Kirkpatrick, "The Army and the A-10: The Army's Role in Developing Close Air Support Aircraft, 1961–1971" (Washington, DC, c. 1987), 35–37; Smith, *Close Air Support*, 161–62.

51. "USA and USAF Agreement on Apportionment and Allocation of Offensive Air Support (OAS)—Information Memorandum," 23 May 1981, reproduced in Romjue, *From Active Defense to AirLand Battle*, 100–108. See also C. Lanier Deal Jr., "BAI: The Key to the Deep Battle," *Military Review* 62:3 (March 1982): 51–54. In the later 1980s, one of my duties as a division staff officer was to request airborne electronic jamming missions from the air force. In order to fit such requests into the centralized air tasking process, I was expected to predict the exact type and location of threat enemy air defense radars thirty-six hours in advance of the battle.

52. Michael J. Mazarr, *Light Infantry Forces and the Future of U.S. Military Strategy* (Washington, DC, 1990), 30–37.

53. Ibid., 49.

54. Many of the details in this discussion may be found in Susan L. Marquis, *Unconventional Warfare: Rebuilding U.S. Special Operations Forces* (Washington, DC, 1997), esp. 62–115.

55. The full story of the Grenada operation exceeds the scope of this book. Interested readers should consult ibid., 98–105; Bruce R. Pirnie, *Operation Urgent Fury: The United States Army in Joint Operations* (Washington, DC, 1986); Mark Adkin, *Urgent Fury: The Battle for Grenada* (Lexington, MA, 1989); Jonathan M. House, *The U.S. Army in Joint Operations, 1950–1983* (Washington, DC, 1992), chap. 7; and Daniel P. Bolger, "Special Operations and the Grenada Campaign," *Parameters* 18:4 (December 1988): 49–61.

56. Bruce W. Watson and Peter M. Dunn, eds., *Military Lessons of the Falklands War: Views from the United States* (Boulder, CO, and London, 1984), 42. For an Argentine view of the air war, see Ruben O. Moro, *The History of the South Atlantic Conflict: The War for the Malvinas* (New York, 1989), esp. 219–37, 254–56, 324–28. Martin Middlebrook has attempted to provide a balanced view from both sides in *The Fight for the 'Malvinas': The Argentine Forces in the Falklands War* (London and New York, 1989).

57. Great Britain, Ministry of Defence, *The Falklands Campaign: The Lessons* (London, 1982), 22, 34.

58. The Argentine 2d Marine Battalion had just completed amphibious exercises with the U.S. Marines. On the invasion, see Middlebrook, *Fight for the 'Malvinas,'* 18–39.

59. This account of Goose Green is based on Watson and Dunn, *Military Lessons,* 69–72; Moro, *History of the South Atlantic War,* 260–66; Scales, *Firepower in Limited War,* 199–203; Middlebrook, *Fight for the 'Malvinas,'* 177–97.

60. Anthony H. Cordesman, *NATO's Central Region Forces: Capabilities/Challenges/Concepts* (London, 1988), 140–47.

61. Ibid., 91–103.

62. Ibid., 199–209.

63. Allen, *Military Helicopter Doctrines,* 195–96.

64. David M. Glantz, "Soviet and Commonwealth Military Doctrine in Revolutionary Times" (Ft. Leavenworth, KS, 1992), esp. 9–10, and Glantz, "Operational Art and Tactics," *Military Review* 68:12 (December 1988): 32–40. I wish to thank Les Grau for his advice concerning the organization of the Russian/Soviet corps.

65. Charles J. Dick, "A Bear Without Claws: The Russian Army in the 1990s," *Journal of Slavic Military Studies* 10:1 (March 1997): 1–10; for Russian force structure, see Terence Taylor, ed., *The Military Balance 1997/98* (London, 1997), 109.

66. Eugeniusz Jendraszczak, "The Restructuring of the Polish Armed Forces in a Defensive Strategy," *Journal of Slavic Military Studies* 8:4 (December 1995): 752–59.

67. The best Western analysis of the Iran-Iraq conflict is Stephen C. Pelletiere and Douglas V. Johnson II, *Lessons Learned: The Iran-Iraq War* (Carlisle Barracks, PA, 1991). See also Efraim Karsh, *The Iran-Iraq War: A Military Analysis* (London, 1987), and Edgar O'Ballance, *The Gulf War* (London, 1988). On Iraqi tactics, see U.S. Army, National Training Center Handbook 100-91, *The Iraqi Army: Organization and Tactics* (Ft. Irwin, CA, 1991), 1, 46, 117–19; and Scales, *Firepower in Limited War,* 235–38.

Chapter 9. War in the 1990s

1. Jeffrey Record, "Why the Air War Worked," *Armed Forces Journal International* (April, 1991): 44–45; similar sentiments appear in Michael J. Mazarr et al., *Desert Storm: The Gulf War and What We Learned* (Boulder, CO, 1993), 4–5. The official (and somewhat optimistic) analysis of Gulf War air operations is the Department of the Air Force's five-volume *Gulf War Airpower Survey* (Washington,

DC, 1993). Williamson Murray's *Air War in the Persian Gulf* (Baltimore, 1995) is based on this official study but includes some notable observations about the limits of airpower. See also Fred Frostic, *Air Campaign Against the Iraqi Army in the Kuwaiti Theater of Operations* (Santa Monica, CA, 1994), and James A. Winnefield et al., *A League of Airmen: U.S. Air Power in the Gulf War* (Santa Monica, CA, 1994).

2. Frostic, *Air Campaign Against the Iraqi Army*, 49–52.

3. Martin S. Kleiner, "Joint STARS Goes to War," *Field Artillery* (February 1992): 25–29, and Scales, *Certain Victory*, 167–70.

4. Jeffrey B. Rochelle, "The Battle of Khafji: Implications for Airpower" (master's thesis, U.S. Air Force School of Advanced Aerospace Studies, 1997), 12, 27. On the battle of Khafji, see also Murray, *Air War in the Persian Gulf*, 252–53; Martin N. Stanton, "The Saudi Arabian National Guard Motorized Brigades," *Armor* 105:2 (March–April 1996): 6–11; Richard M. Swain, *"Lucky War": Third Army in Desert Storm* (Ft. Leavenworth, KS, 1994), 190–91.

5. Swain, *"Lucky War,"* 186–90; Frostic, *Air Campaign Against the Iraqi Army*, 19, 28.

6. Murray, *Air War in the Persian Gulf*, 293.

7. U.S. Department of Defense, *Conduct of the Persian Gulf War* (Washington, DC, 1992), 13–14; see also Winnefield et al., *A League of Airmen*, 132.

8. Royal Koepsall, "The Ultimate High Ground! Space Support to the Army. Lessons from Operations Desert Shield and Storm," Center for Army Lessons Learned Newsletter no. 91-3 (Ft. Leavenworth, KS, 1991).

9. The battle of 73 Easting has been the subject of innumerable popular and serious studies, most of which focus only on the first firefight between the cavalry and the Tawalkalna Division. This discussion is based on a variety of such studies, including Scales, *Certain Victory*, 1–4, 261–65, and 282–91; Swain, *"Lucky War,"* 250–62; and Daniel L. Davis, "Artillerymen in Action—The 2d ACR at the Battle of 73 Easting," *Field Artillery* (April 1992): 48–53.

10. Davis, "Artillerymen in Action," 51–53.

11. Quoted in Swain, *"Lucky War,"* 262.

12. Scales, *Certain Victory*, 287–91.

13. Swain, *"Lucky War,"* 40.

14. U.S. Department of Defense, National Guard Bureau, "Army National Guard After Action Report: Operation Desert Shield, Operation Desert Storm. Executive Summary" (Arlington, VA, n.d.). Upon mobilization, the roundout units underwent a fixed program of training instead of following army doctrine by having the commanders train to correct their own training weaknesses.

15. Vernon Lowry, "Initial Observations by Engineers in the Gulf War," *Engineer* 21 (October 1991): 42–48; Alan Schlie, "Close Up: Engineer Restructuring Initiative," *Engineer* 23 (February 1993): 20–24.

16. See, for example, Mark Newell, "4th ID Pioneers New Division Design," *Armor* 107:6 (November–December 1998): 49–50.

17. Thomas M. Feltey, "The Brigade Reconnaissance Troop," *Armor* 107:5 (September–October 1998): 26–27, 55; Wayne T. Westgaard, "Will the New Bri-

gade Reconnaissance Troop Be Adequately Protected?" *Armor* 108:2 (March–April 1999): 27–29.

18. On the military issues of Operations Other Than War, see Christopher Lord, "Intermediate Deployments: The Strategy and Doctrine of Peacekeeping-Type Operations," Occasional Paper no. 25 (Camberley, England, 1996), especially 21–23.

19. Kelly Orr, *From a Dark Sky: The Story of U.S. Air Force Special Operations* (Novato, CA, 1996), 310–11.

20. U.S. Department of the Army. *Field Manual 100-5: Operations* (Washington, DC, 1993); for a detailed explanation of the changes in U.S. doctrine after 1991, see John L. Romjue, *American Army Doctrine for the Post–Cold War* (Ft. Monroe, VA, 1997).

21. On the Chechen conflict, see the series of articles by Timothy L. Thomas, especially "The Caucasus Conflict and Russian Security: The Russian Armed Forces Confront Chechnya, III: The Battle for Grozny, 1–26 January 1995," *Journal of Slavic Military Studies* 10:1 (March 1997): 50–108, and "The Battle of Grozny: Deadly Classroom for Urban Combat," *Parameters: Journal of the U.S. Army War College* 29:2 (summer 1999): 87–102.

22. Thomas, "The Caucasus Conflict and Russian Security," 52.

23. See, for example, Robert J. Bunker, "Information Operations and the Conduct of Land Warfare," *Military Review* 78:5 (September 1998): 4–17; Ryan Henry and C. Edward Peartree, "Military Theory and Information Warfare," *Parameters: Journal of the U.S. Army War College* 28:3 (autumn 1998): 121–35; and Fred Levien, "Information Warfare: The Plain Truth," *Journal of Electronic Defense* 22:4 (April 1999): 47–53.

BIBLIOGRAPHY

• • • • • • • • • • • • • • • •

Abbreviations

ACSI   U.S. Department of the Army, Assistant Chief of Staff for Intelligence
CARL   Combined Arms Research Library, Ft. Leavenworth, KS
CSI   Combat Studies Institute, U.S. Army Command and General Staff
           College
FBIS   Foreign Broadcast Information Service
HMSO   His/Her Majesty's Stationery Office
MHI   Military History Institute, Carlisle Barracks, PA
OCMH   Office of the Chief of Military History
USACGSC   U.S. Army Command and General Staff College

Unpublished Works

Almond, Lt. Gen. Edward. Letter to Gen. James Van Fleet, Commander, Eighth
    Army, concerning tactical air support, 2 January 1952. Copy in author's files.
Conner, Fox. Letter to Maj. Malin Craig concerning division organization, 24 April
    1920. MHI.
Erickson, John. "Soviet Combined Arms: Theory and Practice." Typescript.
    Edinburgh: University of Edinburgh, 1979.
Fisher, Ernest F. Jr. "Weapons and Equipment Evolution and Its Influence upon
    Organization and Tactics in the American Army from 1775–1963." Washing-
    ton, DC: OCMH, n.d.

France. Ministère de la Guerre. École Superieure de Guerre. "Memorandum on Nuclear Weapons." ACSI translation, no. K-6556, 1976. CARL N16582.519.

Glantz, David M. "Soviet Airborne Forces in Perspective." Typescript. Ft. Leavenworth, KS: CSI, 1983.

———. "Soviet and Commonwealth Military Doctrine in Revolutionary Times." Typescript. Ft. Leavenworth, KS: Foreign Military Studies Office, March 1992.

Great Britain. War Office. Letter 20/GEN/6059 (S.D. 1) concerning reorganization of divisions, dated 1 October 1942. CARL N-6136.

Haworth, W. Blair Jr. "The Bradley and How It Got That Way: Mechanized Infantry, Organization, and Equipment in the United States Army, 1936–1996." 1999 typescript, based on Ph.D. diss., Duke University, 1995.

Kida, Hideto. "Imperial Japanese Army's Operations in New Guinea and the Decisive Battle at Leyte (Shoichigo Operation)." Paper presented at the U.S.–Japanese Military History Exchange Conference, Tokyo, 1984 (copy in author's file).

Lange, H. W. W. "The French Division–Concept of Organization and Operation." ACSI Report no. 2137175, R-217-60, 16 March 1960. Typescript. CARL N-15835.18.

Marshall, Samuel L. A. "CCF in the Attack (Part II)." Operations Research Office Report ORO-S-34 (EUSAK). Typescript. Tokyo: Operations Research Office, 1951. CARL N-16454.59.

McKenney, Janice E. "Field Artillery Army Lineage Series." Typescript. Washington, DC: Center of Military History, 1980.

Peress, Philippe C. "The Combined Arms Battalion Concept in the French Army." Handwritten student study project. Ft. Leavenworth: USACGSC, 1977. CARL N-13423.472.

Pershing, John J. Wrapper Indorsement Forwarding Report of AEF Superior Board on Organization and Tactics. General Headquarters, American Expeditionary Force, 16 June 1920. Copy in MHI, UA 25 C65.

Reinhardt, Hellmuth, ed. "Small Unit Tactics: Infantry, Part I." Typescript. U.S. European Command German manuscript series no. P-060d, 10 November 1950. CARL N-17500.16-A.

Schnabel, James F. "History of the Korean War, Part Two, Volume I, General Headquarters Support and Participation (25 June 1950–30 April 1951)," typescript (no date) in U.S. Army Center of Military History.

United States. Department of the Army. European Command, Historical Division. "Small Unit Tactics, Artillery." Foreign Military Studies, vol. 1, no. 7, 1952. German manuscript series no. P-060h. CARL R-17500.6-A.

———. ACSI. "The Soviet Army: A Department of the Army Assessment, May 15, 1958." Typescript, MHI.

———. 101st Airborne Division (Airmobile). "Final Report—Airmobile Operations in Support of Operation Lam Son 719, 8 February–6 April 1971." 2 vols. Mimeographed, 1971.

———. U.S. Continental Army Command. Letters of Instruction on Pentomic Division. "Changes in ROCID TOE (U)," dated 8 January 1959; "The Armored Division (U)," dated 13 January 1959; and "The Armored Cavalry Regiment (U)," dated 13 January 1959, with supporting documentation in CARL, N-17935.62-U.

———. War Department. Army Ground Forces. "The New Inf, Armd, and A/B Divisions." Typescript explanation and tables, 24 January 1947. CARL N-15338-B.

———. 1st Armored Division. "After Action Report, 1st Armd Div. 25 Jan–31 Dec 43." Typescript with maps. U.S. Army Armor School Library.

———. General Staff. "Provisional Manual of Tactics for Large Units." Typescript. Washington Barracks, DC: U.S. Army War College, 1923–24.

———. General Staff. Military Intelligence Division. Special Bulletin no. 2: "Soviet-Finnish War: Operations from November 30, 1939, to January 7, 1940." G2 document no. 2657-D-1054.

———. Headquarters, 32d Infantry Division. "Report After Action of 32d Infantry Division in Leyte Engagement or K-2 Operation for the Period 16 November–25 December 1944." CARL N-11706.1.

———. Headquarters, 128th Infantry Regiment. "Historical Record, 128th Infantry, from 14 November 1944 to 25 November 1944." U.S. Army Infantry School Library, microfilm D548.

———. 2d Division. "Special Report Based on Field Service Test of the Provisional 2d Division, conducted by the 2d Division, U.S. Army, 1939." Typescript, MHI.

Van Fleet, James, Commander, Eighth Army. Letter to Lt. Gen. Edward Almond concerning tactical airpower, 22 January 1952. Copy in author's files.

Government Regulations

France, Ministère de la Guerre. *Décret du 2 Decembre 1904 portant réglement sur les manoeuvres de l'infanterie.* Paris: Charles-Lavauzelle, 1909.

———. *Décret du 28 Octobre 1913 portant réglement sur la conduit des grandes unités.* Paris: Berger-Levrault, 1913.

———. *Décret du 18 Octobre 1913 portant réglement sur la service des armées en campagne.* Paris: Librairie Chapelot, 1914.

———. *Instruction du 28 Decembre 1917 sur liaison pour les troupes de toutes armes.* With changes 1–3 and annex dated 5 July 1919. Paris: Charles-Lavauzelle, 1921.

———. *Instruction provisoire sur l'emploi tactique des grandes unités, 1921.* Paris: Charles-Lavauzelle, 1922.

———. *Instruction sur l'emploi tactique des grandes unités, 1936.* Paris: Charles-Lavauzelle, 1940.

———. *Instructions for the Offensive Combat of Small Units,* January 2, 1918. Translated and published by Headquarters, American Expeditionary Force, March 1918. No. 160.

———. *Instructions for the Tactical Employment of Large Units, 1937* [*sic,* 1936]. Typescript trans. Richard U. Nicholas, U.S. Army, n.d.

———. *Provisional Field Service Regulations, French Army, Annex no. 1 to Provisional Instructions for the Employment of Large Units.* Trans. U.S. Army General Service Schools. Ft. Leavenworth, KS: General Service Schools Press, 1924.

———. *Réglement des unités de chars légers.* 2 vols. Paris: Charles-Lavauzelle, 1930.

———. Direction de la Cavalerie. *Instruction provisoire sur l'emploi et la manoeuvre des unités d'auto-mitrailleuses de cavalerie.* Vol. 1. Paris: Charles-Lavauzelle, 1921.

———. École Superieure de Guerre. *Études operations 1er cycle, 1953–1954.* Mimeographed. Paris: École Superieure de Guerre, 1953.

———. *Tactique générale.* Rambouillet, FR: Pierre Leroy, 1922.

———. Etat-Major de l'Armée. *Instruction provisoire sur l'emploi des chars de combat comme engins d'infanterie,* 23 March 1920. Paris: Charles-Lavauzelle, 1921.

———. *Instruction sur l'emploi des chars de combat.* Paris: Charles-Lavauzelle, 1930.

Germany. Federal Republic. *TF 59: Truppenfuhrung 1959.* Translated as *Operations Manual of the Federal German Army* by Headquarters, U.S. Military Assistance Advisory Group, 1961.

———. Ministry of Defense. *Army Regulation 100/100: Command and Control in Battle.* Weimar, FRG: Federal Office of Languages, 1973.

———. [Weimar] Ministry of National Defense. Truppenamt. *Command and Control of the Combined Arms.* Trans. U.S. Army General Service Schools. Ft. Leavenworth, KS: General Service Schools Press, 1925.

Great Britain. Ministry of Defence. *The Falklands Campaign: The Lessons.* London: HMSO, 1982.

———. War Office. *Field Service Regulations, 1923–24 (Provisional).* 2 vols. London: HMSO, 1923–1924.

———. *Field Service Regulations,* vol. 2, *Operations (1929).* London: HMSO, 1929.

———. *Field Service Regulations,* vol. 2, *Operations-General (1935).* London: HMSO, 1935.

———. *Field Service Regulations,* vol. 2, *Operations–Higher Formations (1935).* London: HMSO, 1935.

———. *Infantry Training,* vol. 2, *War (1926).* London: HMSO, 1926.

———. *Infantry Training,* vol. 2, *War (1931).* London: HMSO, 1931.

———. *Instructions for the Training of Divisions for Offensive Action.* Reprinted and ed., U.S. Army War College. Washington, DC: Government Printing Office, 1917.

———. *Report of the Committee on the Lessons of the Great War.* London: War Office, 1932.

———. *Tank and Armoured Car Training,* vol. 2, *War (1927).* London: HMSO, 1927.

———. First Army. "German Methods of Trench Warfare." 1 March 1916. London: War Office, 1916.

———. General Staff. "German Armoured Tactics in Libya." In "Periodic Notes on the German Army." London: HMSO, 1942.

———. *German Army Handbook, April 1918.* Reprint. London: Arms and Armour Press, 1977.

————. SS 210. "The Division in Defence." London: HMSO, 1918.

————. SS 135. "The Training and Employment of Divisions, 1918." Rev. ed. London: HMSO, 1918.

————. Middle East Command. "Notes on Main Lessons of Recent Operations in the Western Desert." 10 August 1942, typescript copy by U.S. War Department, General Staff, Military Intelligence Command. N-3915 CARL.

Prussia. Kriegsministerium. Generalstab des Feldherres. *Vorshiften für den Stellungskrieg für alle Waffen.* Teil 14: *Der Angriff im Stellungskrieg.* Berlin: Kriegsministerium, 1918.

Union of Soviet Socialist Republics. Commissariat of Defense. *Field Service Regulations of the Red Army (1942).* ACSI translation; no number. Copy in MHI.

————. *(Draft) Manual on the Breakthrough of Fortified Areas, 1944.* ACSI translation no. F-798.

————. *Field Service Regulations of the Red Army (1944).* ACSI translation; no number. Copy in MHI.

————. *Field Service Regulations of the Soviet* [sic Red] *Army (1936) (Tentative).* Moscow: Commissariat of Defense, 1937. Typescript trans. Charles Berman.

————. *Field Service: Staff Manual (Tentative).* Moscow: Commissariat of Defense, 1935. Typescript trans. Charles Berman.

————. "Instructions for Air Force and Ground Troops on Combined Operations." 1944. ACSI translation no. F-3733.

————. "Manual for Combat Support Aviation." (n.d.) ACSI translation of chapters 6–8, no. F-3732a.

————. *1942 Infantry Tactical Manual of the Red Army.* Decree no. 347, 9 November 1942. ACSI translation no. F-2830.

————. *Service Regulations for the Armored and Mechanized Forces of the Red Army.* Part 1: *The Tank, Tank Platoon, and Tank Company.* 1944. ACSI translation no. F-7176.

United States. Department of the Army. Field Manual 100-5, *Operations.* Washington, DC, 1976.

————. Field Manual 100-5, *Operations.* Washington, DC: Government Printing Office, 1982.

————. Field Manual 100-5, *Operations.* Washington, DC: Government Printing Office, 1993.

————. Field Manual 71-2, *The Tank and Mechanized Infantry Battalion Task Force.* Washington, DC: Government Printing Office, 1977.

————. War Department. Field Manual 100-5: *Tentative Field Service Regulations–Operations.* Washington, DC: Government Printing Office, 1939.

————. Chief of Infantry. *Infantry Field Manual,* vol. 2, *Tank Units.* Washington, DC: Government Printing Office, 1931.

————. General Service Schools. *General Tactical Functions of Larger Units.* Ft. Leavenworth, KS: General Service Schools Press, 1922.

————. *Tactics and Technique of the Separate Branches.* 2 vols. Ft. Leavenworth, KS: General Service Schools Press, 1925.

————. General Staff. *Field Service Regulations, 1914.* With change 7, dated 18 August 1917. Monasha, WI: George Banta Publishing Company, 1918.

————. General Stuff. *Provisional Manual of Tactics for Large Units.* Washington, DC, 1922–1923.

————. *Field Service Regulations: United States Army, 1923.* Washington, DC, 1924.

————. *A Manual for Commanders of Large Units (Provisional).* 2 vols. Washington, DC, 1930–1931.

————. U.S. Army Command and General Staff School. *FM 101-10 (Tentative): Staff Officers' Field Manual.* Undated, with changes to 1943.

Published Works

Abrahamson, James L. *America Arms for a New Century: The Making of a Great Military Power.* New York: Free Press/Macmillan, 1981.

Adan, Avraham. *On the Banks of the Suez: An Israeli General's Personal Account of the Yom Kippur War.* San Rafael, CA: Presidio Press, 1980.

Addington, Larry H. *The Blitzkrieg Era and the German General Staff, 1865–1941.* New Brunswick, NJ: Rutgers University Press, 1971.

Adkin, Mark. *Urgent Fury: The Battle for Grenada.* Lexington, MA: Lexington Books, 1989.

Agar-Hamilton, J. A. I., and L. C. F. Turner. *The Sidi Rezegh Battles, 1941.* Cape Town, SA: Oxford University Press, 1957.

Alferov, S. "Wartime Experience: Breakthrough of Enemy Defenses by a Rifle Corps." *USSR Report: Military Affairs* 1771 (2 June 1983) 36–43. JPRS 83593. Trans. FBIS, from *Voyenno-istoricheskiy zhurnal* [Military History Journal], March 1983.

Alfoldi, Lazlo M. "The Hutier Legend." *Parameters* 5 (June 1976): 69–74.

Allen, Matthew. *Military Helicopter Doctrines of the Major Powers, 1945–1992: Making Decisions About Air-Land Battle.* Westport, CT: Greenwood Press, 1993.

Allen, Robert S. *Lucky Forward: The History of Patton's Third U.S. Army.* New York: Vanguard Press, 1947.

Altham, E. A. *The Principles of War Historically Illustrated.* Vol. 1. London: Macmillan and Company, 1914.

Altmayer, René. *Étude de tactique générale.* 2d ed. Paris: Charles-Lavauzelle, 1937.

Anthérieu, Étienne. *Grandeur et sacrifice de la ligne Maginot.* Paris: Durassie, 1962.

Ardant du Picq, Charles Jean. *Battle Studies: Ancient and Modern Battle.* Trans. John N. Greely and Robert C. Cotton. Harrisburg, PA: Military Service Publishing Company, 1958.

Arendt, Edgar D. "Comparative Analysis of Contemporary Non–U.S. Army Small Infantry Unit Organizations (U)." Washington, DC: Booz-Allen Applied Research, 1 September 1967. Report no. DA-04-495-AMC-845 (X) for U.S. Army Combat Developments Command, Infantry Agency, Ft. Benning, GA. CARL N-18760.57.

Armstrong, Richard N. *Red Army Tank Commanders: The Armored Guards.* Atglen, PA: Schiffer Military/Aviation History, 1994.

Army League. Great Britain. *The Army in the Nuclear Age: Report of the Army League Sub-Committee.* London: Saint Clements Press, 1955.

———. *The British Army in the Nuclear Age.* London: Army League, 1959.

Atkinson, C. T. *The Seventh Division, 1914–1918.* London: John Murray, 1927.

Bailey, Jonathan. "The First World War and the Birth of the Modern Style of Warfare." Occasional Paper no. 22. Camberley, England: Strategic and Combat Studies Institute, 1996.

Baily, Charles M. "Faint Praise: The Development of American Tanks and Tank Destroyers During World War II." Ph.D. diss., Duke University, 1977.

Balck, Wilhelm von. *Development of Tactics—World War.* Trans. Harry Bell. Ft. Leavenworth, KS: General Service Schools Press, 1922.

Baldwin, Hanson W. *The Crucial Years, 1939–1941.* New York: Harper and Row, 1976.

Bank, Aaron. *From OSS to Green Berets: The Birth of Special Forces.* Novato, CA: Presidio Press, 1986.

Barnett, Correlli. *Britain and Her Army, 1509–1870.* New York: Morrow, 1970.

———. *The Desert Generals.* 1st ed. New York: Viking Press, 1961.

Baumann, Robert F. *Russian-Soviet Unconventional Wars in the Caucasus, Central Asia, and Afghanistan.* Leavenworth Paper no. 20. Ft. Leavenworth, KS: USACGSC, 1993.

Beaver, Daniel R. "Politics and Policy: The War Department Motorization and Standardization Program for Wheeled Transport Vehicles, 1920–1940." *Military Affairs* 47 (October 1983): 101–8.

Becker, G. *L'Infanterie d'après-guerre en France et en Allemagne.* Édition de 1924, mise à jour et completée. Paris: Berger-Levrault, 1930.

Beesly, Patrick. *Very Special Intelligence: The Story of the Admiralty's Operational Intelligence Centre, 1939–1945.* London: Hamish Hamilton, 1977.

Belfield, Eversley. *The Boer War.* London: Leo Cooper, 1975.

Bell, Raymond E. "Division Cuirassée, 1940." *Armor* 83 (January–February 1974): 25–29.

Bennett, Ralph. *Ultra in the West: The Normandy Campaign, 1944–45.* New York: Charles Scribner's Sons, 1979.

Bergeron, Frederic A. *The Army Gets an Air Force: Tactics of Insurgent Bureaucratic Politics.* Baltimore: Johns Hopkins University Press, 1980.

Bermudez, Joseph J. Jr. *North Korean Special Forces.* 2d ed. Annapolis, MD: Naval Institute Press, 1998.

Bernhardi, Friedrich von. *The War of the Future in the Light of the Lessons of the World War.* Trans. F. A. Holt. New York: D. Appleton, 1921.

Bidwell, Shelford, and Dominick Graham. *Fire-Power: British Army Weapons and Theories of War, 1904–1945.* Boston: George Allen and Unwin, 1982.

Binckley, John C. "A History of U.S. Army Force Structuring." *Military Review* 57 (February 1977): 67–82.

Binoch, Jacques. "L'Allemagne et le lieutenant-colonel Charles de Gaulle." *Révue Historique* 248 (July–September 1972): 107–16.

Biriukov, Grigoriy F., and G. V. Melnikov. *Tank Warfare*. Mimeographed trans. ACSI, no. J-4398, from the Russian book, *Bor'ba s Tankami*. Moscow: n.p., 1967.

Blair, Clay. *The Forgotten War: America in Korea, 1950–1953*. New York: Times Books/Random House, 1987.

Blau, George E. *The German Campaign in Russia: Planning and Operations (1940–1942)*. Washington, DC: OCMH, 1955.

Blumenson, Martin. *Breakout and Pursuit*. United States Army in World War II. Washington, DC: OCMH, 1961.

———. *Kasserine Pass*. Boston: Houghton Mifflin, 1967.

———, ed. *The Patton Papers*, vol. 2, *1940–1945*. Boston: Houghton Mifflin, 1974.

Bolger, Daniel P. *Scenes from an Unfinished War: Low Intensity Conflict in Korea, 1966–1969*. Leavenworth Paper no. 19. Ft. Leavenworth, KS: USACGSC, 1991.

———. "Special Operations and the Grenada Campaign." *Parameters: U.S. Army War College Quarterly* 18:4 (December 1988): 49–61.

Bolton, E. "Talks on the Soviet Art of War." *Soviet Military Review* 2 (1967): 46–48; no. 6: 42–45; no. 10: 26–29; no. 11: 45–48.

Bond, Brian. *Britain, France, and Belgium, 1939–1940*. 2d ed. London: Brassey's/Maxwell Pergamon, 1990.

———. *British Military Policy Between the Two World Wars*. Oxford: Clarendon Press, 1980.

———. *Liddell Hart: A Study of His Military Thought*. New Brunswick, NJ: Rutgers University Press, 1977.

Bonnal, G. A. B. E. Henri. *La Première Bataille; Le Service de deux ans; du caractère chez les chefs; discipline—Armée nationale; cavalerie*. Paris: R. Chapelot, 1908.

Bradin, James W. *From Hot Air to Hellfire: The History of Army Attack Aviation*. Novato, CA: Presidio Press, 1994.

Braun, ———. "German and French Principles of Tank Employment." Typescript trans. F. W. Merten, U.S. War Department, General Staff, War College Division, January 1938. From the article in *Militar-Wochenblatt*, 22 October 1937.

Brindel, ———. "La Nouvelle organization militaire." *Révue des Deux Mondes* 7th series, 51 (1929): 481–501.

Brodie, Bernard, and Fawn M. Brodie. *From Crossbow to H-Bomb*. Rev. ed. Bloomington, IN: Indiana University Press, 1973.

Brown, Anthony Cave. *Bodyguard of Lies*. New York: Harper and Row, 1975.

Browne, D. G. *The Tank in Action*. Edinburgh: William Blackwood and Sons, 1920.

Bruckmüller, Georg. *The German Artillery in the Break-Through Battles of the World War*. 2d ed. Trans. J. H. Wallace and H. D. Kehm. Berlin: E. S. Mittler and Son, 1922.

Bunker, Robert J. "Information Operations and the Conduct of Land Warfare." *Military Review* 78:5 (September 1998): 4–17.

Burgess, H. *Duties of Engineer Troops in a General Engagement of a Mixed Force*. Occasional Paper no. 32, U.S. Army Engineer School. Washington, DC: Press of the Engineer School, 1908.

Caemmerer, Rudolf von. *The Development of Strategical Science During the Nineteenth Century*. Trans. Karl von Donat. London: Hugh Rees, 1905.

Caidin, Martin. *The Tigers Are Burning.* New York: Hawthorn Books, 1974.

Callahan, Raymond. *Burma, 1942–1945.* Newark: University of Delaware Press, 1978.

Cameron, Robert S. "American Tank Development During the Cold War." *Armor* 107:4 (July–August 1998): 30–38.

Camut, ——. *L'Emploi des troupes du génie en liaison avec les autres armes.* Paris: Librairie Militaire R. Chapelot, 1910.

Cannon, M. Hamlin. *Leyte: The Return to the Philippines.* U.S. Army in World War II series. Washington, DC: OCMH, 1954.

Carell, Paul [Paul Schmidt]. *Stalingrad: The Defeat of the German 6th Army.* Trans. David Johnston. Atglen, PA: Schiffer Military History, c. 1993.

Carter, Donald A. "From G.I. to Atomic Soldier: The Development of U.S. Army Tactical Doctrine, 1945–1956." Ph.D. diss., Ohio State University, 1987.

Carus, W. Seth. *Cruise Missile Proliferation in the 1990s.* Washington Paper no. 156, Center for Strategic and International Studies. Westport, CT, and London: Praeger, 1992.

Carver, Michael. *Dilemmas of the Desert War: A New Look at the Libyan Campaign, 1940–1942.* Bloomington: Indiana University Press, 1986.

——. *Twentieth-Century Warriors: The Development of the Armed Forces of the Major Military Nations in the Twentieth Century.* New York and London: Weidenfeld and Nicolson, 1987.

Carver, Richard M. *The Apostles of Mobility.* New York: Holmes and Meier, 1979.

Chandler, David G. *The Campaigns of Napoleon.* New York: Macmillan, 1966.

Chuikov, Vasili I. *The Fall of Berlin.* Trans. Ruth Kisch. New York: Holt, Rinehart and Winston, 1968.

Clark, Alan. *The Donkeys.* New York: William Morrow, 1962.

Clarke, Jeffrey J. "Military Technology in Republican France: The Evolution of the French Armored Force, 1917–1940." Ph.D. diss., Duke University, 1969.

Clayton, Aileen. *The Enemy Is Listening.* London: Hutchinson, 1980.

Clendenen, Clarence C. *Blood on the Border: The United States Army and the Mexican Irregulars.* New York: Macmillan, 1969.

Cole, Hugh M. *The Ardennes: Battle of the Bulge.* U.S. Army in World War II series. Washington, DC: OCMH, 1965.

——. *The Lorraine Campaign.* U.S. Army in World War II series. Washington, DC: OCMH, 1950.

Cole, Ronald H. "'Forward with the Bayonet!' The French Army Prepares for Offensive Warfare, 1911–1914." Ph.D. diss., University of Maryland, 1975.

Coll, Blanche D., Jean E. Keith, and Herbert H. Rosenthal. *The Technical Services: The Corps of Engineers: Troops and Equipment.* U.S. Army in World War II series. Washington, DC: OCMH, 1958.

Comparato, Frank E. *Age of Great Guns: Cannon Kings and Cannoneers Who Forged the Firepower of Artillery.* Harrisburg, PA: Stackpole, 1965.

Contamine, Henry. *La Revanche, 1871–1914.* Paris: Berger-Levrault, 1957.

Cooling, B. Frank, ed. *Case Studies in the Development of Close Air Support.* Washington, DC: Office of Air Force History, 1990.

Cooper, Bryan. *The Battle of Cambrai.* New York: Stein and Day, 1968.

Cooper, Matthew, and James Lucas. *Panzer: The Armoured Force of the Third Reich.* New York: St. Martin's Press, 1976.

Cordesman, Anthony H. *NATO's Central Region Forces: Capabilities/Challenges/Concepts.* London: Janes Publishing, 1988.

Corum, James S. *The Luftwaffe: Creating the Operational Air War 1918–1940.* Lawrence: University Press of Kansas, 1997.

———. *The Roots of Blitzkrieg: Hans von Seeckt and German Military Reform.* Lawrence: University Press of Kansas, 1992.

Court, Geoffrey D. W. *Hard Pounding: The Tactics and Technique of Antitank Warfare, with Observations on Its Past, Present, and Future.* Washington, DC: U.S. Field Artillery Association, 1946.

Cox, Gary C. "Beyond the Battle Line: U.S. Air Attack Theory and Doctrine, 1919–1941." Maxwell Air Force Base, AL: U.S. Air Force School of Advanced Aerospace Studies, 1996.

Craig, Gordon A. *The Battle of Königgrätz: Prussia's Victory over Austria, 1866.* Philadelphia: J. B. Lippincott, 1964.

Culmann, F. *Tactique générale d'après l'experience de la Grande Guerre.* 4th rev. ed. Paris: Charles-Lavauzelle, 1924.

Cushman, John H. "Pentomic Infantry Division in Combat." *Military Review* 37 (January 1958): 19–30.

Daley, John L. S. "The Theory and Practice of Armored Warfare in Spain." Part 1. *Armor* 108:2 (March–April 1999): 30, 39–43.

Dastrup, Boyd L. *King of Battle: A Branch History of the U.S. Army's Field Artillery.* TRADOC Branch History Series. Ft. Monroe, VA: U.S. Army Training and Doctrine Command, 1992.

Davis, Daniel L. "Artillerymen in Action—The 2d ACR at the Battle of 73 Easting." *Field Artillery* (April 1992): 48–53.

Davis, Richard G. *The 31 Initiatives: A Study in Air Force–Army Cooperation.* Washington, DC: Office of Air Force History, 1987.

Dayan, Moshe. *Diary of the Sinai Campaign.* New York: Harper and Row, 1966.

———. *Moshe Dayan: Story of My Life.* New York: William Morrow, 1976.

Deal, C. Lanier Jr. "BAI: The Key to the Deep Battle." *Military Review* 62:3 (March 1982): 51–54.

Deichmann, Paul. *German Air Force Operations in Support of the Army.* USAF Historical Studies no. 13. Maxwell Air Force Base, AL: USAF Historical Division, Research Studies Institute, 1962.

Deygas, F. J. *Les Chars d'assaut: Leur passé, leur avenir.* Paris: Charles-Lavauzelle, 1937.

Dick, Charles J. "A Bear Without Claws: The Russian Army in the 1990s." *Journal of Slavic Military Studies* 10:1 (March 1997): 1–10.

DiNardo, Richard L. *Germany's Panzer Arm.* Westport, CT: Greenwood Press, 1997.

Doubler, Michael D. *Closing with the Enemy: How the GIs Fought the War in Europe, 1944–1945.* Lawrence: University Press of Kansas, 1994.

Doughty, Robert A. *The Breaking Point: Sedan and the Fall of France, 1940.* Hamden, CT: Archon/Shoe String Press, 1990.

———. "The Enigma of French Military Armored Doctrine, 1940." *Armor* 83 (September–October 1974): 39–44.

———. *The Evolution of U.S. Army Tactical Doctrine, 1946–76.* Leavenworth Paper no. 1. Ft. Leavenworth, KS: USACGSC, 1979.

———. *The Seeds of Disaster: The Development of French Army Doctrine, 1919–1939.* Hamden, CT: Archon Books, 1985.

Doyle, Arthur Conan. *The British Campaign in France and Flanders, 1915.* London: Hodder and Stoughton, 1917.

Drea, Edward J. *Defending the Driniumor: Covering Force Operations in New Guinea, 1944.* Leavenworth Paper no. 9. Ft. Leavenworth, KS: USACGSC, 1984.

———. *MacArthur's Ultra: Codebreaking and the War Against Japan, 1942–1945.* Lawrence: University Press of Kansas, 1992.

———. *Nomonhan: Japanese-Soviet Tactical Combat, 1939.* Leavenworth Paper no. 2. Ft. Leavenworth, KS: USACGSC, 1981.

Dunn, Walter S. *Kursk: Hitler's Gamble, 1943.* Westport, CT: Praeger, 1997.

Dupuy, Trevor N. *Elusive Victory: The Arab-Israeli Wars, 1947–1974.* New York: Harper and Row, 1978.

Dutil, L. *Les Chars d'assaut: Leur création et leur rôle pendant la Guerre, 1915–1918.* Paris: Berger-Levrault, 1919.

Edmonds, James E., and Archibald F. Becke. *History of the Great War: Military Operations France and Belgium, 1914.* Vol. 1, 3d ed. London: Macmillan, 1933.

Edmonds, James E., and Graeme C. Wynne. *History of the Great War: Military Operations, France and Belgium 1915.* London: Macmillan, 1927.

Elble, Rolf. *Die Schlacht an der Bzura im September 1939 aus deutscher und polnischer Sicht.* Freiburg, FRG: Verlag Rombach, 1975.

Ellis, L. F., and A. E. Warhurst. *Victory in the West,* vol. 2, *The Defeat of Germany.* History of the Second World War. London: HMSO, 1968.

Ellis, William D., and Thomas J. Cunningham Jr. *Clarke of St. Vith: The Sergeants' General.* Cleveland, OH: Dillon/Liderbach, 1974.

English, John A. *The Canadian Army and the Normandy Campaign.* New York: Praeger, 1991.

———. *A Perspective on Infantry.* New York: Praeger, 1981.

English, John A., and Bruce I. Gudmundsson. *On Infantry.* Rev. ed. Westport, CT: Praeger, 1994.

Epley, William W. "America's First Cold War Army, 1945–1950." Land Warfare Papers no. 15. Institute for Land Warfare. Arlington, VA: Association of the United States Army, 1993.

Eremenko, Andrei. *The Arduous Beginning.* Trans. Vic Schneierson. Moscow: Progress Publishers, [1966].

Erickson, John. "The Ground Forces in Soviet Military Policy." *Strategic Review* 4 (April 1976): 64–79.

————. *The Road to Berlin: Continuing the History of Stalin's War with Germany.* Boulder, CO: Westview Press, 1983.

————. *The Road to Stalingrad: Stalin's War with Germany.* Vol. 1. New York: Harper and Row, 1975.

————. *The Soviet High Command: A Military-Political History, 1918–1941.* New York: St. Martin's Press, 1962.

Ernest, N. "Self-Propelled Artillery in Offensive Combat by the Combined Arms." Mimeographed ACSI translation, no. F-6592, from *Voyenni vestnik [Military Herald]*, 1945, no. 22.

Essame, Hubert. *The Battle for Europe, 1918.* New York: Charles Scribner's Sons, 1972.

Estienne, Jean-Baptiste Eugène. *Conférence faite le 15 Février 1920 sur les chars d'assaut.* Paris: Librairie de l'Enseignement Technique, 1920.

Evanhoe, Ed. *Darkmoon: Eighth Army Special Operations in the Korean War.* Annapolis, MD: Naval Institute Press, 1995.

Everett, H. W. "The Secret War in the Desert." *British Army Review* 60 (December 1978): 66–68.

Falls, Cyril. *Armageddon, 1918.* Philadelphia: J. B. Lippincott, 1964.

————. *The Battle of Caporetto.* Philadelphia: J. B. Lippincott, 1966.

Farrar-Hockley, Anthony. *The Army in the Air: The History of the Army Air Corps.* Dover, NH, and Stroud, Gloucestershire: Alan Sutton Publishing, 1994.

————. *The Edge of the Sword.* London: Frederick Muller, 1954.

————. *Goughie: The Life of General Sir Hubert Gough, CGB, GCMG, KCVC.* London: Hart-Davis, MacGibbon, 1975.

Fehrenbach, T. R. *This Kind of War: Korea, a Study in Unpreparedness.* New York: Pocket Books, 1963.

Feltey, Thomas M. "The Brigade Reconnaissance Troop." *Armor* 107:5 (September–October 1998): 26–27, 55.

Ferré, Georges. *Le Défaut de l'armure.* Paris: Charles-Lavauzelle, 1948.

Ferris, John. "The British Army and Signals Intelligence in the Field During the First World War." *Intelligence and National Security* 3:4 (October 1988): 23–48.

————. "The British Army, Signals, and Security in the Desert Campaign, 1940–42." *Intelligence and National Security* 5:2 (April 1990): 255–91.

Finnegan, J. P. "U.S. Army Signals Intelligence in World War II." Paper presented at Society for Military History convention, March 1990, Washington, DC (copy in author's files).

Foch, Ferdinand. *The Memoirs of Marshal Foch.* Trans. T. Bentley Mott. London: William Heinemann, 1931.

————. *Des Principes de la Guerre: Conférences faites en 1900 à l'École superieure de Guerre.* 4th ed. Paris: Berger-Levrault, 1917.

France. Ministère de la Defense. État-Major de l'Armée de l'Air. *Le Haut commandement français face au progres technique entre les deux guerres.* Vincennes: Service Historique de l'Armée de l'Air, 1980.

France. Ministère de la Guerre. État-Major de l'Armée. Service Historique. *Les Armées Françaises dans la Grande Guerre.* Vol. 1, 2d ed. Paris: Imprimerie Nationale, 1936.

Frostic, Fred. *Air Campaign Against the Iraqi Army in the Kuwaiti Theater of Operations*. Santa Monica, CA: RAND Corporation, 1994.

Fuller, John F. C. *Armament and History: A Study of the Influence of Armament on History from the Dawn of Classical Warfare to the Second World War*. New York: Charles Scribner's Sons, 1945.

———. *Armoured Warfare: An Annotated Edition of Lectures on F.S.R. III (Operations Between Mechanized Forces)*. Harrisburg, PA: Military Service Publishing Company, 1943.

———. *The Foundations of the Science of War*. London: Hutchinson and Company, 1925.

———. *Memoirs of an Unconventional Soldier*. London: Ivor Nicholson and Watson, 1936.

———. *On Future Warfare*. London: Sifton Praed and Company, 1928.

Gabel, Christopher R. *Seek, Strike, and Destroy: U.S. Army Tank Destroyer Doctrine in World War II*. Leavenworth Paper no. 12. Ft. Leavenworth, KS: USACGSC, 1985.

Gabriel, Richard A. *Operation Peace for Galilee: The Israeli–PLO War in Lebanon*. New York: Hill and Wang, 1984.

Galvin, John R. *Air Assault: The Development of Airmobile Warfare*. New York: Hawthorne Books, 1969.

Gander, Terry. *Encyclopaedia of the Modern British Army*. 2d ed. Cambridge, England: Patrick Stephens, 1982.

Garland, Albert N., and Howard M. Smith. *Sicily and the Surrender of Italy*. U.S. Army in World War II series. Washington, DC: OCMH, 1965.

de Gaulle, Charles. *Vers l'armée du métier*. Paris: Presses Pocket, 1963.

Gavin, James M. *Airborne Warfare*. Washington, DC: Infantry Journal Press, 1947.

Gawrych, George W. *Key to the Sinai: The Battles for Abu Agheila in the 1956 and 1967 Arab-Israeli Wars*. Research Survey no. 7. Ft. Leavenworth, KS: USACGSC, c. 1990.

———. *The 1973 Arab-Israeli War: The Albatross of Decisive Victory*. Leavenworth Paper no. 21. Ft. Leavenworth, KS: USACGSC, 1996.

Gebhardt, James F. *The Petsamo-Kirkenes Operation: Soviet Breakthrough and Pursuit in the Arctic, October 1944*. Leavenworth Paper no. 17. Ft. Leavenworth, KS: USACGSC, 1989.

le Général J. B. Estienne, père des chars. Paris: Imprimerie Union, n.d.

George, Alexander L. *The Chinese Communist Army in Action; The Korean War and Its Aftermath*. New York and London: Columbia University Press, 1967.

Gilbert, Gerald. *The Evolution of Tactics*. London: Hugh Rees, 1907.

Gillie, Mildred H. *Forging the Thunderbolt: A History of the Development of the Armored Force*. Harrisburg, PA: Military Service Publishing, 1947.

———. *Autumn Storm*. Leavenworth Paper no. 7. Ft. Leavenworth, KS: U.S. Army Command and General Staff College, 1984.

Glantz, David M. "Observing the Soviets: U.S. Military Attachés in Eastern Europe During the 1930s." *Journal of Military History* 5:2 (April 1991): 153–83.

————. "Operational Art and Tactics." *Military Review* 68:12 (December 1988): 32–40.

————. *Soviet Military Deception in the Second World War.* London: Frank Cass, 1989.

————. *The Soviet Airborne Experience.* Research Survey no. 4. Ft. Leavenworth, KS: U.S. Army Command and General Staff College, 1984.

————. "Soviet Offensive Military Doctrine Since 1945." *Air University Review* 34 (March–April 1983): 24–35.

————. "Soviet Operational Formation for Battle: A Perspective." *Military Review* 63 (February 1983): 2–12.

————, ed. *The Initial Period of War on the Eastern Front, 22 June–August 1941.* London: Frank Cass, 1993.

Glantz, David M., and Jonathan M. House. *The Battle of Kursk.* Lawrence: University Press of Kansas, 1999.

————. *When Titans Clashed: How the Red Army Stopped Hitler.* Lawrence: University Press of Kansas, 1995.

Gooderson, Ian. *Air Power at the Battlefront: Allied Close Air Support in Europe 1943–45.* London and Portland, Oregon: Frank Cass, 1998.

Gordon, John IV. "Joint Fire Support: The Salerno Experience." *Military Review* 69:3 (March 1989): 38–49.

Graham, Dominick. "Sans Doctrine: British Army Tactics in the First World War." In *Men at War: Politics, Technology and Innovation in the Twentieth Century,* ed. Timothy Travers and Christian Archer. Chicago: Precedent Publishers, 1982.

Grandmaison, Louis L. de *Deux conférences faites aux officiers de l'état-major de l'armée, Février 1911. La notion de sûrété et l'engagement des grandes unités.* Paris: Berger-Levrault, 1913.

————. *Dressage de l'infanterie en vue du combat offensif.* 2d ed. Paris: Berger-Levrault, 1906.

Grau, Lester W., ed. *The Bear Went over the Mountain: Soviet Combat Tactics in Afghanistan.* Washington, DC: National Defense University Press, c. 1996.

Grau, Lester W., and Mohammad Y. Nawroz. "The Soviet Experience in Afghanistan," *Military Review* 75:5 (September–October 1995): 17–27.

Green, Constance M., Harry C. Thomson, and Peter C. Roots. *The Ordnance Department: Planning Munitions for War.* U.S. Army in World War II series. Washington, DC: OCMH, 1955.

Greenfield, Kent Roberts. *Army Ground Forces and the Air-Ground Battle Team, Including Organic Light Aviation.* Army Ground Forces Study no. 35. Washington, DC: Historical Section, Army Ground Forces, 1948.

————, Robert R. Palmer, and Bell I. Wiley. *The Organization of Ground Combat Troops.* U.S. Army in World War II series. Washington, DC: Historical Division, Department of the Army, 1947.

Greenhous, Brereton. "Aircraft versus Armor: Cambrai to Yom Kippur." In *Men at War: Politics, Technology and Innovation in the Twentieth Century,* ed. Timothy Travers and Christian Archer. Chicago: Precedent Publishers, 1982.

————. "Evolution of a Close Ground-Support Role for Aircraft in World War I." *Military Affairs* 39 (February 1975): 22–28.

Greer, Thomas H. *The Development of Air Doctrine in the Army Air Arm, 1917–1941.* USAF Historical Studies no. 189. Reprint. Manhattan, KS: *Aerospace Historian*, n.d.

Griffith, Paddy. *Battle Tactics of the Western Front: The British Army's Art of Attack, 1916–18.* New Haven, CT: Yale University Press, 1994.

Guderian, Heinz. "Armored Forces." *Infantry Journal* 44 (September–October, November–December 1937): 418–21, 522–28.

————. *Panzer Leader.* Trans. Constantine Fitzgibbon. New York: E. P. Dutton, 1952.

Gudmundsson, Bruce I. *On Artillery.* Westport, CT and London: Praeger, 1993.

————. *Stormtroop Tactics: Innovation in the German Army, 1914–1918.* Westport, CT: Greenwood Press, 1989.

Hallion, Richard P. *Strike from the Sky: The History of Battlefield Air Attack, 1911–1945.* Washington, DC: Smithsonian Institution Press, 1989.

Hamilton, Nigel. *Monty: The Making of a General, 1887–1942.* New York: McGraw-Hill, 1981.

Harkavy, Robert E., and Stephanie G. Neuman, eds. *The Lessons of Recent Wars in the Third World.* Vol. 1. Lexington, MA: Lexington/D. C. Heath, 1985.

Harris, J. P. *Men, Ideas, and Tanks: British Military Thought and Armoured Forces, 1903–1939.* Manchester, England: Manchester University Press, 1995.

Harris, J. P., and F. H. Toase, eds. *Armoured Warfare.* New York: St. Martin's Press, 1990.

Harrison, Gordon A. *Cross-Channel Attack.* U.S. Army in World War II series. Washington, DC: OCMH, 1951.

Hawkins, Glen R. *United States Army Force Structure and Force Design Initiatives, 1939–1989.* Washington, DC: U.S. Army Center of Military History, n.d.

Hay, Ian. *Arms and the Men.* The Second World War, 1939–1945: A Short Military History Series. London: HMSO, 1977.

Hazen, David M. "Role of the Field Artillery in the Battle of Kasserine Pass." Master's thesis, USACGSC, 1973.

Heikal, Mohamed. *The Road to Ramadan.* London: William Collins, and New York: New York Times Book Co., 1975.

Heller, Charles E., and William A. Stofft, eds. *America's First Battles, 1776–1965.* Lawrence: University Press of Kansas, 1986.

Henry, Ryan, and C. Edward Peartree. "Military Theory and Information Warfare." *Parameters: Journal of the U.S. Army War College* 28:3 (autumn 1998): 121–35.

Herbert, Paul H. *Deciding What Has to Be Done: General William E. DePuy and the 1976 Edition of FM 100-5, Operations.* Leavenworth Paper no. 16. Ft. Leavenworth, KS: USACGSC, 1988.

Hermes, Walter G. *Truce Tent and Fighting Front.* United States Army in the Korean War. Washington, DC: OCMH, 1966.

Herzog, Chaim. *The War of Atonement: October 1973.* Boston: Little, Brown, 1975.

Hightower, Charles D. "The History of the United States Air Force Airborne Forward Air Controller in World War II, the Korean War, and the Vietnam Conflict." Master's thesis, USACGSC, 1984.

Hinsley, F. H., E. E. Thomas, C. F. G. Ransom, and R. C. Knight. *British Intelligence in the Second World War: Its Influence on Strategy and Operations.* Vol. 1. London: HMSO, 1979, and New York: Cambridge University Press, 1981.

Hofman, George F. "Doctrine, Tank Technology, and Execution: I. A. Khalepskii and the Red Army's Fulfillment of Deep Offensive Operations." *Journal of Slavic Military Studies* 9:2 (June 1996): 283–334.

Hogan, David W. Jr. *Raiders or Elite Infantry? The Changing Role of the U.S. Army Rangers from Dieppe to Grenada.* Westport, CT: Greenwood Press, 1992.

Hogg, Ian V. *Barrage: The Guns in Action.* New York: Ballantine Books, 1970.

———. *The Guns, 1914–18.* New York: Ballantine Books, 1971.

Holder, L. Donald. "Maneuver in Deep Battle." *Military Review* 62:5 (May 1982): 54–61.

Holder, L. Donald, and Huba Wass de Czege. "The New FM 100-5." *Military Review* 62:7 (July 1982): 53–70.

Horne, Alistair. *To Lose a Battle: France, 1940.* Boston: Little, Brown, 1969.

House, Jonathan M. "The Decisive Attack: A New Look at French Infantry Tactics on the Eve of World War I." *Military Affairs* 40 (December 1976): 164–69.

———. *The U.S. Army in Joint Operations, 1950–1983.* Washington, DC: U.S. Army Center of Military History, 1992.

———. "Waiting for the Panther: Kursk, 1943." In *The Limits of Technology in Modern Warfare,* ed. Andrew J. Bacevich and Brian R. Sullivan. Cambridge: Cambridge University Press, forthcoming.

Howard, Michael. *The Franco-Prussian War: The German Invasion of France, 1870–1871.* New York: Collier Books, 1969.

Howze, Hamilton H. "The Howze Board." *Army Magazine* 24 (February, March, April 1974): 8–14, 18–24, 18–24.

———. "Tactical Employment of the Air Assault Division." *Army Magazine* 13 (September 1963): 36–53.

Hughes, Thomas A. *Over Lord: General Pete Quesada and the Triumph of Tactical Air Power in World War II.* New York: Free Press, 1995.

Hunnicutt, R. P. *Abrams: A History of the American Main Battle Tank.* Vol. 2. Novato, CA: Presidio Press, 1990.

Ingles, Harry C. "The New Division." *Infantry Journal* 46 (November–December 1939): 521–39.

International Institute for Strategic Studies. "The Military Balance, 1977–78." *Air Force* 60 (December 1977): 59–126.

Isby, David C. *War in a Distant Country—Afghanistan: Invasion and Resistance.* London: Arms and Armour Press, 1989.

———. *Weapons and Tactics of the Soviet Army.* 3d ed. London: Jane's Publishing, 1988.

Isby, David C., and Charles Kamps Jr. *Armies of NATO's Central Front.* London and New York: Jane's Publishing, 1985.

Isely, Jeter A., and Philip A. Crowl. *The U.S. Marines and Amphibious War: Its Theory, and Its Practice in the Pacific.* Princeton, NJ: Princeton University Press, 1951.

Jackson, Lawrence M. II. "Force Modernization—Doctrine, Organization, and Equipment," *Military Review* 62:12 (December 1982): 2–12.

Jackson, Robert. *Air War over France, May–June, 1940.* London: Ian Allen, 1974.

Jacobs, W. A. "Air Support for the British Army, 1939–1943." *Military Affairs* 46 (December 1982): 174–82.

Jacomet, Robert. *L'Armement de la France, 1936–1939.* Paris: Editions LaJeunesse, 1945.

Jalili, Ahmad, and Lester W. Grau. *The Other Side of the Mountain: Mujahideen Tactics in the Soviet-Afghan War.* Quantico, VA: U.S. Marine Corps University, 1995.

Jencks, Harlan W. *From Muskets to Missiles: Politics and Professionalism in the Chinese Army, 1945–1981.* Boulder, CO: Westview Press, 1982.

Jendraszczak, Eugeniusz. "The Restructuring of the Polish Armed Forces in a Defensive Strategy." *Journal of Slavic Military Studies* 8:4 (December 1995): 752–59.

Joffre, Ellis. *The Chinese Army After Mao.* Cambridge: Harvard University Press, 1987.

Johnson, Hubert C. *Breakthrough! Tactics, Technology, and the Search for Victory on the Western Front in World War I.* Novato, CA: Presidio Press, 1994.

Karsh, Efraim. *The Iran-Iraq War: A Military Analysis.* Adelphi Papers no. 220. London: International Institute for Strategic Studies, 1987.

Kauffmann, Kurt. *The German Tank Platoon: Its Training and Employment in Battle.* 2d ed. *Panzer Kampfwagenbuch.* Berlin: Verlag Offene Worte, n.d. Mimeographed translation by U.S. Army, Headquarters Armored Force.

Kemp, Anthony. *The Unknown Battle: Metz, 1944.* Briarcliff Manor, NY: Stein and Day, 1981.

Kennedy, Robert M. *The German Campaign in Poland (1939).* Department of the Army Pamphlet 20-255. Washington, DC: OCMH, 1956.

King, Michael J. *Rangers: Selected Combat Operations in World War II.* Leavenworth Paper no. 11. Ft. Leavenworth, KS: U.S. Army Command and General Staff College, 1984.

Kintner, William R. "Who Should Command Tactical Air Forces?" *Combat Forces Journal* 1:4 (November 1950): 34–37.

Kipp, Jacob. "The Russian Military and the Revolution in Military Affairs: A Case of the Oracle of Delphi or Cassandra?" Ft. Leavenworth, KS: Foreign Military Studies Office, 1995.

Kireyev, N., and N. Dovbenko. "From the Experience of the Employment of Forward Detachments of Tank (Mechanized) Corps." *USSR Report: Military Affairs* 1736 (17 January 1983): 25–35. FBIS translation from *Voyenno-istoricheskiy zhurnal [Military History Journal],* September 1982.

Kirkpatrick, Charles E. "The Army and the A-10: The Army's Role in Developing Close Air Support Aircraft, 1961–1971." Washington, DC: U.S. Army Center of Military History, c. 1987.

Kleiner, Martin S. "Joint STARS Goes to War." *Field Artillery* (February 1992): 25–29.

Klotz, Helmut. *Les Leçons militaires de la Guerre Civile en Espagne.* 2d ed. Paris: privately published, 1937.

Klyuchkov, A. "The Tank Battalion in the Forward Element." *Zhurnal bronetankovykh i mekhanizirovannykh voisk [Journal of Armored and Mechanized Forces]* 7 (1946): 9–14. ACSI translation no. F-6585-B.

Kobrin, N. "Encirclement Operations." *Soviet Military Review* 8 (1981): 36–39.

Koepsell, Royal. "The Ultimate High Ground! Space Support to the Army. Lessons from Operations Desert Shield and Storm." Center for Army Lessons Learned Newsletter no. 91-3. Ft. Leavenworth, KS: Center for Arms Lessons Learned, 1991.

Krepinevich, Andrew F. Jr. *The Army and Vietnam.* Baltimore: Johns Hopkins University Press, 1986.

Lackland, Frank D. "Attack Aviation." Student paper written at the Command and General Staff College, 1931, reproduced in USACGSC Reference Book 20-18, *Selected Readings in Military History: Evolution of Combined Arms Warfare.* Ft. Leavenworth, KS: USACGSC, 1983.

Lafitte, Raymond. *L'Artillerie d'assaut de 1916 à 1918.* Paris: Charles-Lavauzelle, 1921.

Leeb, Wilhelm von. *Defense.* Trans. Stefan T. Possony and Daniel Vilfroy. Harrisburg, PA: Military Service Publishing Company, 1943.

de Leon, Peter. "The Laser-Guided Bomb: Case History of a Development." RAND Report no. R-1312-1-PR. Santa Monica, CA: RAND Corporation, 1974.

Lester, J. R. *Tank Warfare.* London: George Allen and Unwin, 1943.

Lestringuez, Pierre. *Sous l'armure: Les Chars d'assaut Français pendant la guerre.* Paris: La Renaissance du Livre, 1919.

Levien, Fred. "Information Warfare: The Plain Truth." *Journal of Electronic Defense* 22:4 (April 1999): 47–53.

Liddell Hart, Basil H. "French Military Ideas Before the First World War." In *A Century of Conflict,* ed. Martin Gilbert. New York: Atheneum, 1967.

———. *Foch: The Man of Orleans.* London: Eyre and Spottiswoode, 1931.

———. *The German Generals Talk.* New York: William Morrow, 1948.

———. *History of the Second World War.* New York: G. P. Putnam's Sons, 1971.

List, Single [pseud.]. *The Battle of Booby's Bluffs.* Washington, DC: U.S. Infantry Association, 1922.

Lord, Christopher. "Intermediate Deployments: The Strategy and Doctrine of Peacekeeping-Type Operations." Occasional Paper no. 25. Camberley: Strategic and Combat Studies Institute, 1996.

Losik, C. A. *Stroitel'stvo i boevoe primenie sovetskikh tankovykh voisk v gody velikoi otechestvennoi voiny.* [Construction and combat use of Soviet tank forces in the years of the Great Patriotic War.] Moscow: Voennoe izdatel'stvo, 1979; portions trans. David M. Glantz.

Lowry, Vernon. "Initial Observations by Engineers in the Gulf War." *Engineer* 21 (October 1991): 42–48.

Lucas, James. *Kommando: German Special Forces of World War II.* New York: St. Martin's Press, 1985.

Lucas, Pascal Marie Henri. *Evolution des idées tactiques en France et en Allemagne pendant la Guerre de 1914–1918* [Evolution of tactical ideas in France and Germany During the 1914–1918 War]. 3d ed. Paris: Berger-Levrault, 1932.

Luttwak, Edward, and Dan Horowitz. *The Israeli Army.* New York and Evanston: Harper and Row, 1975.

Macksey, Kenneth J. *Afrika Korps.* New York: Ballantine Books, 1968.

———. *Armoured Crusader: A Biography of Major-General Sir Percy Hobart.* London: Hutchinson, 1967.

———. *Guderian: Creator of the Blitzkrieg.* New York: Stein and Day, 1976.

———. *Rommel: Battles and Campaigns.* London: Arms and Armour Press, 1979.

———. *The Tank Pioneers.* London: Jane's Publishing, 1981.

Madill, Donald L. "The Continuing Evolution of the Soviet Ground Forces." *Military Review* 62:8 (August 1982): 52–58.

Maier, Klaus A., Horst Rohde, Bernd Stegemann, and Hans Umbreit. *Germany's Initial Conquests in Europe*, vol. 2, *Germany and the Second World War.* Trans. Dean S. McMurry and Ewald Osers. New York and Oxford: Oxford University Press, 1991.

Malinovskiy, R. Y., and O. Losik. "Wartime Operations: Maneuver of Armored and Mechanized Troops." *USSR Report: Military Affairs* 1559 (29 December 1980): 18–26. JPRS 77064. FBIS translation from *Voyenno-istoricheskiy zhurnal [Military History Journal]*, September 1980.

Mark, Edward. *Aerial Interdiction: Air Power and the Land Battle in Three American Wars.* Washington, DC: Office of Air Force History, 1994.

Marquis, Susan L. *Unconventional Warfare: Rebuilding U.S. Special Operations Forces.* Washington, DC: Brookings Institution, 1997.

Marshall, George C. *Memoirs of My Services in the World War, 1917–1918.* Boston: Houghton Mifflin, 1976.

Marshall, Samuel L. A. *Commentary on Infantry Operations and Weapons Usage in Korea, Winter of 1950–51.* Operations Research Office Study ORO-R-13. Chevy Chase, MD: Operations Research Office, 1951.

———. *Operation Punch and the Capture of Hill 440, Suwon, Korea, February 1951.* Technical Memorandum ORO-T-190. Chevy Chase, MD: Operations Research Office, 1952.

———. *Pork Chop Hill: The American Fighting Man in Action; Korea, Spring 1953.* New York: Morrow, 1956.

———. *The River and the Gauntlet: Defeat of the Eighth Army by the Chinese Communist Forces, November 1950, in the Battle of the Chongchon River, Korea.* New York: Time Reading Program, 1952.

———. *Sinai Victory.* New York: Morrow, 1967.

———. *World War I.* New York: American Heritage Press, 1971.

Marshev, M. "Action of a Tank Platoon During Combat in the Depths of the Enemy Defenses." *Voyenni vestnik [Military Herald]* 23 (1948): 28–33. ACSI translation, no. F-4488B.

Martel, Giffard LeQ. *In the Wake of the Tank.* 2d ed. London: Sifton Praed and Company, 1935.

———. *Our Armoured Forces.* London: Faber and Faber, 1945.

———. *An Outspoken Soldier: His Views and Memoirs.* London: Sifton Praed and Company, 1949.

Martin, Michel L. *Warriors to Managers: The French Military Establishment Since 1945.* Chapel Hill: University of North Carolina Press, 1981.

Marty-Lavauzelle, I. *Les Manoeuvres du Bourbonnais en 1909.* Paris: Charles-Lavauzelle, n.d.

———. *Les Manoeuvres de l'ouest en 1912.* Paris: Charles-Lavauzelle, 1913.

———. *Les Manoeuvres de sud-ouest en 1913.* Paris: Charles-Lavauzelle, 1913.

Marzloff, Jean. "The French Mechanized Brigade and its Foreign Counterparts." *International Defense Review* 6 (April 1973): 176–80.

Masters, John. *The Road Past Mandalay.* London: Michael Joseph, 1961.

Mataxis, Theodore C., and Seymour L. Goldberg. *Nuclear Tactics, Weapons, and Firepower in the Pentomic Division, Battle Group, and Company.* Harrisburg, PA: Military Service Publishing Company, 1958.

Mazarr, Michael J. *Light Infantry Forces and the Future of U.S. Military Strategy.* AUSA Institute of Land Warfare. Washington, DC: Brassey's/Macmillan, 1990.

Mazarr, Michael J., Don M. Snider, and James A. Blackwell Jr. *Desert Storm: The Gulf War and What We Learned.* Boulder, CO: Westview Press, 1993.

McCormick, Robert R. *The Army of 1918.* New York: Harcourt, Brace, and Howe, 1920.

McIntosh, Malcolm. *Juggernaut: A History of the Soviet Armed Forces.* New York: Macmillan, 1968.

McKenna, Charles D. "The Forgotten Reform: Field Maneuvers in the Development of the United States Army, 1902–1920." Ph.D. diss., Duke University, 1981.

Meller, R. "Federal Germany's Defense Potential, Part I: The Armed Forces." *International Defense Review* 7 (April 1974): 167–74.

Messenger, Charles. *The Blitzkrieg Story.* New York: Charles Scribner's Sons, 1976.

———. *Trench Fighting 1914–18.* New York: Ballantine Books, 1972.

Middlebrook, Martin. *The Fight for the 'Malvinas': The Argentine Forces in the Falklands War.* London and New York: Viking/Penguin, 1989.

Milner, Samuel. *Victory in Papua.* U.S. Army in World War II series. Washington, DC: OCMH, 1957.

Monash, John. *The Australian Victories in France in 1918.* New York: E. P. Dutton, n.d.

Montross, Lynn, and Nicholas A. Canzona. *U.S. Marine Operations in Korea, 1950–1953,* vol. 2, *The Inchon-Seoul Operation.* Washington, DC: Headquarters, U.S. Marine Corps, 1955.

Morelock, Jerry D. *Generals of the Ardennes: American Leadership in the Battle of the Bulge.* Washington, DC: National Defense University Press, 1993.

Moro, Ruben C. *The History of the South Atlantic Conflict: The War for the Malvinas.* Trans. Michael Valeur. New York: Praeger Publishers, 1989.

Mortensen, Daniel R. *A Pattern for Joint Operations: World War II Close Air Support, North Africa.* Washington, DC: Office of Air Force History, 1987.

Moulton, J. L. *A Study of Warfare in Three Dimensions: The Norwegian Campaign of 1940.* Athens: Ohio University Press, 1967.

Mrazek, James E. *The Fall of Eben Emael.* Washington, DC: Luce, 1971.

Murray, Williamson. *Air War in the Persian Gulf.* Baltimore, MD: Nautical and Aviation Publishing Company of America, 1995.

Négrier, François-Oscar de. "Quelques enseignemen[t]s de la Guerre Sud-Africaine." *Révue des Deux Mondes,* 5th ser., 9 (1909): 721–67.

Neillands, Robin. *In the Combat Zone: Special Forces Since 1945.* New York: New York University Press, and London: Butler and Tanner, 1998.

Nenninger, Timothy K. "The Development of American Armor, 1917–1940." Master's thesis, University of Wisconsin, 1968.

Newell, Clayton R., and Michael D. Krause, eds. *On Operational Art.* Washington, DC: U.S. Army Center of Military History, 1994.

Newell, Mark. "4th ID Pioneers New Division Design." *Armor* 107:6 (November–December 1998): 49–50.

Ney, Virgil. *Evolution of the U.S. Army Division 1939–1968.* Combat Operations Research Group Memorandum M-365. Ft. Belvoir, VA: U.S. Army Combat Developments Command, 1969.

Nikolaieff, A. M. "Secret Causes of German Successes on the Eastern Front." *Coast Artillery Journal* (September–October 1935): 373–77.

Novitskiy, K. "Coordination Between Medium and Heavy Tanks in Offensive Combat." *Tankist* 9 (1947): 40–43, ACSI translation no. F-6602A.

O'Ballance, Edgar. *The Gulf War.* London: Brassey's/Pergamon, 1988.

———. *No Victor, No Vanquished: The Yom Kippur War.* San Rafael, CA: Presidio Press, 1978.

———. *The Third Arab-Israeli War.* Hamden, CT: Archon Books, 1972.

Odom, William O. *After the Trenches: The Transformation of U.S. Army Doctrine, 1918–1939.* College Station: Texas A&M Press, 1999.

Ogorkiewicz, Richard M. *Armoured Forces: A History of Armoured Forces and Their Vehicles.* New York: Arco Publishing Company, 1970.

———. *Design and Development of Fighting Vehicles.* Garden City, NY: Doubleday, 1968.

Orenstein, Harold S., trans. *The Evolution of Soviet Operational Art, 1927–1991; The Documentary Basis,* vol. 2, *Operational Art, 1965–1991.* London: Cass, 1995.

Orr, Kelly. *From a Dark Sky: The Story of U.S. Air Force Special Operations.* Novato, CA: Presidio Press, 1996.

Osadchiy, N. "Cooperation of Tanks with Motorized Infantry." *Zhurnal brone-tankovykh i mekhanizirovannykh voisk [Journal of Armored and Mechanized Forces]* 7 (1946): 15–17. ACSI translation no. F-6585C.

Osterhoudt, Henry J. "The Evolution of U.S. Army Assault Tactics, 1778–1919: The Search for Sound Doctrine." Ph.D. diss., Duke University, 1986.

Owens, Elmer G., and Wallace F. Veaudry, "Control of Tactical Air Power in Korea," *Combat Forces Journal* 1:9 (April 1951): 19–21.

Palat, Barthelemy Edmond. "Les Manoeuvres du Languedoc en 1913," *Révue des Deux Mondes*, 6th ser., 17 (1913): 799–817.

Palazzo, Albert P. "The British Army's Counter-Battery Staff Office and Control of the Enemy in World War I." *Journal of Military History* 63:1 (January 1999): 55–74.

Palmer, Dave R. *Summons of the Trumpet: U.S.–Vietnam in Perspective.* San Rafael, CA: Presidio Press, 1978.

Pelletiere, Stephen C., and Douglas V. Johnson II. *Lessons Learned: The Iran-Iraq War.* [Unclassified version.] Carlisle Barracks, PA: U.S. Army War College, 1991.

Perre, Jean Paul. *Batailles et combats des chars Français: La Bataille defensive Avril–Juillet 1918.* Paris: Charles-Lavauzelle, 1940.

Perrett, Bryan. *A History of Blitzkrieg.* New York: Stein and Day, 1983.

———. *Knights of the Black Cross: Hitler's Panzerwaffe and Its Leaders.* New York: St. Martin's Press, 1986.

Pershing, John J. *My Experiences in the World War.* 2 vols. New York: Frederick A. Stokes, 1931.

Peterson, Bernard W. *Short Straw: Memoirs of Korea by a Fighter Pilot/Forward Air Controller.* Scottsdale, AZ: Chuckwalla Publishing, 1996.

Piekalkiewicz, Janusz. *The Cavalry of World War II.* New York: Stein and Day, 1980.

———. *Operation "Citadel": Kursk and Orel: The Greatest Tank Battle of the Second World War.* Novato, CA: Presidio Press, c. 1987.

Pirnie, Bruce R. *Operation Urgent Fury: The United States Army in Joint Operations.* Washington, DC: U.S. Army Center of Military History, 1986.

Pitt, Barrie. *The Crucible of War: Western Desert 1941.* London: Jonathan Cape, 1980.

Playfair, I. S. O., G. M. S. Stitt, C. J. C. Molony, and S. E. Toomer. *The Mediterranean and the Middle East*, vol. 1, *The Early Successes Against Italy (to May 1941).* History of the Second World War. London: HMSO, 1954.

Playfair, I. S. O., F. C. Flynn, C. J. C. Molony, and S. E. Toomer. *The Mediterranean and the Middle East*, vol. 2, *The Germans Come to the Help of Their Allies (1941).* History of the Second World War. London: HMSO, 1956.

Playfair, I. S. O., F. C. Flynn, C. J. C. Molony, and T. P. Gleave. *The Mediterranean and the Middle East*, vol. 3, *British Fortunes Reach Their Lowest Ebb.* History of the Second World War. London: HMSO, 1960.

———. *The Mediterranean and the Middle East*, vol. 4, *The Destruction of the Axis Forces in Africa.* History of the Second World War. London: HMSO, 1966.

Postgate, Malcolm. *Operation Firedog: Air Support in the Malayan Emergency, 1948–1960.* London: HMSO, 1992.

Pratt, Edwin A. *The Rise of Rail-Power in War and Conquest, 1833–1914.* Philadelphia: J. B. Lippincott, 1916.

Radzievsky, A. I. *Tankovyi udar* [Tank Blow]. Moscow: Voennoe Izdatel'stvo, 1977. Portions trans. David M. Glantz.

Raney, George H. "Tank and Anti-Tank Activities of the German Army." *Infantry Journal* 31 (February 1927): 151–58.

Record, Jeffrey. "Why the Air War Worked." *Armed Forces Journal International* (April 1991): 44–45.

Reznichenko, Vasiliy G., ed. *Tactics*. Moscow: Military Publishing House, 1966. ACSI Translation no. J-1731.

Ripperger, Robert M. "The Development of the French Artillery for the Offensive, 1890–1914." *Journal of Military History* 59:4 (October 1995): 599–618.

Ritter, Gerhard. *The Sword and the Scepter: The Problem of Militarism in Germany*, vol. 1, *The Prussian Tradition, 1740–1890*. Trans. Heinz Horden. Coral Gables, FL: University of Miami Press, 1969.

Rochelle, Jeffrey B. "The Battle of Khafji: Implications for Airpower." Master's thesis, U.S. Air Force School of Advanced Airpower Studies, 1997.

Rockenbach, Samuel D. "Tanks." Mimeographed. Ft. Meade, MD: U.S. Tank School, 1922.

Romjue, John L. *American Army Doctrine for the Post–Cold War*. Ft. Monroe, VA: U.S. Army Training and Doctrine Command, 1997.

———. *From Active Defense to AirLand Battle: The Development of Army Doctrine 1973–1982*. Ft. Monroe, VA: U.S. Army Training and Doctrine Command, 1984.

Rommel, Erwin. *Infantry Attacks*. Trans. G. E. Kidde. Washington, DC: Infantry Journal, 1944.

———. *The Rommel Papers*. Ed. B. H. Liddell Hart, trans. Paul Findlay. New York: Harcourt, Brace, 1953.

Ross, G. MacLeod. *The Business of Tanks, 1933–1945*. Ilfracombe, England: Arthur H. Stockwell, 1976.

Rotmistrov, Pavel. "Cooperation of Self-Propelled Artillery with Tanks and Infantry," *Zhurnal bronetankovykh i mekhanizirovannykh voisk [Journal of Armored and Mechanized Forces]* 7 (1945): 8–13. ACSI translation no. F6583B.

———. "Tanks—The Decisive Power of the Attack." *Voyennaya mysl' [Military Thought]* 8 (1946). ACSI translation no. F-6089A.

———. *Stal'naia gvardiia* [Steel guards]. Moscow: Voenizdat, 1988.

Rowan-Robinson, H. *Further Aspects of Mechanization*. London: William Clowes and Sons, 1929.

Ryan, Stephen. *Pétain the Soldier*. New York: A. S. Barnes, 1969.

Ryazanskiy, A. "The Creation and Development of Tank Troop Tactics in the Pre-War Period." *Voyenni vestnik [Military Herald]* 11 (1966): 25–32. ACSI translation no. J-1376, 35–46.

———. "Land Forces *Podrazdeleniye* [battalion] Tactics in the Postwar Era." *Voyenni vestnik* 8 (1967): 15–20. ACSI translation no. J-2745.

———. "Tactics of Tank Forces During the Great Patriotic War." *Voyenni vestnik [Military Herald]* 5 (1967): 13–20.

Samuels, Martin. *Command or Control? Command, Training and Tactics in the British and German Armies, 1888–1918.* London: Frank Cass, 1995.

Scales, Robert H. Jr. "Artillery in Small Wars: The Evolution of British Artillery Doctrine, 1860–1914." Ph.D. diss., Duke University, 1976.

———. *Certain Victory: The U.S. Army in the Gulf War.* 1993. Rpt., Ft. Leavenworth, KS: USACGSC, 1994.

———. *Firepower in Limited War.* Rev. ed. Novato, CA: Presidio Press, 1995.

Schlie, Alan. "Close Up: Engineer Restructuring Initiative." *Engineer* 23 (February 1993): 20–24.

Schmidt, Paul Karl [Paul Carell, pseud.]. *The Foxes of the Desert.* Trans. Mervyn Savill. New York: E. P. Dutton, 1961.

———. *Stalingrad: The Defeat of the German 6th Army.* Trans. David Johnston. Atglen, PA: Schiffer Military History, c. 1993.

Schnabel, James F. *Policy and Direction: The First Year.* United States Army in the Korean War. Washington, DC: OCMH, 1972.

Scott, Harriet Fast, and William F. Scott. *Soviet Military Doctrine: Continuity, Formulation, and Dissemination.* Boulder, CO: Westview Press, 1988.

Seaton, Albert. *Stalin as Military Commander.* New York: Praeger, 1976.

Serré, C. (Rapporteur). *Rapport fait au nom de la commission chargée d'enquête sur les événements survenues en France de 1933 à 1945.* 8 vols. Paris: Imprimerie de l'Assemblée Nationale, 1947–1951.

Sheppard, E. W. *Tanks in the Next War.* London: Geoffrey Bles, 1938.

Showalter, Dennis E. *Railroads and Rifles: Soldiers, Technology, and the Unification of Germany.* Hamden, CT: Archon Books, 1976.

Shtrik, S. "The Encirclement and Destruction of the Enemy During Combat Operations Not Involving the Use of Nuclear Weapons." *Voyennaya mysl'* 1 (1968): 279–92. Defense Intelligence Agency translation no. FPD 0093/68.

Simkin, Richard E. *Mechanized Infantry.* Oxford, England: Brassey's Publishers, 1980.

———. *Tank Warfare: An Analysis of Soviet and NATO Tank Philosophy.* London: Brassey's Publishers, 1979.

Skillen, Hugh. *Spies of the Airwaves: A History of Y Sections During the Second World War.* Pinner, England: privately printed, 1989.

Slim, William. *Defeat into Victory.* Abrgd. ed. New York: David McKay, 1961.

Smith, Bradley F. *The Shadow Warriors: O.S.S. and the Origins of the C.I.A.* New York: Basic Books, 1983.

Smith, Peter C. *Close Air Support: An Illustrated History, 1914 to the Present.* New York: Orion Books, 1990.

Snyder, Jack. *The Ideology of the Offensive: Military Decision Making and the Disasters of 1914.* Ithaca, NY: Cornell University Press, 1984.

Snyder, James M., ed. *History of the Armored Force, Command and Center.* U.S. Army Ground Forces Study no. 27. Washington, DC: Army Ground Forces Historical Section, 1946.

Sokolovsky, Vasili D., ed. *Soviet Military Strategy.* 3d ed. Trans. and ed. Harriet Fast Scott. New York: Crane, Russak, 1975.

Spiller, Roger J., ed. *Combined Arms in Battle Since 1939.* Ft. Leavenworth, KS: USACGSC, 1992.

Stanton, Martin N. "The Saudi Arabian National Guard Motorized Brigades." *Armor* 105:2 (March–April 1996): 6–11.

Stanton, Shelby L. *America's Tenth Legion: X Corps in Korea, 1950.* Novato, CA: Presidio Press, 1989.

Steadman, Kenneth A. "A Comparative Look at Air-Ground Support and Practice in World War II, with an Appendix on Current Soviet Close Air Support Doctrine." CSI Report no. 2. Ft. Leavenworth, KS: USACGSC, 1982.

———. "The Evolution of the Tank in the U.S. Army." Typescript. CSI Report no. 1. Ft. Leavenworth, KS: USACGSC, 1982.

Stolfi, R. H. S. "Equipment for Victory in France in 1940." *History* 55 (February 1970): 1–20.

Stone, Norman. *The Eastern Front, 1914–1917.* New York: Charles Scribner's Sons, 1975.

Strokov, Alesandr A. *Vooruzhennye sily i voennoe iskusstvo v Pervoi Mirovoi Voine* [The Armed Forces and Military Art in the First World War]. Moscow: Voenizdat, [1974]. Portions trans. David M. Glantz.

Strong, Kenneth. *Men of Intelligence: A Study of the Roles and Decisions of Chiefs of Intelligence from World War I to the Present Day.* London: Cassell, 1970.

Subbotin, A. "Concerning the Characteristics of the Direction of Present-Day Offensive Operations." *Voyennaya mysl' [Military Thought]* 6 (1946): ACSI translation no. F-6123A.

Sunderland, Riley. "Massed Fires and the FDC." *Army* 8 (May 1958): 56–59.

Suvenirov, O. F. "Vsearmeiskaia tragediia" [An armywide tragedy], *Voyenno-istoricheskii zhurnal* 3 (March 1989): 42ff.

Swain, Richard M. *"Lucky War": Third Army in Desert Storm.* Ft. Leavenworth, KS: USACGSC, 1994.

Sweet, John J. T. *Iron Arm: The Mechanization of Mussolini's Army, 1920–1940.* Westport, CT: Greenwood Press, 1980.

Tarrant, V. E. *Stalingrad: Anatomy of an Agony.* London: Leo Cooper, 1992.

Taylor, Terrence, ed. *The Military Balance 1997/98.* International Institute for Strategic Studies. London: Oxford University Press, 1997.

Terraine, John. *To Win a War—1918, the Year of Victory.* Garden City, NY: Doubleday, 1981.

Terrett, Dulany. *The Technical Services: The Signal Corps: The Emergency (to December 1941).* U.S. Army in World War II. Washington, DC: OCMH, 1956.

Teveth, Shabtai. *The Tanks of Tammus.* New York: Viking, 1968–1969.

Thomas, Timothy L. "The Battle of Grozny: Deadly Classroom for Urban Combat." *Parameters: Journal of the U.S. Army War College* 29:2 (summer 1999): 87–102.

———. "The Caucasus Conflict and Russian Security: The Russian Armed Forces Confront Chechnya, III: The Battle for Grozny, 1–26 January 1995." *Journal of Slavic Military Studies* 10:1 (March 1997): 50–108.

Thompson, George R., and Dixie P. Harris. *The Signal Corps: The Outcome (Mid-*

1943 Through 1945). U.S. Army in World War II. Washington, DC: OCMH, 1966.

Thompson, Paul W. *Engineers in Battle.* Harrisburg, PA: Military Service Publishing Company, 1942.

Times (London). "The Insight Team." In *The Yom Kippur War.* Garden City, NY: Doubleday, 1974.

Tolson, John J. *Airmobility, 1961–1971.* Vietnam Studies. Washington, DC: Department of the Army, 1973.

Travers, Tim. *The Killing Ground: The British Army, the Western Front and the Emergence of Modern Warfare, 1900–1918.* London: Allen and Unwin, 1987.

Trupener, Ulrich. "The Road to Ypres: The Beginnings of Gas Warfare in World War I." *Journal of Modern History* 47 (September 1975): 460–80.

Union of Soviet Socialist Republics. Commissariat of Defense. "Collection of Materials for the Study of the War Experience, no. 2 (September–October 1942)." 1942. ACSI translation no. F-7565.

———. Institut Marksizma-Leninizma. *History of the Great Patriotic War,* vol. 3, *Radical Turning Point in the Course of the Great Patriotic War.* Moscow: Military Publishing House of the Ministry of Defense, 1961. ACSI translation, no number.

United States. Department of the Air Force. *Gulf War Airpower Survey.* 5 vols. Washington, DC: Office of the Secretary of the Air Force, 1993.

———. Department of the Army. The Armored School. "Armored Division Organizational and Manning Charts, TO&E 17N." Instructional Pamphlet no. CS-2. Ft Knox, KY: U.S. Army Armored School, 1949.

———. "Armor in the Exploitation: The Fourth Armored Division Across France to the Moselle River." Student Research Project, mimeographed, 1949. CARL N-2146.74.

———. General Staff. *Soviet Army Organization: Mechanized Division (Wartime).* Washington, DC: ACSI, 1954.

———. *Soviet Army Organization: Tank Division (Wartime).* Washington, DC: ACSI, 1954.

———. "Soviet Tank and Motorized Rifle Division." Intelligence Research Project no. A-1729. Washington, DC: ACSI, 1958.

———. Intelligence and Threat Analysis Center. *How They Fight: Armies of the World.* ATC-RA-2600-147-92. Washington, DC: Government Printing Office, 1992.

———. National Training Center Handbook 100-91. *The Iraqi Army: Organization and Tactics.* Ft. Irwin, CA: U.S. Army National Training Center, 1991.

———. OCMH. Department of the Army Pamphlet no. 20-230. *Historical Study—Russian Combat Methods in World War II.* Washington, DC: OCMH, 1972.

———. U.S. Army Forces, Far East. *Japanese Night Combat.* 3 vols. Japanese Research Division report, mimeographed, Tokyo, 1955. CARL N-17807.28A.

———. USACGSC. Reference Book 100-2, vol. 1. *Selected Readings in Tactics: The 1973 Middle East War.* Ft. Leavenworth, KS: USACGSC, 1976.

———. Department of Defense. *Conduct of the Persian Gulf War: Pursuant to Title V* of the Persian Gulf Conflict Supplemental Authorization and Personnel Benefits Act of 1991 (Public Law 102-25). Washington, DC: Government Printing Office, 1992.

———. Defense Intelligence Agency. *Handbook on the Chinese Armed Forces.* Washington, DC: Government Printing Office, 1976.

———. National Guard Bureau. "Army National Guard After Action Report: Operation Desert Shield, Operation Desert Storm. Executive Summary." Arlington, VA: n.d.

———. Weapons System Evaluation Group. "A Historical Study of Some World War II Airborne Operations." WSEG Staff Study no. 3. Mimeographed, 1951. CARL R-17309.1.

———. "A Study on Tactical Use of the Atomic Bomb." WSEG Staff Study no. 1. Mimeographed, 1949. CARL N-16687.2.

———. War Department. Technical Manual *TM-E30-451: Handbook on German Military Forces,* 15 March 1945. Washington, DC: War Department, 1945.

———. Technical Manual *TM-E30-480: Handbook on Japanese Military Forces,* 15 September 1944. Washington, DC: War Department, 1944.

———. European Theater of Operations. The General Board. "Army Tactical Information Service." Study no. 18. Mimeographed, n.d.

———. "Chemical Mortar Battalions." Study no. 70. Mimeographed, n.d.

———. "The Control of Tactical Aircraft in the European Theater of Operations." Study no. 55. Mimeographed, n.d.

———. "Organization, Equipment, and Tactical Employment of the Airborne Division." Study no. 16. Mimeographed, n.d.

———. "Organization, Equipment, and Tactical Employment of the Armored Division." Study no. 48. Mimeographed, n.d.

———. "Organization, Equipment, and Tactical Employment of Tank Destroyer Units." Study no. 60. Mimeographed, n.d.

———. "Organization, Operations, and Equipment of Air–Ground Liaison in all Echelons from Division Upwards." Study no. 21. Mimeographed, n.d.

———. "Tactical Employment, Technique, Organization, and Equipment of Mechanized Cavalry Units." Study no. 49. Mimeographed, n.d.

———. "Types of Divisions—Postwar Army." Study no. 17. Mimeographed, n.d.

———. General Staff. "Lessons from the Tunisian Campaign, 15 October 1943." Washington, DC: Government Printing Office, 1943.

———. *Order of Battle of the United States Land Forces in the World War: American Expeditionary Forces; Divisions.* Washington, DC: Government Printing Office, 1931.

———. ACSI. *Determination of Fighting Strength, USSR.* 2 vols. Washington, DC: ACSI, 1942.

———. "The Organization and Tactical Employment of Soviet Ground Formations." Mimeographed, July 1944. Study no. C5859.

———. Military Intelligence Service. Special Bulletin no. 36. *The Battle of Salum, June 15–17, 1941.* Washington, DC: War Department, 1941.

——. Special Series no. 6. *Artillery in the Desert.* Washington, DC: War Department, 1942.

——. Special Studies no. 8. *German Tactical Doctrine.* Washington, DC: Government Printing Office, 1942.

——. Special Studies no. 26. *Japanese Tanks and Tank Tactics.* Washington, DC: Government Printing Office, 1942.

——. Information Bulletin no. 12, MIS 461. "Japanese Warfare: A Summary." Washington, DC: War Department, 1942.

——. Information Bulletin no. 16, MIS 461. "Japanese Warfare: A Summary." Washington, DC: War Department, 1942.

——. War Plans Division. *A Survey of German Tactics, 1918.* Monograph no. 1. Washington, DC: War Department, 1918.

——. The Infantry School. Special Text no. 265. *Infantry in Defensive Combat.* Ft. Benning, GA: Army Field Printing Plant, 1926.

——. Third U.S. Army. Letters of Instruction nos. 1, 2, and 3, 1944. Reprinted in *Third Army After Action Report,* Third U.S. Army, 1945.

Ushakov, D. "The Role of Engineer Troops in Modern War." *Voyenno-inzhenerny zhurnal [Military Engineer Journal]* 5–6 (1946): 9–13. ACSI translation no. F-1622a.

Van Creveld, Martin. *Supplying War: Logistics from Wallenstein to Patton.* Cambridge and New York: Cambridge University Press, 1977.

Viner, Joseph W. *Tactics and Techniques of Tanks.* Ft. Leavenworth, KS: General Service Schools Press, 1920.

Volodin, N. "Assignment of Combat Engineers in the Breakthrough and Development of the Offensive." *Voyenno-inzhenerny zhurnal [Military Engineer Journal]* 5–6 (1946): 21–25. ACSI translation no. F-1522a-b.

Voroushenko, V., and A. Pozdnyshev. "Heavy Tanks in Combat with Counterattacking Tanks of the Enemy." *Zhurnal bronetankovykh i mekhanizirovannykh voisk [Journal of Armored and Mechanized Forces]* 1945. ACSI translation no. F-6606C.

Wakebridge, Charles. "The Egyptian Staff Solution." *Military Review* 55 (March 1975): 3–11.

Walthall, Melvin. *We Can't All Be Heroes.* Hicksville, NY: Exposition Press, 1975.

Wass de Czege, Huba, and L. Donald Holder. "The New FM 100-5." Military Review 62:7 (July 1982): 53–70.

Watson, Bruce W., and Peter M. Dunn, eds. *Military Lessons of the Falklands War: Views from the United States.* Boulder, CO: Westview Press, and London: Arms and Armour Press, 1984.

Wavell, Archibald. *Allenby: A Study in Greatness.* New York: Oxford University Press, 1941.

——. *The Palestine Campaigns.* 2d ed. London: Constable and Company, 1929.

Weeks, John. *Men Against Tanks: A History of Anti-Tank Warfare.* New York: Mason/Charter, 1975.

Weigley, Russell F. *The American Way of War: A History of United States Military Strategy and Policy.* New York: Macmillan, 1975.

——. *History of the United States Army.* New York: Macmillan, 1967.

———. "Shaping the American Army of World War II: Mobility Versus Power." *Parameters, Journal of the U.S. Army War College* 11 (September 1981): 13–21.

Weinberg, Gerhard L. *A World at Arms: A Global History of World War II.* Cambridge: Cambridge University Press, 1994.

Weller, Jac. "Armor and Infantry in Israel." *Military Review* 55 (March 1975): 3–11.

Westgaard, Wayne T. "Will the New Brigade Reconnaissance Troop Be Adequately Protected?" *Armor* 108:2 (March–April 1999): 27–29.

Wheldon, John. *Machine Age Armies.* London: Abelard-Schuman, 1968.

Wieder, Joachim, and Heinrich von Einsiedel. *Stalingrad: Memories and Reassessments.* Trans. Helmut Bogler. London: Arms and Armour Press, 1995.

Wilson, Dale E. *Treat 'Em Rough! The Birth of American Armor, 1917–20.* Novato, CA: Presidio Press, 1989.

Wilson, John B. *Maneuver and Firepower: The Evolution of Divisions and Separate Brigades.* Army Lineage Series. Washington, DC: U.S. Army Center of Military History, 1998.

Winnefield, James A., Preston Niblack, and Dana J. Johnson. *A League of Airmen: U.S. Air Power in the Gulf War.* Santa Monica, CA: RAND Corporation, 1994.

Winton, Harold R. "General Sir John Burnett-Stuart and British Military Reform, 1927–1938." Ph.D. diss., Stanford University, 1977.

Woodward, David. *Armies of the World, 1854–1914.* New York: G.P. Putnam's Sons, 1978.

Wray, Timothy A. "Standing Fast: German Defensive Doctrine on the Russian Front During the Second World War." Master's thesis, USACGSC, Ft. Leavenworth, KS, 1983.

———. *Standing Fast: German Defensive Doctrine on the Russian Front During World War II: Prewar to March 1943.* Research Survey no. 5. Ft. Leavenworth, KS: USACGSC, 1986.

Wynne, Graeme C. "The Development of the German Defensive Battle in 1917, and Its Influence on British Defence Tactics." *Army Quarterly* 24 (April, July 1937): 15–32, 248–66.

———. *If Germany Attacks: The Battle in Depth in the West.* London: Faber and Faber, 1940.

Yamaguchi, Jiso, ed. *Burma Operational Record: 15th Army Operations in Imphal Area and Withdrawal to Northern Burma.* Rev. ed. Japanese Monograph no. 134. Washington, DC: OCMH, 1957.

Yarchevsky, P. "Breakthrough of the Tactical Defense," *Voyennaya mysl' [Military Thought]* 9 (1946): ACSI translation no. F-6090B.

Yekimovskiy, A. "Tactics of the Soviet Army During the Great Patriotic War," *Voyenni vestnik [Military Herald]* 4 (1967): 12–20. ACSI translation no. J-2507.

Yekimovskiy, A., and N. Makarov. "The Tactics of the Red Army in the 1920s and 1930s." *Voyenni vestnik [Military Herald]* 3 (1967): 8–13. ACSI translation no. J-2141.

Yunker, Stephen F. "'I Have the Formula': The Evolution of the Tactical Doctrine of General Robert Nivelle." *Military Review* 54 (June 1974): 11–25.

Zakharov, M. V., ed. *50 let vooruzhennykh sil SSSR* [50 Years of the Soviet Armed Forces]. Moscow: Voennoe Izdatel'stvo, 1968. Portions trans. David M. Glantz.

Zaloga, Steven J. "Soviet Tank Operations in the Spanish Civil War." *Journal of Slavic Military Studies* 12:3 (September 1999): 134–62.

———. *Target America: The Soviet Union and the Strategic Arms Race, 1945–1964.* Novato, CA: Presidio Press, 1993.

———. "Technological Surprise and the Initial Period of War: The Case of the T-34 Tank." *Journal of Slavic Military Studies* 6:4 (December 1993): 634–46.

Zaloga, Steven J., and James Grandsen. *Soviet Tanks and Combat Vehicles of World War II.* London: Arms and Armour Press, 1984.

Zaloga, Steven J., and Peter Sarson. *T-34/76 Medium Tank, 1941–1945.* London: Osprey, 1994.

Ziemke, Earl F. *Stalingrad to Berlin: The German Defeat in the East.* U.S. Army Historical Series. Washington, DC: OCMH, 1968.

Zubok, Vladislav, and Constantine Pleshakov. *Inside the Kremlin's Cold War: From Stalin to Khrushchev.* Cambridge: Harvard University Press, 1996.

INDEX